Human Sciences

SUNY series in

Science, Technology, and Society

Sal Restivo and Jennifer Croissant, editors

HUMAN SCIENCES

Reappraising the Humanities through History and Philosophy

JENS HØYRUP

STATE UNIVERSITY OF NEW YORK PRESS

Published by
State University of New York Press, Albany

Printed in the United States of America

For information, address State University of New York Press,
State University Plaza, Albany, NY 12246

Production, Laurie Searl
Marketing, Anne Valentine

Library of Congress Cataloging-in-Publication Data

Høyrup, Jens.
Human sciences : reappraising the humanities through history and philosophy / Jens Høyrup.
p. cm.
Includes bibliographical references and index.
ISBN 0-7914-4603-4 (alk. paper) — ISBN 0-7914-4604-2 (pbk. : alk. paper)
1. Humanities—Philosophy. 2. Humanities—History. 3. Science and the Humanities.
I. Title.

AZ103 .H69 2000
001.3′01—dc21
99-088563

10 9 8 7 6 5 4 3 2 1

Ludovica in memoriam

"Ubi caritas et amor,

Deus ibi est"

CONTENTS

Acknowledgements

A number of publishers and authors or representatives of deceased authors have kindly permitted use of the following poems:

Cecil Bødker, "Generationer", quoted from Cecil Bødker, *Samlede digte.* København: Hasselbalch, 1964. © 1964 Cecil Bødker. Permission (including permission to translate) granted by Cecil Bødker.

Bertolt Brecht, "General, dein Tank ist ein starker Wagen", quoted from Bertolt Brecht, *Gesammelte Werke in 20 Bänden,* vol. IX. Frankfurt am Main: Suhrkamp, 1967. © 1967 Suhrkamp Verlag, Frankfurt am Main. Permission granted by Suhrkamp Verlag.

Bertolt Brecht, "Lob des Zweifels", excerpt quoted from Bertolt Brecht, *Gesammelte Werke in 20 Bänden,* vol. IX. Frankfurt am Main: Suhrkamp, 1967. © 1967 Suhrkamp Verlag, Frankfurt am Main. Permission granted by Suhrkamp Verlag.

Paul Celan, "Ich Höre, die Axt hat geblüht", quoted from Paul Celan, *Gesammelte Werke,* vol. II. Frankfurt am Main: Suhrkamp, 1986. © 1983 Suhrkamp Verlag, Frankfurt am Main. Permission granted by Suhrkamp Verlag.

Otto Gelsted, "Døden", quoted from Otto Gelsted, *Udvalgte digte.* København: Gyldendal, 1957. © 1957 Gyldendalske Boghandel Nordisk Forlag. Permission (including permission to translate) granted by Gyldendalske Boghandel Nordisk Forlag and Otto Gelsted Mindefond.

Ivan Malinowski, "Kritik af kulden", quoted from Ivan Malinowski, *Vinterens hjerte.* København: Borgen, 1980. © 1980 Ivan Malinowski and Borgens Forlag. Permission (including permission to translate) granted by Ruth Malinowski and Borgens Forlag.

In the main, the transformation of a set of lecture notes into the first version of a book manuscript was made during my stay as a visiting scholar at the Max-Planck-Institut für Wissenschaftsgeschichte, Berlin in Autumn 1994. It is a pleasant duty to express my gratitude for the hospitality I enjoyed.

It is also a pleasure to thank Marianne Marcussen for providing me with a photo from the original edition of Dürer's *Unterweysung der Messung.*

INTRODUCTION

The present book grew out of a course on the "theory of the humanities" held at Roskilde University in Denmark, the participants in which were first- and second-year students of the humanities.[1] The title of the course may sound odd, the very notion of a 'theory of the humanities' being unfamiliar within the Anglo-American tradition. In German, it would be much more regular, namely *Wissenschaftstheorie der Geisteswissenschaften.* The subject is related to the traditional philosophical discipline *philosophy of science,* but with two important differences. Firstly, since the nineteenth century English *science* is narrower than German/Latin *Wissenschaft/scientia,* and often it encompasses only the exact and natural sciences to the exclusion of other scholarly pursuits; secondly, *Wissenschaftstheorie* may draw more on empirical (historical, sociological, and psychological) foundations than standard twentieth-century philosophy—and even standard philosophy *of science.* On the other hand, it is less prone than certain types of "science studies" to forget that science is a practice concerned with *knowledge.*

Any approach to the theory of the humanities must apply a double perspective. If the humanities are to be understood as *sciences* in the German/Latin sense, they must share properties that characterize many if not all other sciences as well: a "theory of the humanities" must ask what can be said about the humanities *qua* sciences. Yet if it is meaningful to single out "the humanities" as a particular and somehow coherent area, the "theory" must also be able to tell what distinguishes the humanities from other scientific fields, that is, to tell the *distinctive characteristics* of the humanities.

The book consists of three parts. Parts I and II concentrate (each in its own way) upon the second perspective; part III is mainly devoted to the first issue.

As a historian of science I find it natural to approach the problem of the humanities through their genesis and development. Part I is therefore devoted to a presentation of select episodes and developments from the history of the humanities, *not*

1. More about this outlandish institution will be told in chapter 21.

only as a field of knowledge but also as a sequence of social practices. In our own world, indeed, the humanities are not only a type of scholarly work supported by teaching and popularization. They are also a *profession securing a living* for the social group of humanists, which entangles them in a particular social and societal context—and one of the insights gained by the history of science over the last thirty years is that there is an intimate connection between the professional setting of a field, the types of insight at which it aims, and its mode of thinking and of organizing the insights which it gains. Discussions of this interplay in nonfamiliar historical settings may, firstly, awaken our appreciation of similar relations between the intellectual aspect and the social and professional situation of *the humanities today;* secondly, the presentation of central ideas and characteristic problems and methods of the humanities in the context where they were created and once put to use will often give essential information about their meaning and carrying capacity.[2]

Etymology, however, is rightly claimed to "tell what words don't mean any longer." The humanities are no longer found in the settings where they developed. If their value (or some value of theirs) remains, this cannot be due to their origin (in the philosophy of science, this problem is spoken of as the difference between *genesis* and *validity*). Even though Copernicus *may* have found the mental courage to remove the Earth from the center of the universe because of the breakdown of the medieval social and ecclesiastical world order, his theory (as recast by Kepler, Newton, and Einstein) now serves to send planetary sondes successfully to the outer planets; similarly, the validity of psychoanalysis does not depend upon Sigmund Freud's personal frustrations and imagined mother fixation (as claimed by some of those who do not like the "Godless Jew," as he called himself). Taken alone, a historical approach to a body of ideas may give clues to their meaning but does not explain or demonstrate their general validity and coherence. Part II therefore switches to a *systematic* approach to the different "anthropologies"—that is, fundamental notions about the distinctive nature of human beings and human society—that may be presupposed in the human sciences.[3] It may be difficult to sum up in a simple formula what constitutes the object of the humanities. So much is certain, however, that they cannot be defined simply as "sciences concerned with human beings." The law of gravitation also deals with human beings, and so does biological

---

2. The presentation may seem unduly culturo-centric. However, while it is impossible to trace the development of modern natural sciences without taking the developments of at least the Islamic and Indian world into account, this is on the whole not true concerning the humanities. The humanities themselves have indeed been strongly culturo-centric since the Hellenistic epoch—first "Greek," then "Christian," then "European," now "Western."

3. The term *anthropology* is thus used as when we speak of "philosophical anthropology," which has only indirect connections to the concepts of 'cultural' or 'social anthropology' (cf. p. 191).

science. However much it makes use of guinea pigs and investigates bacteria, medicine is even *applied human biology* in its very essence. If we insist on setting up an easy (possibly facile) delimitation, the humanities will rather deal with those aspects of human existence which distinguish, or seem to distinguish, human beings from the entities considered by other sciences, and which therefore also enforce other theoretical approaches: the use of language, the production of symbols, the possibility of reflexive thought, the presence of culture.[4] Philosophical anthropologies try to specify or formulate—perhaps even to explain—these distinctive characteristics.

Such anthropologies may be *deterministic* in tendency; if we explain human behaviour or find the real meaning of human communication in terms of human biology or sociology or in the structure of language, little seems to be left to human freedom. Or they may (like original Sartrean existentialism) declare that everything which is explained is thereby nonhuman, because *human nature is freedom* aware of itself. Ultimately, the former kind of anthropologies assert that the apparently distinctive characteristics are illusive, and that they can be derived from and reduced to levels of reality considered by other sciences (be it systems theory or biology); the latter kind, by contrast, moves in a closed circle, *defining* so to speak the distinctively human as that which is irreducibly and thus distinctively human.

Quite apart from this logical fallacy, neither determinism nor the postulate of abstract freedom gives a meaningful account of the complexities of human existence, human communication, and human history. Therefore, the final pages of part II attempt a synthesis under the headline "human nature as dialectic and history."

---

4. These aspects of human existence are neither fully distinct nor identical; in part they extend, in part they explicate and explain, in part they condition each other, in a way which allows us to regard them as aspects of that elusive *specifically human* which we are after. To the same complex belong features like the production and understanding of art, theoretical knowledge, and religion; the sense of humor; and the consciously planned production of tools.

It is certainly possible to find additional characteristics that distinguish human beings from other animals. According to an anecdote told in Diogenes Laërtios's *Lives of Eminent Philosophers* (VI, 40, ed. Hicks 1925: II, 42), Plato once defined Man as "a featherless, biped animal"—to which the further qualification "with broad nails" was added, as Diogenes of Sinope presented the Academy with a plucked chicken (which is, of course, parodic and was always meant to be, but which also illustrates the problem). There is no reason to deny that the choice of language, symbols, reflexive thought, and culture is inspired by the actual interests of the humanities and meant to exhibit the inner coherence of a field which extends from theoretical grammar to the history of literature and social psychology.

We observe that even Plato's second definition holds for females no less than males. In Plato's language, *man* (*anthrōpos*) is the human being in general (German *Mensch*, etc.). Thus also, in order to avoid extreme linguistic clumsiness, everywhere in the following (even in all quotations but two). Every reader is asked to ascribe to the abstract person in question her favorite gender of the moment—be it her own or the complement.

Part III presents a general philosophy of science, that is, investigates those features which the humanities share with the natural and social sciences, and applies some of the insights gained to philosophical problems of particular relevance for the humanities, namely, moral philosophy and the relation between art and cognition. The detailed presentation is best postponed. Some general observations on the character of the volume as a whole may be useful, however.

Firstly, the exposition is meant throughout to be *read at different levels:* readers who have never heard of "scholasticism" or "normal science" may miss some of the more delicate points in the discussion without much damage to their increased basic understanding. Those who are already familiar with the fundamental notions will be offered particular interpretations, connections to parallel phenomena and related discussions, and adjustments of the coarser summary statements at the basic level; such adjustments and qualifications often occur as open-ended or elusive invitations to further thought. Many of them will appear in the footnotes.

Quite a few names occur in the text. For those who already know (for instance) about Thales of Miletus, his name should serve to anchor an argument with respect to what they already know; those who do not know him but are curious will find his date in the name index. If further information about him had been relevant (which it happens not to be), it would have been given in the text.

Secondly, and for this reason, *footnotes are not peripheral* but as important as the main text. They often contain further reflections, objections, and qualifications, or they serve as a device that allows a branching of the argument. Some of them contain material which is essential in subsequent parts of the text (in which case cross-references will be made); some of the longer notes are meant to suggest open-ended historiographic or philosophical lines of thought which cannot be pursued in full consequence. Notes should therefore *not* be skipped—but readers who approach the topic for the first time may find it useful to concentrate on the main text.

Much of the philosophical argument in parts II and III refers to a rather narrow array of authors: Aristotle, Immanuel Kant, Jürgen Habermas, Jean-Paul Sartre, Jean Piaget, Karl Popper, Imre Lakatos, Thomas Kuhn, Robert Merton. They are not chosen solely because of their undeniable general importance but rather because they provide convenient stepping stones for the actual discussion. Nor does my selection imply that those who do not appear are irrelevant; some of the points I make in connection with discussions of Piaget and Kuhn might (for instance) have been made instead with reference to Ludwig Wittgenstein—but less conveniently, I believe.

The aim of the volume is *not* to give an encyclopedic survey of the opinions of select philosophers (etc.) but to develop a certain view on the subject matter. On the premise that nothing is detracted from (say) Aristotle, Sartre, or Merton if a book from the 1990s reads them with a particular purpose in view, I have therefore abstained from any attempt to cover globally the views of the authors to whom I refer.

Similar "productive readings" of more recent authors would be more disputable, in particular if they were based on disagreement. Even though part III can be seen as an alternative reading of facts and structures which also interest constructivist and other "post-Kuhnian" and "post-Mertonian" science studies, I have therefore intentionally avoided explicit confrontation as liable to be unjust if brief and unduly extensive if doing justice to the approaches under discussion.

References are made in most cases according to the author/editor–date system. A few standard encyclopedias, such as the *Dictionary of the History of Ideas,* however, are represented by abbreviations; these abbreviations are listed in the bibliography. The other exceptions are authors for whom a standard reference system or a standard edition exists (Plato, Aristotle, Kant, etc.); here I have followed established conventions and omitted a reference to the specific edition, unless it is quoted.

As a consequence of the different characters of the three parts, the use of references is uneven. In part I it would be impossible to give references for every point I make. In cases where I draw on relatively well-known secondary literature, I have omitted the text reference. Instead, the bibliographic essay (chapter 9) refers to essential works on the main topics. Quotations and specific points drawn from recent or less well-known publications are always provided with a reference; since the boundary lines between the specific and the general, between the recent and the less recent, and between the well-known and the less familiar are blurred, I have certainly erred on quite a few occasions, omitting references that were due and including others that might safely have been omitted.

Part II is much more of a personal synthesis, and the need for references is correspondingly smaller. I have attempted to include references for all specific points of importance, but much of what is said concerns general views and widespread attitudes, for which it would be meaningless to give a reference unless a particular example is discussed in some depth. Much the same could be said concerning part III.

All translations into English are mine and made from the original language if nothing else is stated. If a translation into another language is referred to, I am responsible for the retranslation.

Some Fundamental Concepts

Firstly, although this usage is not current in English (or not yet quite current—things may be gradually changing), the term science is used in the following as an equivalent of German *Wissenschaft,* that is, in the sense of *socially organized and systematic search for and transmission of coherent knowledge* in any domain. On this account chemistry, the study of law, sociology, and literary history are sciences in full right. Theology may be so, if founded on critical discussion and hence on the search for intellectual coherence, whereas the mere teaching of traditional dogma in a fundamentalist preachers' school is not. Nor are "pseudo-sciences" like chiromancy or astrology to be counted as sciences: they contain a body of coherent knowledge (which may be false, but sciences also err), and they hand it down systematically through a socially organized network; but they do not involve systematic search for extension or critical revision of that knowledge. Nor does the mere collection of information, for example for the compilation of a telephone guide, constitute a science: it lacks both the aim of intellectual coherence and the character of a continuous endeavor bound together by systematic transmission. Finally, the knowledge gathered by some Robinson Crusoe isolated on his island and forgotten with him constitutes no science, even if it should happen to be eminently true and coherent: it differs from the sciences by the lack of *social* organization and by the absence of systematic (indeed, any) transmission links.

This definition implies no value judgement. Nobody will blame our Robinson because he works in isolation, and the higher or lower moral value of fundamentalist theology does not depend on its being a science. The definition is first of all intended to be descriptive, telling the characteristics of the actual undertaking of science in our modern world. It also concentrates on features which influence each other, and influence in particular the character of the cognitive contents achieved by the undertaking: *social organization* of the search for knowledge makes each worker dependent on the work and the critical judgement of others. The aim of creating *new* (in some sense "certified" and reliable) *knowledge* conditions the sociology and the norms of the institution creating and transmitting the knowledge; and so forth.[1]

The use of the above definition does not mean that those ventures which are excluded are without interest. Quite the opposite, indeed. If we are to gain insight into the distinctive characteristics of scientific activities, a privileged method is to compare with those undertakings which in some respects are similar yet in others not. For example, which are the differences that explain the dissimilar outcomes of science and organized crime, two not wholly unlike types of social activity, as it has been pointed out (see p. 333)? What distinguishes science from technology? And from pseudo-science?

Secondly, another ancient anecdote about Plato and Diogenes of Sinope:

> One day Plato the philosopher met his fellow philosopher Diogenes, who, as so often, made a teasing comment on Plato's philosophy. "Good Friend," he said, "Table and cup I see; but your Tablehood and Cuphood, Plato, I can nowhere see." "That's readily accounted for, Dear Diogenes," replied the other. "You have that which is needed to see the table and the cup: that's the eyes. But you lack what is required to grasp Tablehood and Cuphood: namely the intellect."
>
> (Slightly paraphrased from Hicks 1925: II, 55)

This story illustrates the contents of some of the fundamental concepts of the philosophy of science. Diogenes is a *positivist* and an *empiricist:* only that which can be *positively seen* or otherwise registered by direct experience is real. And he is a *materialist:* experience of reality has to be sense experience; the direct and immediate insights of the intellect are irrelevant, since *reality is out there,* outside our mind. Plato, on the other hand, is an *idealist:* supreme reality is the reality of ideas (Cuphood); that material reality (the cups) to which our senses have access is secondary only, a pale and distorted reflection. Moreover, he is an *objective* idealist: for him, the world of ideas exists *out there,* in the *Universal Intellect,* beyond the control

---

1. Since the notion of the *institution* will turn up often in the following, an explanation already at this point may be useful. An institution is not necessarily (and never primarily) *an organization* or a social arrangement provided with its own building(s) or payroll, as shown by concepts like the 'family institution' and the 'parliamentary institution'—the latter is certainly more than the actually elected parliament and the parliamentary building, and the former certainly not to be understood as the sum total of actual families, nor as the mere set of laws ruling their existence as families. Summed up in a few words, an institution is rather to be understood as *a socially stabilized pattern of rules, expectations and habits.* Often, of course, an institution in this sense is also coupled to the presence of a particular organization—the existence of which may even be implied by the pattern of rules.

The *institutionalization* of a social phenomenon, correspondingly, is its development into a socially stabilized pattern of rules, expectations and habits.

of our mind and our fancies; our individual intellects have access to this world of ideas because they partake in the universal intellect (and, according to Plato, sense experience only serves to *remind us* of the knowledge once possessed but now deeply buried in our minds).

These concepts are still essential to discussions of the philosophy of science and knowledge, and we shall return to them in part III. At present I shall use them to characterize my approach to the history of the humanities. This approach does not postulate or look for the transhistorical and unchanging existence of the Humanities across all epochs and cultural borders, that is, it does not believe in *the Humanities as Cuphood.* It is empiricist, and presupposes that the humanities can only be approached in their appearance as actual, historically specific undertakings and vocations. It is not positivist, however, but founded on the conviction that the grouping of these undertakings—the individual cups—under a common headline is inherently meaningful, reflecting real similarities and relationships, and thus more than a mere device dependent solely upon historical accident or on our arbitrary choices and whims.

PART I

Institutions, Professions, and Ideas

Approaching the Humanities through their History and Settings

For educational purposes, the past is a bank to be raided, not a church to worship in; but it is also not a useless museum.
—Ivor Grattan-Guinness

It should be obvious, I hope, that only a small section of the following builds directly upon my own research. The rest is based in part on selective reading of sources and original literature; just as important, however, has been thoughtful reading of the secondary literature, a modest segment of which is listed in the bibliographic essay which follows. However, as everybody knows who has been engaged in research in a specific domain, secondary literature cannot always be relied upon; at times it contains downright errors, at times it makes use of generalizing formulations which the author would be able to interpret so as to make them agree with the sources, but which nevertheless mislead the innocent reader. Unguided reading of select sources within an unfamiliar field may, however, be equally misleading. The two methods may supplement each other and thus help avoid many errors, and only the universal or divine genius can do without them; but errors will still remain in any interdisciplinary work of broad scope. So also, no doubt, in the following. CAVEAT LECTOR! In other words: READER BEWARE!

CHAPTER 1

A Bronze Age Scribal Culture: A Sociological Fable with an Implicit Moral

BRAIN WORK AND STATE FORMATION

Humanists are brain workers, and hence specialists of a particular kind. It is therefore meaningless to look for an environment of a "humanistic" character in so-called *egalitarian societies* as exemplified by Bushmen or Inuit. In such societies as everywhere, of course, individuals differ. But status differences depend on the gender and age and on the personal abilities of individuals, not on inherited position or social class; in principle, everybody belonging to the same gender and the same age group makes a living in the same way. There is little or no room for specialization in the sphere of intellectual work; the closest we come is the possible existence of a shaman or a priestess, but even they will normally not be full-time specialists[1] and in any case not be part of *an environment* or profession.

Chiefdoms, another main category of pre-state society distinguished by political anthropology, are characterized by differentiation along several dimensions: socio-economic division, implying unequal access to basic resources (for instance land for agriculture), and maybe a genuine class division into a ruling class, commoners, and slaves belonging to members of the ruling class; and political division between the chief supported by his associates and retinue on one side, and the general population on the other.[2] Priestly functions may be the privilege of the chief and his associates;

1. A fitting example is the Ibo priestess Chielo in Chinua Achebe's novel *Things Fall Apart* (1986), who from time to time is possessed by the spirit of her God but on all other occasions fulfils the same social roles as other women.

2. It hardly needs to be said that these descriptions of pre-state societies simplify the variability of real societies in the extreme.

but still one finds no group specializing in intellectual work, neither of religious character nor associated with the chief's political control (cf. Goody 1977: 19–35).

The division between manual and intellectual work, the precondition for the emergence of anything approaching, however vaguely, a stratum of "humanists," is thus a consequence of the organization of society as a state. As a minimal definition of the early state we may borrow the following, current in one or the other form in contemporary political anthropology and socioarchaeology: a state is a society provided with

1. a structure of control with at least three levels ("vertical" specialization);
2. division of labor, both in the productive and in the controlling sphere ("horizontal" specialization);
3. precedence of social over kinship divisions;
4. reasonable temporal permanency and stability;
5. possession of a reasonably well-defined fixed territory.

Apart perhaps from number 5, all these features are necessary prerequisites for the creation of a social stratum (3) of people specializing in intellectual work (2) yet not belonging to the ruling class (1), and understood (both by its own members and by others) as a coherent group (3 and 4). But they are of course not sufficient. "Control" may be of many sorts, and so may division of labor. Only where *writing* developed together with the state has a stratum of professional intellectuals emerged—and even writing has not always been sufficient.[3]

Three cases where the step was actually taken are ancient Mesopotamia, ancient Egypt, and ancient China. In Mesopotamia and Egypt the stratum of scribes carried literate culture for millennia before eventually collapsing in the Hellenistic era; in China the classical "mandarin culture" has survived continuously until this century, and only future developments will tell whether it has been shattered or only shaken by the double revolution and by the modernization process. China could therefore be chosen to demonstrate the importance of scribal "humanism" even in recent times; but since the early beginnings are poorly documented and I myself am not conversant with the relevant details of Chinese culture, I shall concentrate instead on Mesopotamia, with which I am fairly familiar (familiar enough to discern with some precision what I do not know), and where the permanency of clay has guaranteed the survival of early written sources.

---

3. This is born out by the Mycenaean society of the second millennium BCE. This early Greek culture was a bureaucratic "Palace economy" administered by scribes. Not the slightest written evidence for intellectual interests can be traced, however, nor is there any indication that the script developed for accounting purposes survived the downfall of the Palaces.

In Mesopotamia, the earliest social system fulfilling conditions (1) through (5) (with some reservations for condition 3) arose between c. 3500 and 2800 BCE in Southern Mesopotamia ("Sumer"), which by this time became suited for irrigation agriculture and could thus yield a surplus large enough to allow certain population groups to specialize in activities other than food production. According to all evidence, this is the earliest statal system in the world. Its center was a temple institution, where the body of priests took care of a number of societally important functions: long-range trade; exchange of the produce of different groups of producers specializing in agriculture, herding, fishing, and other trades; presumably organization of handicraft work and in any case of huge temple building projects; perhaps even reparation of the consequences of climatic or other catastrophes. As in other state formation processes, the carriers of central functions took advantage of their positions and made themselves masters of society—to judge from the favorite pictorial motifs of cylinder seals, not by peaceful means alone.

None the less, the fundament for the newly emerging state organization of Mesopotamian society was functional and theocratic-economical, *at least as far as legitimization is concerned.* War and similar organized violence played a much smaller role than in other state formation processes; economic class division, moreover, appears to have been a consequence rather than a cause of the process. This legitimization of the state through purported functionality will probably have been one reason for the systematic development of tools for bureaucratic management—another one being the availability of an age-old system for primitive accounting which easily lent itself to refinement.

One of the newly developed tools was *writing*—at first with a purely ideographic script, that is, a script where each sign stands for a specific word or conceptually related group of words, as "to eat" and "food ration"; the other was *mathematics* used for accounting and in metrology. Together they were used by the priesthood acting collectively as a civil service to keep track of taxation and of the temple economy. They were thus the instruments for the emergence of a class of intellectual workers separate from but controlling manual work.

So far, this seems to have little to do with the humanities. The operation of any specialized tool, however, has to be learned, and in the case of the Mesopotamian scribal tools this was done by institutionalized schooling.[4] Here the script and the mathematical notations and techniques were taught—the script by means of systematic sign lists. Indeed, these lists and a few mathematical exercises constitute the only evidence left by the school institution. We can therefore not say very much about the organization of the school, but still something about its intellectual impact.

---

4. In Egypt, on the other hand, scribes were taught as apprentices "on the job" until the end of the third millennium BCE. This may be a main reason for the difference between Egyptian and Mesopotamian scribal culture—another being a different legitimization of the Pharaonic state as the result of conquest and as the upholder of cosmical order and stability.

This follows from a comparison of the organization of the sign lists with the results of an investigation of the "psychology of literacy and modernization" made by the psychologist Alexander Luria in the 1930s in Soviet Central Asia.[5] He distinguishes "categorical classification" from "situational thinking," in a way which can be illustrated by this dialogue:

> *Luria,* explaining a psychological test: "Look, here you have three adults and one child. Now clearly the child doesn't belong in this group" [categorical classification].
>
> *Rakmat,* an illiterate peasant: "Oh, but the boy must stay with the others! All three of them are working, you see, and if they have to keep running out to fetch things, they'll never get the job done, but the boy can do the running for them. . . ." [situational thinking].
>
> Luria 1976: 55

Situational thinking was found to be "*the* controlling factor among uneducated, illiterate subjects," whereas both modes were applied (with situational thinking dominating) among "subjects whose activities were still confined primarily to practical work but who had taken some courses or attended school for a short time." On the other hand, young people with only a year or two of schooling, but who were actively engaged in building up collective farms, employed the principle of categorical classification "as their chief method of grouping objects." In other words, people living in a stable world dominated by a restricted number of apparently ever-fixed situations presuppose this stability in their mode of thought, whereas those accustomed to change (and perhaps engaged in planned change) arrange theirs in a more abstract and less directly applicable but precisely therefore more flexible pattern.

Now, the sign lists are arranged according to the categorical principle. One list enumerates professions in a hierarchical scheme; one records wooden objects; one inventories vessels; one is a catalogue of place names; and so on. None of them groups together the ox, the seed plough which it draws, and the grain that is sown. Apart from teaching the pictographic signs, the lists can thus be said to convey in a double way a *world view:* firstly, they teach that the world is to be understood in terms of general categories; secondly, they tell the categories involved. Being an intellectual worker in the early Mesopotamian temple state not only implied social segregation but also existence in *a mental cosmos wholly different* from that of illiterate peasants and workers.

Another perspective on the early scribal culture is provided by a scheme proposed by Jürgen Habermas (*Erkenntnis und Interesse,* 1973) in a discussion of the different incentives for the quest for knowledge; since we shall return repeatedly to

5. Similar consequences could be drawn from analysis of the mathematical texts and techniques.

the *motives for knowing*, for which the scheme provides an adequate starting point, it is worthwhile to introduce it already at this point (a more thorough and critical discussion will be found in chapter 10). Habermas distinguishes three distinct *Erkenntnisinteressen* or interests motivating the pursuit of knowledge.[6] One is the *technical interest*, which looks for means to achieve given ends, and which is identified by Habermas as the incentive motivating the natural sciences. The other is *interpretive* or *legitimizing*, leading to *understanding* of why things (in particular society, culture, and cultural products) are as they are and hence potentially to acceptance that they are thus; Habermas identifies it as the incentive motivating the humanities. The third, finally, is emancipation, which *ought to be* the distinctive motivation of social science; in reality, Habermas observes, the real incentive of much actual social science is technical, which he regards as a philosophical mistake of categories, namely as a treatment of fellow human beings as objects to be manipulated (cf. below, p. 185, n. 7).

Obviously, the primary motive of the priestly managers for their construction of a coherent system of knowledge was *technical:* their aim was to *know how* to manage the temple estate and that society in which they had acquired a pivotal position. This position was not legitimized by their *possession of knowledge*—instead, legitimacy followed from actual or pretended societal functions, to which knowledge was subservient. Nor can we ascribe more than a secondary role to the emancipatory interest; however, the liberation from the bonds of traditional modes of thought suggested by the organization of the lexical lists makes it conceivable to speak precisely of emancipation as a *secondary motivation* for a social stratum ridding itself of the restrictions imposed by functional service to a formerly more egalitarian society while perpetuating the functional ideology.

THE FIRST INTELLECTUALS

During the following millennium, the character of the Mesopotamian state underwent a series of changes. At first, a system of city states in perpetual conflict (mainly, it seems, over basic resources like water) developed. In these states, a secular supreme power (a "king") came to overshadow the traditional theocracy. The most striking illustration of this is provided by the so-called "royal tombs of Ur" (c. 2600 BCE), where up to eighty servants, maidens, and soldiers were killed in order to follow their royal master to the nether world. Clearly, statal power was now firmly established on its own and no longer to be understood as an expression of social functions alone.

---

6. Anthony Giddens (1985: 127) translates the untranslatable German phrase as "knowledge-constitutive interests." At the cost of precision, I shall use the less suffocating expression "cognitive interests" in the following.

Epics originating during the same epoch (though only written down around 2100 BCE) also show us a social structure where the king is protector of the shrine of the city but not subject to the priesthood. King Gilgameš of Uruk is no absolute ruler; but the bodies which he must consult are the council of "men" (able to bear arms, one may presume) and the council of "elders" (powerful citizens, presumably, like the Roman Senate, the highest-ranking members of leading families). The leading priests may well have belonged to the latter body; but the epic is only interested in their role as first-rank citizens.

Yet in spite of its absence from the epic horizon, the literate tradition was conserved and expanded. Royal inscriptions begin to turn up—the very first we know is due to a king known from the Gilgameš story. What is more, the functional ideology originally connected with the beginnings of writing was carried on by the literate environment. Nobody would believe without the evidence provided by city walls and by the originally oral epic tradition that the king was first of all a military leader: the literate environment would for centuries depict him as a builder of temples and of canals, that is, as a representative of functional and theocratic power. Nobody would guess from contemporary written evidence that servants were slaughtered in honour of the king—only the oral tradition as reflected in later written epics gives some ambiguous hints. "Early Dynastic" Mesopotamia was thus a truly *dual* society, a society with Janus face. As seen by epic poetry it was a "military democracy";[7] as reflected in written sources—and hence as seen by the literate and by literate culture—it was still a bureaucratic-functional state supposedly providing for communal needs.

In the longer run, of course, the dual society was unstable. Around the middle of the third millennium (the so-called Fara period), the primitive-democratic components of society were increasingly incorporated into an overall structure where written contracts and monetary relations were important. At the same occasion, however, literacy itself stopped being bound up exclusively with its bureaucratic function.

Firstly, the increased use of writing for many socially important functions called for an increase in the number of literate functionaries and hence—given the complex character of writing—for professional specialization. For the first time an organized group of *scribes* distinct from the stratum of priestly managers turns up in the sources. The scribes were *professionally literate and numerate,* not just professional managers who had to be literate and numerate in order to attend to their business.

Secondly, this profession also became aware of itself as a particular group. The scribes started investigating the capacity of their distinctive professional tools: writing

---

7. A concept originating in nineteenth-century anthropology and mainly known today from Friedrich Engels's *Origin of the Family, Private Property and the State.*

and computation. They started writing down *literary texts* (hymns, epics, and proverb collections), and they produced the earliest examples of *pure mathematics,* that is, mathematical problems chosen for their inherent interest and in spite of lack of direct relevance for everyday scribal practice.[8]

Thirdly, the students in the Fara scribe school were enthusiastic about the newly invented idea of intellectual work. They do not inform us so directly but they have left more telling evidence: the empty corners of many of the school tablets made by this first generation are filled in by nice and occasionally really artistic decorations, in a way not seen in any other period of Mesopotamian history, where the cane was always the chief teaching aid. In Fara it was apparently great fun to go to school—at least when you were looking back in mature age (as pointed out by the Assyriologist Aage Westenholz (personal communication), the most beautiful "school tablets" look rather like *de luxe* editions).

Novelties do not remain new, nor will a subordinate social group be left in possession of its distinctive subculture if this culture can serve those in power—this was as true in Mesopotamia as in our century. After the twenty-fourth century, when a centralizing tendency replaced the city states by larger "territorial states," literary texts were no longer made in the scribal school in order to explore the possibilities of professional tools; they had become the product of a "court chancellery" and were made as a vehicle for royal propaganda. Enḫeduanna, the first poet of world history known by name, was a princess, and her hymns are clearly meant to reshape mythology in a way that would suit the territorial state created by her father.

During the twenty-first century BCE, scribal intellectual autonomy reached a paradoxical low point. In this century, southern Mesopotamia formed a single state (the so-called "Third Dynasty of Ur" or "Ur III"), which presumably constituted one of the most highly bureaucratized systems in history until the advent of the modern corporation. Most land was concentrated in royal estates (some of them formally temple estates, but realities were the same), and most of the rural population worked here as serfs, supervised by scribes who were accountable for the performance of their crew calculated in units of twelve minutes (one sixtieth of a work-day). Textile and other industries as well as foreign trade were organized according to similar principles, and of course *the scribe,* the central pivot of the

---

8. A modern mathematician, of course, would not easily recognize the problems as "pure mathematics"—a favorite problem type was division of phantasmagorically large round numbers by divisors which were as irregular as possible with respect to the number systems in use. Nor will a modern literary scholar perhaps fathom the historical importance of a collection of proverbs. But on a background where nobody had ever used writing and numbers beyond the purely utilitarian realm, both are important steps towards the formation of a sphere of autonomous intellectual work.

whole machine, stands out as the culture hero of the era, for instance in various royal hymns.

At the same time, however, we possess a number of descriptions of the curriculum of the Ur III scribe school, as well as texts that were used to implant a suitable ideology in the minds of future scribes. It turns out that the education of the ordinary scribe was strictly utilitarian. Most of the Sumerian literature, it is true, was written down precisely during Ur III; according to all evidence, however, the focus for this activity was the royal court, and the purpose was propagandistic (as it can be seen from the way old stories were twisted so as to fit the political conditions of the day). The rank-and-file scribe was to be an *engineer* and an *accountant* in *civilized society,* and he was to be proud of that. He was a *trusted* and *privileged* subject of the state but still a subject and not a member of an autonomous profession with its own values and interests. In this respect his situation was quantitatively but not qualitatively different from that of the enslaved former peasant, once a member of a free clan but now reduced to being a mere subject.

SCRIBAL "HUMANISM"

Ur III was a swing of the pendulum in one direction. It was soon followed by a reverse swing, both as concerns socioeconomic conditions and regarding scribal culture.

In the long run, the costs of the bureaucracy that was needed for running the Ur III system and keeping the workers busy had been too high for the yield of the land. Breakdown followed, and after an intermediate period (the twentieth century BCE) a new, highly individualized socioeconomic structure emerged in the "Old Babylonian" period (which lasted until 1600 BCE).

Firstly, the economy itself was individualized. Much land was still held by the king, by temples, and by the richest citizens. Estates run by enslaved workers, however, had become rare or disappeared, and land was instead cultivated by tenants on contract or by wage-earning workers. Trade had been taken over by private merchants, and so had industry. Banking of a sort developed, and monetary economy flourished.[9]

These changes had a strong impact on culture and ideology, where the emergence of the *individual* can be observed. In earlier times, only official letter writing had existed; now, the private letter turns up (and even in the official correspondence

9. This description may sound almost like modern capitalism. That would be to overstate things, however. Commodities were produced for the market, it is true, and even land could be bought and sold. The exchange of land, however, did not take place on real market terms but depended on the social standing and kinship affiliation of the parties involved. As land was still the all-important economic asset, it is therefore meaningless to speak of a real market economy and hence of capitalism.

of King Hammurapi, the king stands out as an individual character); the *cylinder seal,* until then an attribute of the official, now becomes a token of private identity; and so on. Society no longer consists of mere *subjects,* but of *private human beings*—in a few unique cases (the merchant communities in certain cities), it seems, even of *citizens,* that is, of persons possessing political co-responsibility.

In this new context, the role and self-awareness of the scribe changed, too. Maybe 80 percent of the scribe school candidates still went into traditional scribal occupations (engineering, accounting, surveying, cancellarian functions at court or in the service of wealthy families). But the scribe-school now provided an *education of individuals aware of themselves as such,* and no longer a mere training of officials and overseers. This double situation explains the distinctive character of the Old Babylonian scribal culture.

Firstly, the ideology that legitimized the state in whose service the scribe could expect employment was still the traditional "social-democratic" idea of the functional state securing affluence *and* justice.[10] In so far as this idea was believed (and it was easier to believe in now than during Ur III), the scribe could be proud of his service to the state – for who but the scribe took care of the functions that actually secured affluence and justice by means of accounting, surveying, letter writing for the illiterate King, and so forth?

Secondly, however, the scribe was taught in school to be proud of being *somebody special.* The mathematics of normal accounting and surveying was fairly trivial, and so was writing in the current Babylonian language.[11] The abilities *actually required* to procure affluence and justice were thus no adequate basis for professional pride. Pride, however, would be fully justified if the scribe mastered the non-trivial (yet, alas, useless!) levels of scribal cunning: solving second-degree equations (occasionally even third to eighth degree!); reading, writing, and speaking the dead Sumerian language understood by nobody but other scribes; knowing the three different occult meanings of cuneiform signs; and so on.

In contrast to his Ur III colleague, the Old Babylonian scribe was thus no mere technician but a virtuoso—and in contrast to his fellow countrymen he was *culturally polished.* He was very much aware of this and had a name for this specific characteristic of his: nam-lú-ulù, (of course Sumerian for) *humanity.* Like the humanist of more recent times he was aware of being a human being *par excellence,* and he was proud of that.

---

10. One of the most clear expressions of this ideology is found in the prologue and epilogue of "Hammurapi's Law-Code."

11. Babylonian could be written adequately by means of some eighty syllabic cuneiform signs, as demonstrated by the members of an Assyrian merchant colony in Cappadocia in the nineteenth century BCE. They wrote their own letters without scribal assistance. The scribes took care that this would never happen again by making the script more complex.

Superficially, this reminds us of the joy of the Fara scribes to be the first intellectuals in history, and some scribe students have certainly experienced the pleasures of Sumerian poetry or of mathematical dexterity. To judge from surviving texts, however, the overall climate of the Old Babylonian scribe school was as repressive as that seventeenth to nineteenth-century Latin school which inculcated "Latinity" or "Graecity" into the sore backs of future priests and officials. Like the humanists produced by the latter institution, the brood of the Old Babylonian school would usually be full of arrogance toward both aristocrats and commoners yet fully uncritical and submissive to the existing social order (cf. also p. 92, n. 18).[12]

Returning to Habermas's cognitive interests, we may say that the glimpse of emancipatory interest which can be read into the early lexical lists is indubitably present in the Fara emergence of autonomous intellectual activity, though even here only as a by-product. The training of ordinary Ur III scribes, on the other hand, was apparently driven by purely technical interest, whereas that of the contemporary court chancellery was legitimizing. Legitimization and understanding of the proper cultural tradition of the scribal profession was also the driving force behind Old Babylonian "humanism"—no wonder, in fact, since Habermas locates the interpretive-legitimizing interest precisely in that humanist tradition which sprang from the post-Renaissance Latin school, and which impregnated the Wilhelmian German mandarinate.

The fall of the Old Babylonian state was the end of the age-old idea of the functional state; of the culture of individualism; and of the scribe school. As one should expect, it was also the end of the scribal culture which had grown out of this well-blended soil. Scribal humanism disappeared; for the next one and a half millennia the scribes, now trained as apprentices inside a "scribal family," would posture as members of a venerable and patriarchal *tradition* and would mix up as much priestly mysticism as possible into their secular business. This phase of Mesopotamian scribal history is therefore no longer interesting as an elucidating parallel to the professional situation of the humanities in the contemporary world, even though it may throw oblique light on twentieth-century fringe mysticism and occultism.

---

12. A striking example of this: in 1600 BCE, the Hittites raided Babylon, thereby putting a violent end to the Old Babylonian era. In reaction to the ensuing social chaos, the population of the city rose in rebellion, only to be crushed by the conquering Kassite warrior tribes, who imposed a new military state on the region—a state that was totally devoid of "social-democratic" legitimation. A contemporary scribe has left us a description of this sequence of events, metaphorically transformed into the shape of a myth. The popular rising is compared to a plague, and the foreign conquerors are seen as deliverance sent by benevolent gods (cf. Brentjes 1966: 30–40).

A question close at hand is of course whether the earlier period is a real elucidation, or whether I have just used the opportunity to tell some more or less amusing anecdotes. Is the presentation of the Babylonian scribal culture really "a sociological fable with an implicit moral," as claimed in the title? An answer will be provided by the following chapters, toward which it will repeatedly be possible to hold up the models presented here. For the moment I shall just oppose three counter-questions to my own rhetorical question: firstly, is it fully excluded that the interest of modern humanists in literature and philosophy fulfils the same functions (irrespective of individual pleasure in the endeavour, which is a different matter) as Old Babylonian scribal "humanity" with regard to legitimation of their social identity and position? Secondly, can we trust our conviction that we serve general interests while engaged by the public authority as teachers, librarians, researchers, TV commentators, and so forth, or is this Hammurapian ideology nothing but a convenient veil hiding quite different realities? Thirdly, is the complete technicalization of Ur III intellectual work without parallels in the contemporary age?

Whoever answers "yes" to these three questions without hesitation will see no moral in the fable. Others may well see it.

CHAPTER 2

Classical Antiquity

The preceding chapter dealt, firstly, with some very general preconditions for the rise of anything akin to the humanities considered as a social undertaking—namely the segregation of intellectual work as a particular occupation and the emergence of literate culture; and secondly, with *sociological parallels* to the modern business of the humanities as they could be found in the various phases of Mesopotamian scribal culture. The present chapter, devoted to classical antiquity, will still offer some such parallels. Its chief aim, however, is to introduce some of the *roots* of the humanities—those very roots to which the intellectual culture of Europe has preferably referred itself ever since.

In that perspective four main periods can be distinguished. The first ("pre-Socratic") period covers the span from c. 600 BCE to c. 430 BCE. The second ("from Socrates and the Sophists to Aristotle") comprises the time until 320 BCE. The third ("Hellenism"), finally, until c. 300 CE. A fourth period ("late Antiquity") beginning around 300 CE, can be defined for instance as the period when Christianity became important for the interpretation of the cultural heritage.

These distinctions not only reflect intellectual currents (though that is the way they are defined here) but also are demarcated by important socioeconomic and political changes. At the same time, however, decisive continuities make it meaningful to speak of the whole span as one epoch. Fundamental among these continuities is the lasting importance of slavery.

It is in fact common to speak of classical society as a slave society. As with all simplifications, this characterization can be challenged. It is not true, as one might believe from the term, that slave labour was all-important throughout the period and in the whole Mediterranean world, though on the other hand it remained more important than in most other historical periods, earlier as well as later. Decisive is the quality implied by the corresponding German term *Sklavenhaltergesellschaft*, "slave holders' society"—namely that the ideological and cultural perspective of the social stratum of slave holders was hegemonic, that is, influenced even the culture and perspective of other social strata decisively.[1] An important strain in that hegemonic thought was contempt for manual work (which *ought to* be the domain of slaves), extending even to contempt for the supervision of such work, with the one exception that a gentleman might legitimately engage in the management of his own *rural* estates.

Supervision and management of practical work, we remember, had been the *raison d'être* of the scribal profession. The hegemony of the slave-holders' perspective therefore had as its natural consequence that the scribal function stopped being culturally productive.

THE RISE OF PHILOSOPHY

From our present perspective, the decisive characteristic of the first, "pre-Socratic" period is the emergence of *philosophy as critical and systematic investigation, as organized curiosity*—not too far from the definition of "science" given above. The change in relation to earlier times is fittingly delineated by Aristotle in the following passage from the *Metaphysics* (981^b14–982^a1, trans. W. D. Ross 1928):

> At first he who invented any art whatever that went beyond the common perceptions of man was naturally admired by men, not only because there was something useful in the inventions, but because he was thought wise and superior to the rest. But as more arts were invented, and some were directed to the necessities of life, others to recreation, the inventors of the latter were naturally always regarded as wiser than the inventors of the former, because their branches of knowledge did not aim at utility. Hence when all such inventions were already established, the sciences which do not aim at giving pleasure or at the necessities of life were discovered, ...
>
> So . . . , the theoretical kinds of knowledge [are thought] to be more the nature of Wisdom than the productive

The knowledge of the first phase is the age-old technological wisdom of the human race, which in Aristotle's times had been degraded to being the cunning of the unfree (legally unfree, like slaves, or morally unfree, like those who had to work manually in the likeness of slaves and perhaps together with slaves). The distinctive knowledge of the second phase can be compared to the virtuosity of Old Babylonian scribal humanism, although its scope is somewhat broader. The third phase, finally, is that in which philosophy (*philo-sophia,* meaning "love of wisdom") emerged, as a quest for knowledge not conditioned by considerations of material utility.

Though not aiming at material utility, early Greek philosophy was still directed at the material world. Early sixth-century philosophy (embodied by Thales,

1. The term is borrowed from the Italian Marxist Antonio Gramsci, who uses it to describe the general acceptance of the ideology and perspective of the bourgeois class in capitalist society.

Anaximander, and Anaximenes, all from Miletus on the coast of Minor Asia) was *natural philosophy*, that is, it aimed at understanding the phenomena of the material world through the proper nature (*physis*) of things and not by reference to external divine forces. Thunder and lightning, rain and earthquakes—until then numinous manifestations of the Gods—could be explained as occurrences on a larger scale but not qualitatively different from what could be observed in the artisan's workshop. Through their *theoretical* investigations (from *theōréō*, to look at, to inspect, to regard) the natural philosophers thus brought the incomprehensible and awe-inspiring within the reach of human understanding. Nothing, on the other hand, suggests that the philosopher's results were brought back to the artisan. The aim of understanding the material world was not technological cunning or mastery—power in the Greek city state or *polis* was power over fellow beings, not technical command of nature. In terms of Habermas's scheme, the cognitive interest behind early Greek natural philosophy was emancipation, namely from traditional authority and from the authority of tradition, and nothing but that.[2]

Why? Was this a product of a particular Greek mind? Hardly. As observed above, the Mycenaean (that is, Greek) mid-second-millennium palace scribes had been unable to create a scribal culture similar to that of Babylonia. The best explanation of the "Greek miracle" appears to be sociological rather than racial. Seventh- to sixth-century Greece, indeed, happened to be at the crossroads of a multiplicity of mutually fecundating currents, tendencies, and opportunities.

One of these was the vigor of primitive-democratic institutions. Anybody reading Homer's *Iliad* (c. eighth century BCE) will discover the importance of the council

---

2. The one exception to this blunt statement is to be found in the field of medicine. The improvement of cures can of course be claimed to be a technical matter, and the fifth-century medical works of Hippocrates and his followers are indubitably in debt to contemporary natural philosophy (and *vice versa*). Yet the exception is not really one: the aim and result of "philosophical medicine" was not simply to cure the sick but quite as much to get them out of the grasp of the Asclepian medical temples and their priests; it was, furthermore, to understand the reasons of sickness on *natural terms*—whence the subject and title of one of the Hippocratic works, *Airs, Waters, Places,* and the oft-quoted insistence that "the sacred disease" epilepsy is not "any more divine or more sacred than other diseases, but has a natural cause" (trans. Jones 1972: II, 139).

Another variation on the pattern of natural philosophy is the writing of *critical history*, which took its beginnings in the late sixth and reached an early high point with Herodotos in the mid-fifth century BCE. Once again, the aim was to procure orientation in and reasoned understanding of the actual world.

Language shows that Herodotos and the Hippocratic writers themselves saw their undertaking as a continuation of the tradition that had been established by the natural philosophers (however much some of the medical writers criticized the tendency to borrow the dogmas of natural philosophy and neglect clinical observation): Although their own dialect is Doric, they write in Ionic, the dialect of the natural philosophers (Longrigg 1989: 6).

and the popular assembly. At closer investigation it turns out to be an ever-recurrent topic that the aristocrat truly worth that title possesses equal excellence in battle and in the assembly and earns his honour indiscriminately in both domains. The importance of rhetorical skill and argument is also made clear by the way prayers to the gods are formulated: not "Do *Z* in my favor" but "If I have done *X* for you [implying: which you cannot deny], then do *Z* in my favor."

Such conciliar institutions are not specifically Greek; we remember that Gilgameš consulted two similar bodies back in the early third millennium BCE, and historical anthropologists can point them out in many parts of the world. But in the Greek city states of the seventh to sixth century BCE they happened to be still alive at a moment when the Greek city states were plagued by endemic class struggles (not between masters and slaves but between rich and poor citizens) often evolving into open civil war, and when old aristocratic constitutions were replaced by democracy.[3]

Traditional primitive-democratic institutions are mostly organs of approval and nothing more. The Germanic aristocrat would make his speech, and the commoners would strike their shields with their swords to express consent—or two aristocrats would make different proposals, and the noise level would decide. In the sixth-century Greek assembly, on the other hand, truth was expected to be found "in the Middle," that is, to result from debate among equals in the open squares of the city, and to be decided neither in a closed council of "Elders" nor in the secret deliberations of the counsellors of the ruler, nor by a high priest (cf. Vernant 1982: 47, 53).

The application of reason in realms hitherto governed by tradition and authority thus took its beginning in the political domain. An older contemporary of Thales is the Athenian statesman Solon, who in the 590s gave Athens a new constitution constructed by reason and political insight and meant to check those abuses and instabilities which threatened to destroy the state from within. But the cutting edge of reason was soon taken over by the philosophers, who make clear their debt to the sphere of political discussions through the metaphors they use to describe their endeavor—among which was the idea that truth is found "in the Middle."

---

3. The class struggles resulted from the increasing role of monetary relations, loss of land on the part of the poorer citizens, and accelerating destabilization due to institutions like debt slavery. *Democracy* followed not from the mere goodwill of those in power, nor from popular rebellions alone. It had become a historical possibility through a seventh-century reform of military tactics which gave the key role to *hoplites,* heavily armed infantry. Hoplites were recruited from the stratum of moderately wealthy citizens who could afford the equipment, but who then also could (and did) claim influence in that *polis* which needed them. In fifth-century Athens, which built its military power on a fleet manned by penniless citizens, even these could, in a similar manner, claim their voice in the chapter, which provided the foundation for the radical democracy of Athens.

The happy collision between primitive and developed democracy provided the background and the occasion, and actual political life contributed the metaphors. The *tools* for understanding the world, however, had to come from elsewhere and were provided by another happy collision. Greece, still equipped with the institutions of primitive democracy, was the neighbor of civilizations which had lost them millennia ago but which, on the other hand, had accumulated technical and other knowledge during an even longer time span. The Greek city states were in intensive trading connection with these older civilizations; Greeks went as mercenaries to Egypt and elsewhere; and Greek cities and rulers employed foreign technicians when they needed them. The Greeks learned the alphabet from the Phoenicians, and in the process of adapting it to their own language (by introducing letters for vowels) they developed something much better suited to the diffusion of rudimentary literacy than the original Semitic invention was when transferred to a non-Semitic linguistic environment.[4]

Beyond intellectual tools, the foreign contacts also provided new questions (which the source civilizations had never asked themselves as "theoretical" questions). Questions of thunder and earthquakes and of the origin of the world had been asked in Greece before the advent of philosophy and answered at first in religious terms.[5] But the natural philosophers went further, asking also for the reasons underlying the practices which had been learned from abroad: *Why* do the methods of surveyors and accountants work? These methods were borrowed from Near Eastern and Egyptian scribal mathematicians. *Why* are the Sun and the Moon eclipsed at those points of the ecliptic where it happens? That these eclipses were subject to regularity was known to Babylonian astrologer-priests at least since the

4. Because of the particular character of the Semitic languages, understanding is not impeded by the omission of vowels (the reason that Arabic and Hebrew are mostly written without short vowels even nowadays) (cf. Strohmaier 1991). The introduction of vowels is thus *not* the reason that the Greek city states could bring forth philosophy whereas the Phoenician city states could not, as sometimes maintained. If other preconditions had been present, the lack of vowel writing would not have prevented the Phoenicians from creating the "miracle"—but without vowels, the communication network that came to carry *Greek* philosophy from the middle of the sixth century onward would have been inadequate for the task.

Even with vowels, writing was still a deficient tool as compared with what we know, lacking both word separation and punctuation. None the less, literacy was certainly important if not for the very first beginnings then for the further development of Greek thought. Although the reading ability of the majority of citizens remained severely restricted until the beginning of the fifth century (and in many places outside Athens probably considerably longer—see Harris 1989: 57–65), the very fact that writing was no longer the preserve of a closed professional group made it possible for *some* laymen (and the first philosophers were such laymen) to transform it into an instrument for their own undertakings.

5. For instance by Hesiod in his seventh-century *Theogony*—which, by the way, already borrows from Near Eastern sources, but from mythologies and epics.

seventh century. *Why* are the positions of planets so important for epidemics? The assumption that this was the case was part of the ground for Babylonian astrology.

A final coincidence was needed to make Greece come out of the collision alive. World history is full of confrontations between developed civilizations and cultures on an organizational level comparable to that of Homeric Greece. As a rule the latter cultures have been crushed. Greece, like eleventh- to twelfth-century Western Europe and nineteenth-century Japan, was militarily and economically vigorous enough to survive the confrontation and to incorporate foreign elements into its own culture as inspiration. But whereas Western Europe and Japan adopted the foreign inspiration into a relatively stable structure, Greece received it while itself in the midst of a deep-going transformation. This quadruple coincidence appears to be the ultimate root of the "Greek miracle."

Such grand explanations can always be subjected to doubt. Whichever its roots, however, Greek philosophy began as natural philosophy and as a kind of *enlightenment* aiming at emancipation from the authority of tradition. Thereby it had defined the battleground; since this battleground was the politically decisive city square, the partisans of aristocratic values and of tradition were forced to take up the challenge and to formulate their points of view *as philosophy*—a process not unlike the late eighteenth-century appearance of the Counter-Enlightenment and of Conservatism as an explicit philosophy as attacks on the French revolution and progressive political currents in terms that were congruent with Enlightenment philosophy. The very first person to have adopted the title of "philosopher" may actually be characterized as a *counter-philosopher* in an analogous sense, namely Pythagoras (fl. c. 530 BCE to c. 500 BCE).

Because Pythagoras was regarded in Hellenistic and late ancient philosophy as the Prophet of True Wisdom, we possess a large stock of reports and legends on his life and opinions—and for precisely the same reason it is often impossible to distinguish what may remain of reliable report in the heap of legends. In spite of the prevailing opinion in popular histories of philosophy and science, it seems most safe to disregard the incoherent accounts of Pythagoras the mathematician and Pythagoras the experimental physicist. He appears to have been much more of a guru, a spiritualized teacher, than a "scientist" or philosopher. His fame as a mathematician and scientist seems to be due to a predilection (shared with Ron Hubbard and other neoreligious gurus of our own times) for using as much as he understood of contemporary philosophy and mathematics to illustrate and support his teachings and to impress the public.[6] Like many recent gurus, he also founded a brotherhood

6. Whereas the appraisal of Pythagoras as a shamanic figure can be found in part of the general literature (in particular Burkert 1962, English translation Burkert 1972), the latter statement will not be found in standard textbooks on the history of Greek science and philosophy. I build it partly on a fragment of an early comedy presupposing general knowledge

where his doctrines were taught: for Pythagoras and his followers, truth was certainly not to be found, and not even to be divulged "in the Middle." On the contrary, the initiates of the brotherhood were to use their acquaintance with True Wisdom to grasp or keep political power in their cities (eventually, revolutions in most of those Greek cities in Southern Italy where the order had its strongholds put a violent end to its political dominion).

The predilection for knowledge that was not accessible to everybody led one branch of the Pythagorean order to concentrate its teachings on mathematics—*mathemata,* like Latin *doctrinae,* means "matters to be taught"—and at some point in the later fifth century BCE these so-called "mathematikoi" had created a curriculum based on four *mathematical arts,* which was to remain an important ingredient of any European higher education for 2000 years:

1. Arithmetic
2. Geometry
3. Astronomy
4. Music

The first two subjects are self-explanatory—if only one remembers that arithmetic has as little to do with practical computation as geometry with the measurement of land and distances. Astronomy was first of all a mathematical theory of how the sacred heavenly system *ought* to behave; music was the mathematical theory of musical harmony (considered a model for social harmony, based as the social order should be on correct proportions and not on equality).

Pythagoreanism was apparently a morally-politically motivated reaction against the implications of natural philosophy, but a reaction which at the same time stuck to the central themes of natural philosophy. Mathematics was seen by the Pythagoreans as a way to gain *better* knowledge of the material world than could be obtained by everybody by means of the senses; *number,* to the Pythagoreans, was not *superior to* material reality but *the essence of* the material world. Another countercurrent—carried by figures like Parmenides (possibly c. 515 to c. 450 BCE) and Zeno (c. 490 to c. 425 BCE), both from Elea in Southern Italy—may have been less directly motivated by politics while at the same time differentiating itself more thoroughly from the tenets of natural philosophy. Distinguishing sharply between the thought of the intellect and that reality which is experienced by the senses, the Eleatic thinkers claimed that sensual experience is fallible and even misleading, and the intellect thus the only source for certain knowledge; more than any other philosophical current in history, they deserve the label "rationalists." Movement and

---

of supposedly secret Pythagorean mathematical teachings, partly on evidence from a medieval Arabic book catalogue that Pythagoras's most glorified discoveries in musical theory may have been committed to writing by the musician and composer Sacadas already around the time when Pythagoras was born (Høyrup 1994: 285 n.36).

change, they argued, were illusions, and by a series of famous thought experiments[7] Zeno set out to prove that movement was logically impossible, all common sense experience notwithstanding. Whereas the Counter-Enlightenment of the 1790s declared reason impotent or dangerous (cf. below, p. 144), the Pythagorean and Eleatic reactions to natural philosophy thus exalted it—but in a way which had the same ultimate consequence: the futility of the reason of everyday and every enlightened person, that reason which is the foundation of any enlightenment venture.

Natural philosophy, Pythagoreanism, and Eleatic rationalism were all important for the formation of European philosophy, and thus for the humanities. But were they themselves constituents of a "humanistic community"?

Yes and no. In a loose sense it is possible to speak of "schools" around the philosophers; Zeno, for instance, was a follower of Parmenides. But the usage is only permissible in the loose sense which corresponds to the Greek expression "those around Parmenides." The philosophers were in mutual, critical communication, and they were of course specialized as intellectuals. But the philosophical environment was quite loosely knit, too loosely perhaps to justify the term *community*.

The Pythagorean order was certainly more firm in structure, but probably too firm to be reckoned a primarily *intellectual* community—and, in particular, religious and political in orientation rather than directly intellectual. Again, the differences probably outweigh the similarities.

At a less ambitious intellectual level one might perhaps expect a more positive answer. After all, a system of general education for free boys seems to have existed. On the other hand, the ambitions of the curriculum were probably too low to allow us to speak of more than rudiments of a humanistic environment: gymnastics, meant as preparatory military training; spelling and counting; and *music,* meant in this context as "the domain of the Muses," that is, recitation of religious hymns and Homeric passages meant to strengthen the character and to convey a moral message.

FROM THE SOPHISTS TO ARISTOTLE

Around the middle of the fifth century, the new social system of the *polis* had become (relatively!) stable, and the roles of rich and poor within the social fabric had become institutionalized and more or less a habit. This is especially true of Athens, where even the different political roles of the different social classes and

7. One of them is the "paradox of Achilles and the tortoise," another one the "arrow paradox." The latter goes as follows: Think of an arrow flying from point *A* to point *B*. Before reaching *B* it must pass through the mid-point *B′*, and before that through the mid-point *B′′* between *A* and *B′*. This argument can be continued ad libitum. Before getting anywhere at all the arrow must thus go through a process consisting of an infinite number of steps, which is "obviously" impossible. Therefore the arrow will get nowhere.

groups had been organized within an astonishingly well-performing direct democracy.[8] Thanks in part to this, in part to the wealth and political preeminence of Athens between the Persian and the Peloponnesian wars (500–479 and 431–404 BCE, respectively), Athens became the intellectual meeting place of Greece and the focus of new developments.

One new development was a "technical" utilization of philosophy, brought about by the "Sophists." This did not involve any use of the insights of natural philosophy in the productive sphere—natural philosophy had neither sought nor produced any insights fit for that. In agreement with the principle that "power in the Greek city state or *polis* was power over fellow beings, not technical command of nature," *use* was *political use.*

The problem which the Sophists helped to solve was the common problem of economic elites in a democracy: how to safeguard also the position as a *political* elite. In Greek direct democracy, the road to political influence went through the popular assembly. What you could do with your money was (apart from supporting or favoring your clients, thus making them vote as you wanted) to buy an education that would permit you to perform well on this stage. That meant, firstly, that you needed rhetorical skill—political life was oral; secondly, that you had to be versed in the favorite themes of philosophy, philosophy having developed in close interaction with the political stage, as a sort of meta-political discourse; and thirdly, that you had to be fairly well versed in that traditional literature (first of all Homer) which was part of the upbringing of every citizen and hence common cultural heritage.

The Sophists were thus professional intellectuals who made a living from *teaching what upper class youth needed* in order to achieve political success. In its beginnings, the Sophist movement descended directly from natural philosophy: Gorgias, one of its greatest early names,[9] was a disciple of Empedocles, who devised the theory of the Four Elements; the Sophists were thus the first to live *off, not for* philosophy,

---

8. Democracy was established in Athens in 509–507 BCE after a period of tyrannic rule. Truly, the abuses of Athenian democracy has been a favorite theme for anti-democrats from Plato onwards. As pointed out by the American historian Moses Finley (1973), however, few political regimes in history, democratic or authoritarian, have managed to make so few political blunders per decennium as the Athenian Popular Assembly.

Evidently, Athenian democracy was *democracy for the citizens,* which means that it did not include slaves, women, children, and foreigners, and for a while Athens built up an empire which it treated as its legitimate sphere of influence. This nonobservance of twentieth-century political ideals, however, changes nothing in the fact that the Athenian political system performed well, both as regards efficiency and concerning compliance with its own established moral norms.

9. So great, indeed, that Plato spares him the direct attacks in the dialogue carrying his name, using instead minor or caricatured followers as scapegoats when he is to demonstrate the mutual discordance between Gorgias's two claims: that rhetoric is technically effective, and that perfection in rhetoric is conducive to moral perfection.

to borrow Arthur Schopenhauer's sarcasm. The alliance between natural philosophy and the preparation for a political career on the conditions of democracy was no accident. Already a generation or so before the arrival of the Sophists, Pericles's acquaintance with Anaxagoras and his familiarity with Anaxagoras's natural philosophy enhanced the impetus of his political career—so it is told at least by Plato's Socrates, in a discussion of the rhetorical and persuasive art of this father of radical Athenian democracy:[10]

> All great arts demand discussion and high speculation about nature; for this loftiness of mind and effectiveness in all directions seem somehow to come from such pursuits. This was in Pericles added to his great natural abilities; for it was, I think, his falling in with Anaxagoras, who was just such a man, that filled him with high thoughts and taught him the nature of mind and of lack of mind, subjects about which Anaxagoras used chiefly to discourse, and from these speculations he drew and applied to the art of speaking what is of use to it.
>
> (Plato, *Phaedrus* 269e–270a, trans. Fowler 1977: 547)

In its beginnings, the Sophist movement also presupposed that truth is to be found "in the Middle": Protagoras, the other main character of fifth-century Sophism, is famous for having formulated that "man is the measure of all things"—that is, moral truth derives from neither tradition, nor authority, nor religion, but only from human utility and free decision.

Around the end of the fifth century, the need to bolster up political aspirations with familiarity with natural philosophy seems to have vanished.[11] As Sophist teaching crystallized, the program therefore came to consist of these three parts:

1. Grammar
2. Rhetoric
3. Dialectic

"Grammar" can be explained as *rules for correct and tools for good use of language.* This includes what the term may mean for us ("correct use"), but also knowledge of

---

10. Plutarch, when discussing the same matter, sees the utility of philosophy more technically; Pericles "often made an auxiliary string of Anaxagoras, subtly mingling, as it were, with his rhetoric the dye of natural science" (*Lives* 156,1, trans. Perrin 1967: 21).

11. One reason that the tie between natural philosophy and politics was torn will have been Socrates's influence (cf. below). But Aristophanes's comedy *The Clouds* from 423 BCE, where he ridicules Socrates as a typical Sophist proponent of natural philosophy, shows that Socrates was not the only one in his times to find it socially and morally irrelevant whether wasps produce their buzzing with one or the other end of their body. It even suggests that he was not the first; if the comedy could be expected to be funny, it will have had to portray Socrates in a period of his life where he still believed that natural philosophy contained a clue to existential questions (as he claims once to have hoped, if we are to believe Plato's version of his *Apology*).

literature, in particular poetry ("fitting and agreeable use"); even "correct use" was trained pragmatically with reference to literature—schemes and grammatical systems were inventions of the Hellenistic age. "Dialectic" derives from "dialogue," and is thus the art of persuasion in discussion; the term was used with somewhat changing meanings during antiquity and the Middle Ages but mostly as a near-equivalent of *logic.*[12] But as it is suggested by our own term "sophistry," the dialectical art taught by the Sophists would also embrace the use of pseudo-logical fallacies and tricks, notwithstanding Gorgias's ideals but in agreement with the needs of the customers—who pays the piper chooses the tune.

This well-known principle is also reflected in a reinterpretation of Protagoras's maxim that was made by his followers, at least if we are to believe Plato's polemical but not implausible portraits. Protagoras's critical utilitarianism, speaking in the abstract of "man," had presupposed a shared and general human measure of what is good; when exposed to the realities of political life and leaving to the individual to decide the measure, it changed into moral relativism or nihilism: that which is *good for the strongest* is *good simpliciter.*

In the nineteenth century, this consequence of naïve moral relativism was summed up by Dostoevsky in the statement that "if God does not exist, then everything is permitted." In the fifth century BCE, Socrates (c. 470 to 399 BCE) reacted to the moral breakdown inherent in and represented by the Sophist teachings as Dostoevsky would later do. Socrates lived for, not off, philosophy, and he changed the course of philosophy radically and lastingly.[13] In his youth he had been interested in natural philosophy, believing to find there the answers to essential questions concerning the question of *the good life* (no wonder, indeed, if we recall the enlightenment role of natural philosophy). But natural philosophy did not meet his expectations—its theories about what is were, and could only be, irrelevant to the essential questions, which in Socrates's view were *practical.*[14] Since the roundabout

---

12. To Plato, dialectic was the supreme stage of philosophy, leading to true knowledge; but already Aristotle used it as the kind of logical reasoning which merely "aims at gaining credence" (*Analytica Posteriora* $81^{b}17$), and opposes it to *analytic,* that reasoning from necessary truths which characterizes the sciences (cf. below, p. 42).

13. A. N. Whitehead has claimed all Western philosophy to be but a series of footnotes to Plato. Plato, however, presents most of his works as footnotes to Socrates, and is indeed our main source for Socrates's teachings (even though Xenophon and Aristotle say enough about him to permit us to extricate an approximate picture of the historical Socrates from the diverse distortions of the three authors; Socrates himself never wrote a line). Allowing for obvious exaggeration on the part of Whitehead and for literary stratagems on that of Plato, we may still conclude that the word *lasting* understates Socrates's influence.

14. The term *practical* as used in philosophical contexts is easily misunderstood if interpreted from common usage. The word is borrowed from Greek *praxis,* which originally means

way over natural philosophy had shown itself to be nothing but a dead end, the practical questions had to be tackled directly. This was hence what Socrates attempted to do in his famous, obstinate discussions with whoever dared expose his own inveterate conventional wisdom to Socrates's questions: what is VIRTUE in the absolute, what is THE GOOD, what is THE BEAUTIFUL? The answers, according to Socrates, could not depend on subjective and arbitrary choice, as claimed by the Sophists. Nor could they any longer be taken over from a tradition whose credibility had been undermined both by the Sophists and by tragic authors like Euripides.[15] The investigation had to start from scratch, from the principle of "Socratic doubt": "The only thing I know is that I know nothing." Until his execution Socrates seems not to have attained any conclusive results beyond this, but to all evidence his way to enforce upon others the same awareness of ultimate ignorance was considered far from ineffective by those whom it made his enemies.

Socrates declared (in Plato's version of his *Apology*, 33a–b) never to have been the teacher of anybody. A teacher, in Socrates's opinion, poured ready-made doctrines and opinions into the defenseless minds of his private, paying disciples (*ibid.* 33b; *Protagoras* 314b). Yet through the questions and advice offered publicly to rich and poor alike he had certainly made Plato (c. 428 to 348/47 BCE) as much of a disciple as anybody has ever been. Plato continued Socrates's quest for absolute moral knowledge, not only with the aim of *knowing*, however, but as part of an effort to improve society morally through education.

Plato's early works still reflect the global doubt that had been Socrates's concluding position. Later on, however, he developed that philosophical doctrine

---

"doing," "transaction," "business," "action," or "dexterity," etc., but which in Greek philosophy took on the specialized sense of that *action in itself* which fits the free citizen (in contrast to *materially productive* or *technical* action, ideally the chore of the unfree). *Practical philosophy* therefore means analysis of the question *how one should act* absolutely, in contrast to the instrumental (and nonphilosophical) question "how should one act in order to attain a given goal effectively."

Practical philosophy, encompassing *moral* and *political* philosophy, is thereby one of the main branches of philosophy—the others being *epistemology* (how we attain knowledge, and what is the status of knowledge); *aesthetic philosophy;* and, in some enumerations, *philosophy of religion.*

15. Even though he was the first to submit the ethical questions to philosophical inquiry, Socrates was indeed not the first to raise them publicly and critically; nor were Sophists like Protagoras (whose maxim can hardly be labelled "philosophical inquiry"). The precedence goes to literature, more precisely to the dramatists from Aeschylos to Euripides and Aristophanes. As Socrates's greatness can be attributed to his ability to formulate the urgent dilemmas of his times (most of his mature activity falls within the thirty years of the Peloponnesian war between Athens and Sparta) in a way which was open-ended enough to express and interpret similar dilemmas in later ages, so can much of the majesty of the tragic authors.

("Platonism" of course, or the "doctrine of forms/ideas"[16]) which was hinted at in the anecdote about Plato, Diogenes, and Cuphood (p. 8) and which was meant to solve Socrates's problem.

Starting points for Plato's doctrine were provided by the Pythagorean mathematical and the Eleatic rationalist traditions. Unlike the Pythagoreans, Plato did not claim that supreme reality was mathematical (nor, as the Pythagorean *mathematikoi* may possibly have done, see mathematics as knowledge that should be kept secret). In Plato's view, mathematics served as a symbol and as a preparation of the intellect for the perception of paramount truth. According to Plato's understanding, all the dissimilar, more or less imperfect triangles which we can draw in the sand, with sides never quite rectilinear and never infinitely narrow, are only deficient reproductions of the *ideal triangle,* the *triangle in itself* or the *form* of the triangle ("Trianglehood," in the idiom of the anecdote). The theorem that the sum of the angles of the triangle equals two right angles never holds perfectly for our drawn triangles, and it should not. It holds for the triangle in itself, in the world of ideas.

THE BEAUTIFUL, still according to Plato, is also a form, and the many beautiful things in this material world—temples, statues, young boys, musical harmonies—are only imperfect reflections of the form or idea. When we enjoy the naked boys in the stadium or the sculptures of Phidias, it is therefore not through a process of abstraction that we derive a concept (or an "idea" in the modern, subjective sense) of beauty, as modern thinking would mostly have it, or by extracting the quality shared by all the beautiful things (according to Aristotle, this was Socrates's opinion). The reason is instead that our own minds originated as parts of the Universal Intellect, like sparks from a bonfire. The imperfect beauty of boys and sculptures therefore *reminds us* of what we once knew, before our material incarnation.

The doctrine of forms was no piece of pure epistemology to Plato—no "academic pursuit," in modern terms. In his large dialogue on the construction of the perfect state (*The Republic*) Plato delineates a grand educational program which will allow the best minds of the state to be led to perception of THE GOOD—the supreme form from which all other forms derive, and knowledge of which will solve the ethical problem; having attained that stage of wisdom these true philosophers will have to take on the task of governing the state, since they—and they alone—are capable of discerning which course of action must be chosen as THE GOOD POLICY.

The Republic, of course, is a piece of utopian writing. As a beginning, however, Plato organized an actual educational institution, located close to and borrowing

16. *Idea* corresponds to Greek *idea,* originally "shape," "appearance," "sort"; *form* translates Greek *eidos,* originally "shape," "figure," "sort." Plato used the words without distinction, but with a certain preference for *idea,* for which reason the Platonic notion is habitually referred to as the "idea." Aristotle preferred to speak of his own concept as *eidos.* When Plato's usage is referred to, it should be remembered that his *idea* is not subjective but supremely objective reality.

the name of an athletic ground ("the Academy") much frequented by adolescents. Here, firstly, he used the opportunity offered by the location to get into contact with Athenian youth; this was the trick of the location. Secondly, mature philosophers and mathematicians worked there "making their investigations in common."[17] (Plato's Academy can thus be claimed to be the prototype of the modern university institution with its association of teaching and research).

Plato's philosophy was hence part of a broader political, moral, and educational program. It reflected Plato's dismay at the indubitable crisis of the city state of his time. In Plato's later years, Macedonia under Philip II had already begun the conquest of Athens's allies; thirteen years after Plato's death, Alexander the Great had subdued all city states in mainland Greece. According to Plato's diagnosis, the root of the crisis was the individualistic egoism of the rulers—be they the common people as in the democracies, the rich and powerful in oligarchic states, or a single ruthless individual (a "tyrant"). Paradoxical as it may seem, Plato attempted to save the city state—if anything the state of the citizens—from the blindness of these citizens by transforming them into citizens subject to the state.

> There is a striking parallel—almost certainly not fully accidental—between this subordination of all individuals under the state (as a theoretical construct) and the idea that all "cups" are nothing but defective copies of a "Cuphood" (no less a theoretical construct). No accident, either, that conservative ideologies (emphasizing the primacy of the state and existing social structures over merely ephemeral individuals) have mostly been supported by some sort of objective idealism, and that Stalinist policies were accompanied by a corresponding transfiguration of Marxist philosophy (as pointed out by Sartre in his *Critique de la raison dialectique*—cf. p. 217).
>
> Whereas Plato for the most part has been the idol of European philosophy from the Renaissance onwards, some of the "footnotes" (especially those of liberal but also of many Marxist philosophers and historians of ideas) are strongly critical and negative. Plato is one of the main villains, for instance, in Karl Popper's *The Open Society and its Enemies.*

Thanks to his literary gifts, Plato succeeded in discrediting the Sophists for all times to come. But literary gifts alone are not sufficient to change the real needs of a time. Plato was enough of a genius to go beyond immediate needs, and even enough to inspire others (not only Aristotle but also all of the major mathematicians of the fourth century BCE participated, for a shorter or longer period, in the research at the Academy). But his genius was also sufficient (and of a sort) to permit

17. As told by Proclos Diadochos (c. 410 to 485 CE), one of the last heads of the Academy, which was closed in 529 by the Byzantine Emperor Justinian (In *Primum Euclidis . . . commentarii,* 67).

him to propose strategies and solutions without foundation in real life.[18] Not only were his plans to save the city state therefore without consequence; so was his attempt at moral rearmament through education, of which only some formal structures survived. This can be seen in the "cycle of Liberal Arts" that came, from Plato's mature years onwards, to constitute the canon of good education for the free citizen belonging to the well-to-do and hence cultured classes:[19]

1. Grammar
2. Rhetoric
3. Dialectic
4. Arithmetic
5. Geometry
6. Astronomy
7. Music

The first three (later to be known as the "trivium"), we notice, are identical with the core of Sophist education. The presence of the last four (the "quadrivium" in later ages) may be due to Plato's influence; but their root may also be directly Pythagorean or, more likely, have been inspired by some Sophists (several of these tried to prove their omniscience by muddling with modish geometric problems like the squaring of the circle).

Though devoid of the moral messages which had been intended by Plato, the cycle of Liberal Arts still looks quite encyclopedic.[20] But the all-round character is delusive: in practice only the arts of the trivium were dealt with seriously,[21] and in real depth only grammar and rhetoric. Average education for the upper classes was thus, from the time it was generalized, *almost exclusively a literary education.* Moreover, since this development took place at a time when city state democracy was on the wane (soon to disappear altogether), rhetoric no longer aimed at participation in political life but at the law-courts, or it was simply an *art pour l'art.* In practice, education in the Liberal Arts was thus not only purely humanistic but also

---

18. It may thus be due to political common sense quite as much as to tolerance that democratic Athens never tried to disturb Plato's attempt to overthrow its institutions through education—the danger to social order presented by oppositional scholars and *literati* depends critically on their ability to hit the frequencies of the social resonator.

19. These arts are "liberal," that is, "free" (Greek *eleutheros*) in the sense that they were thought suitable for the free citizen, in contradistinction to the "mechanical arts" used in material production.

20. In fact, the term *encyclopedic* derives from the Greek name for the cycle (*enkyklios paedeia*).

21. This is obvious from some of the introductory compendia in the quadrivial arts for students in the "university level" philosophical schools. They must be presumed to start at the level attained by the brighter pupils from the "high school" level of Liberal Arts (only these would get the idea to attend a philosophical school)—and they start from scratch.

solely directed at procuring *cultural polish,* and hence that self-assurance which comes from being part of the right culture and possessing the best cultural values. There is, of course, a direct line backward from this system of "secondary education" to the *music* of traditional general education (meaning "the domain of the Muses," we remember, and encompassing recitation of religious hymns and Homeric passages). Precisely this backward connection, identifying the Liberal Arts (= grammar + rhetoric) as a qualitatively higher level of the acknowledged right thing, is what would make the new system a support for upper-class class consciousness.[22] Habermas's identification of the humanities with legitimization of existing culture and of the existing social order, therefore, has its roots back in the fourth century BCE—not in Plato's program, however reactionary some will find him to be,[23] but in his failure.

Philosophy itself can be claimed to have undergone a similar change at the same time as it reached a high point not to be surpassed for very long time. This culmination was brought about by Aristotle (384 to 322 BCE), a one-time fellow in Plato's Academy and later educator of Alexander the Great (who, as is known, did not share his teacher's view on the pre-eminence of the city state). Aristotle created a complete system encompassing not only almost every subject which had been treated before him in Greek philosophy but also some where he was the absolute pioneer: a *Metaphysics* where he comes to grips with the doctrine of forms, with Eleatic rationalism, with the Pythagorean belief that everything is numbers, with the problem of causality, and other core problems of existing philosophy, while at the same time writing a history of philosophy[24]; several large works on epistemology, logic, and Sophist dialectic; many more on natural philosophy, beginning with the

22. Maybe not quite without intention. The Greek historian Polybios, who passed seventeen years as a hostage in Rome around 150 BCE, tells with enthusiasm of the skill with which the Roman ruling classes used superstition to keep in check the lower classes: "as every multitude is fickle, full of lawless desires, unreasoned passion and violent anger, the multitude must be held in by invisible terrors and suchlike pageantry. For this reason I think, not that the ancients acted rashly and at haphazard in introducing among the people notions concerning the gods and beliefs in the terrors of Hell, but that the moderns are most rash and foolish in banishing such beliefs" (*Histories* VI,56, trans. Paton 1922: III, 395). Forging and using ideology intentionally was nothing to be ashamed of for the ancient upper classes.

23. Plato, indeed, was utterly distrustful of traditional "Muses' music" as a part of education. So he was in general of poetry: firstly, it was normally immoral (just think of the behavior of the Gods in the Homeric epics: betraying, fornicating, and the Gods know what!); secondly and worse, poetry was qua imitation of (material) real life one step further remote from the higher reality of ideas than material reality itself. An irony of history, then, that Plato's own fine literary style has made him the favorite philosopher of beaux-esprits since Petrarca.

24. To be precise, Aristotle never wrote the *Metaphysics* as one work, but a number of treatises on these topics which some hundred years later were combined into one work by commentators (see Randall 1960: 107–109).

theory of change and motion (the *Physics*) and the description of the heavenly system, and continuing with a many-volume natural history largely built on his own observations (Aristotle was the son of a physician) and with works "on the soul" ("psychology"); and finally several works on ethics, political theory, rhetoric, and poetics. Among the subjects on the periphery of philosophy, mathematics and medicine proper are lacking, and history is only represented in the surviving corpus by an investigation of the constitutional history of Athens, serving as part of the empirical foundation for the treatise on political theory (many studies of other states by his disciples have been lost).

A point which was decisive in making Aristotle's philosophy a better basis for scientific thinking than Platonism was his transformation of the doctrine of forms. Whereas Plato had been an "extreme realist," claiming that the forms had an independent existence and a more real existence than the individual instances representing them, Aristotle was a "moderate realist": forms exist, it is true, and a shared form is what brings a number of individual instances (for example all human beings) together as members of one species (in casu, humankind); but forms exist only through their participation in individuals, just as the form of a statue only possesses real existence as the form (in the everyday sense) imposed upon the bronze; families of forms, on their part, may share features, through which species (such as humankind, catkind, horsekind, and crocodilekind) are brought together in a common genus (*in casu,* animals).[25] It is the task of each particular science to induce from

---

25. This *hierarchy of forms* is only possible in moderate realism where, in some sense, forms are *shared features.* The geometric example of the triangle demonstrates the dilemma of Platonism and of extreme realism in general: how are we to explain that everything which holds for THE TRIANGLE also *by necessity* holds for THE RIGHT TRIANGLE if these two are independently existing entities (more on this on p. 259, n. 4)? No wonder that Plato preferred to take as his example THE CIRCLE, which allows no such subdivision.

If one finds it difficult to grasp what one or the other variant of the doctrine of forms has to do with the possibility of actual research, one may borrow an oft-used analogy referring to more recent science: is *gravitation* nothing but a way to abbreviate our records of the movement of bodies which is convenient but which refers to nothing real (so to speak Diogenes's point of view)? Does gravitation exist in isolation (the "Platonic" persuasion)? Or is it something real, but only as a quality possessed by material bodies in interaction (the "Aristotelian" view, to which most physicists will adhere if only for the psychological reason that it is difficult to engage seriously in the study of something in which you don't believe)?

In one respect Aristotle's conception of the form was different from the pattern in which we would think today. Not every acorn will develop the perfect form of the oak tree—if growing under unkind conditions, it might be crippled. Rather than being the mere set of shared properties of all oak trees, the Aristotelian form is something toward which the acorn will develop if its growth is not obstructed; it is the *final cause* for the development, that which is best for an oak tree and toward which its development strives. In that way the form is a version of THE GOOD, but a specific GOOD for oaks.

experience the forms of the entities falling within its domain (in modern translation, the laws governing the domain) and to make its further conclusions from these *necessary truths.* Poetics has to know what distinguishes a tragedy, namely the arousal of pity and fear (*De poetica* 1453^{b}12); politics as a science has to start out from the form of man, "a political animal" (an animal that should live in and be a citizen of a *polis*), and deduce from this starting point the correct form of the city state. Below the level of shared metaphysical principles (not least the doctrine of forms and the types of causation), knowledge is thus compartmentalized, each discipline dealing with its specific subject-matter according to its own distinctive principles.

But it was also an important characteristic of Aristotle's scientific thinking (positively as well as negatively) that he never stuck dogmatically to these metaphysical doctrines. When absolute certainty about forms could not be attained (and it rarely could, even according to Aristotle's own standards), he would still try to find out as much as possible about the single fields from experience submitted to educated common sense, and he would tell when the outcome did not allow any firm conclusion; when strict application of the metaphysical principles would lead to conclusions which contradicted his own common sense too strongly, he would introduce *ad hoc* modifications (cf. n. 27).

Aristotle's system is impressive, and it is no wonder that he came to be known simply as "the Philosopher" from the thirteenth through the sixteenth centuries CE. It marks the emergence of professional scholarly philosophy, and the point where a general scientific enterprise can be spoken of with some reason. But it also marks the final retreat of philosophy from its enlightenment pretensions. Philosophy (like literature, rhetoric, and mathematics) had become a field for educated leisure,[26] where the gifted dilettante (if gifted and hard-working enough) might produce scholarly works of high quality without endangering the social "peace for the palaces" (and hardly that of the Christian Church when it eventually rose to power).[27] The

---

26. Cf. the quotation from the *Metaphysics* in the beginning of this chapter.

27. The disappearance of critical political aims is confirmed by a paradoxical conflict between Aristotle's metaphysics and natural philosophy and his political theory. According to his variant of the doctrine of forms, all individual representatives of a species share *the same* nature or form. This could easily undermine the moral legitimation of slavery, since everybody (included Aristotle) would agree that slaves were human beings, and thus should be "political animals" who ought to live in and be citizens of a constitutional *polis.* In order to avoid such unhealthy inferences, Aristotle introduced the pragmatic *ad-hoc* postulate that slaves possess a specific nature through which their servile role is predetermined (for instance, *Politica* 1254^{b}20, ed. Jowett 1921; cf. Jürß & Ehlers 1982: 38). He chooses this more radical solution even though his general system would have allowed him to claim that those who grew up under the unfavorable conditions of a despotic or barbarian society were impeded from developing the full human nature as "political animals" (cf. n. 25).

motive force may still be characterized as "organized curiosity," but legitimization (of the social preeminence of the leisured class) had taken the place of emancipation.

THE EPOCH OF HELLENIZATION

Christian power, of course, was still well below the horizon when Alexander the Great and Aristotle died. Palaces, on the other hand, were not. Alexander had conquered the world from Indus to the Libyan Desert and the Adriatic. Within a few years after his death his leading generals would divide this empire among themselves; in the west, Carthage and Rome were soon to begin their fight for supremacy, with the well-known end result that the Roman Empire came to reach from the Scottish to the Iranian border and from Vienna to Nubia. This whole region was soon Hellenized, in the sense that upper- and middle-class culture imitated the Greek model as far as possible. But Hellenization reached still further geographically, as may be illustrated by the death of the Roman politician and general Crassus. In 53 BCE his troops were defeated, and Crassus was killed by the Parthian army (whose capital was in present-day Afghanistan). The Parthians carried off his severed head as a trophy—and eventually, as Plutarch tells, the head ended up as a requisite in a local private performance of Euripides's tragedy *Bacchae.*

Not least in post-Alexandrian Egypt, the adaptation of Hellenization to local culture (and vice versa) was a deliberate policy of the Greek conquerors; in Rome it was no less a policy of the indigenous elite. In both cases the policy certainly had to overcome much greater obstacles than in the initiation of Athenian upper-class adolescents to the choice products of their own mother culture. In several places, therefore, specialized institutions were erected in order to overcome the obstacles, the most famous and prestigious of which are the *Museum* (another term derived from the Muses) and the Library in Alexandria. Here, among other things, literary scholarship took its beginning with the establishment of critical editions of the literary classics, which had attained the status of classics precisely by being transferred from the place where they were living culture to one where they had to be studied.[28] In the Alexandrian and similar institutions, philosophy also changed its character, namely from being the *study of the moral or natural world* or of the *conditions for knowing* the world, to *the study of the doctrines of the great (whence classical) philosophers* on these subjects.

---

28. Paradoxically, we know the Hippocratic medical writings much better than the (much more knowledgeable) Alexandrian physicians: the Hippocratic writings were classics and dutifully collected and copied as such; works written in Alexandria were not, and were therefore lost, apart from quotations and references in treatises from the Roman age—not least those of the second-century physician Galen, who became the paramount medical authority of the subsequent centuries and throughout the Middle Ages, and whose works were therefore also conserved.

Through the establishment of textual criticism, an important branch of humanistic studies can be rightfully claimed to have reached a scientific level. The problems of getting access to the literature of a foreign language (and, in Rome, to develop a corresponding literature in one's own language) were spurs for the development of more sophisticated studies of grammar and semantics. The general *use* of the humanities, however, followed the trend established in fourth-century Athens throughout antiquity: lip service to the complete cycle of Liberal Arts and in practice restriction to grammar and rhetoric (including as much knowledge of literature and philosophy as was needed to give color and substance to your speeches) together with some dialectic. Globally, humanistic culture remained legitimizing when not subservient to the techniques of rhetorical persuasion.[29] What was legitimized, moreover, was power and not mere status: discussing the situation in the Roman Empire of the third to fourth centuries CE, Peter Brown (1992:37) observes that

> in every major province with which the imperial administration had to deal, its representatives met a group of persons who claimed, on the basis of their high culture [namely, training in grammar and rhetoric], to be the natural leaders of society. . . . The few cases of successful careers by lowborn, uneducated persons in the higher reaches of bureaucracy . . . should not blind us to the cumulative success of educated civic notables in obtaining governorships and higher offices. . . .

In one important respect, the situation of humanistic teaching in the Hellenistic empires and Rome was different from that of the beginnings in fourth-century Athens: the scale and degree of professionalization. State interest in the spread of culture led to public employment of teachers (*professores*) in the Liberal Arts (carefully pointed out in Roman Law to be an honorable business, in contradistinction to the teachers of vulgar useful crafts like practical calculation), whereas the most

---

29. It is thus anything but an accident that the main Latin work on education (at least until St. Augustine) is Quintilian's *Institutio Oratoria,* the *Teaching of the Art of Speaking* (first century CE), where the preeminence of rhetoric over all other subjects is proclaimed (whence also that ethics is a subdiscipline of the art of speaking rather than a branch of philosophy—I, Preface, 10–11).

The lack of substance in the teaching of the quadrivial subjects is illustrated by Quintilian's explanation of the utility of geometry for rhetoric (I.x,34–37, ed. Butler 1920: I, 174–179).Firstly, the subject is told to deal *both* with numerical computation and with figures. Numbers are necessary because it makes a bad impression when the speaker "fumbles with his fingers in a way which contradicts the calculation which he states in words." (Roman calculators had a complex system for representing numbers by variously bent fingers). Geometrical figures are needed because lawsuits often regard boundaries and measurement of landed property. On this background it seems questionable whether personal experience is involved in the ensuing assertion that geometry also serves to train formal logical argumentation which "sometimes though rarely" is used by the rhetor.

wealthy families could employ their own educators (or buy them!—we are in a slave holders' society, and slaves could be found on any cultural level). Like Old Babylonian scribe-school teaching but unlike early Greek philosophy and even Sophist activity, the teaching of culture had become a standardized career—characterized, one must presume, by the customary sociology of such careers. Directly dependent as these teachers were on their municipal or private employers, we may surmise that the professionalization of the cultural career contributed to the lasting moral and political domestication of Hellenistic humanism.

Three very important exceptions to the general identification of literate culture with legitimization and cultural polish existed: Stoicism, Epicureanism, and Neoplatonism, all being philosophical currents with mainly moral or quasi-religious implications.[30] The former two philosophical currents arose already in the fourth century BCE and reacted upon the breakdown of political life, not by offering reform proposals but by preaching retreat into private life. Epicureanism especially was an enlightenment movement critical of established religion and superstition and rarely in favor with the rulers; in 173 BCE, Epicurean philosophers were thus expelled from Rome (Farrington 1965: 165)—they were indeed those "moderns" that in Polybios's opinion (see p. 40, n. 22) were "rash and foolish" enough to banish "notions concerning

---

30. These implications are obvious among other places in Epicuros's exhortation that "no one be slow to seek wisdom when he is young nor weary in the search thereof when he is grown old"—because "no age is too early or too late for the health of the soul" (*Letter to Meneceos*, trans. Hicks 1925: II, 649).

The primarily moral character of the Epicurean and Stoic messages did not prevent their doctrines from dealing with topics that carry an only implicit moral message. Atomism had originally been devised by natural philosophers as an answer to the Eleatic dilemma: it was not Nature as a whole which was immutable and changeless but its minutest, indivisible parts (*a-tomos* means *in-divisible*). Epicuros adopted it as a way to dispense with religion and superstition. He did not deny the existence of the gods but made them so distant that his opponents would consider him an atheist without being much mistaken. The Stoics, intent to prove the harmony of the cosmos, developed a physical doctrine involving a match between microcosm and macrocosm. In connection with investigations of semantics (and ultimately with the question "What is Reality composed of?"—the Stoics held that there is an intimate connection between the *names* of things and their real nature), they also developed the earliest grammatical *theory* (that is, earliest in Greek tradition—Sanskrit grammarians precede them by at least two centuries, and Babylonian teachers of Sumerian may have done so by 1500 years).

The categories of the "logical grammar" of the Stoics (word classes—case—tense mixed up with aspect—mood—and voice) are still with us today, and they dominated European grammar until early nineteenth-century linguists got acquainted with the Sanskrit tradition (see Ellegård 1973: 664a). Epicurean atomism, as we shall see, became important both in twelfth-century naturalism and in the seventeenth century.

the gods and beliefs in the terrors of Hell," thereby jeopardizing the social order. Stoicism, a philosophy of resignation, was more acceptable: one Roman Emperor (Marcus Aurelius) was himself a Stoic writer, and a major Stoic philosopher (Seneca) was the educator of Nero (who eventually disliked his moral preaching so much that he ordered him to commit suicide, which the obedient teacher did). Neoplatonism was a selective reinterpretation of Plato, basing itself among other things on the theory of Love set forth in the *Symposium* and making use of Aristotelian metaphysical concepts; its intention can be summed up as an attempt to make philosophical sense of the mystical experience. It was formulated by Plotinus around the mid-third century CE and soon became very influential, though often in diluted form where miracles and occultism were more important than genuine mysticism; its prominence can be seen as an expression of the same moods as those that gave rise to the acceptance of Christianity among the educated classes. A characteristic theme is the "Great Chain of Being" through which influence emanates from the wholly ineffable "One" through the Divine Universal Mind and an increasingly complex hierarchy of intermediaries ("angels" when the doctrine was taken over by Christians and Muslims) and further via Man to the lower, animate and (at bottom) inanimate orders of Nature. It was very influential in Christian philosophy from St. Augustine to the seventeenth century, and also in medieval Islamic philosophy.[31]

THE IMPACT OF CHRISTIANITY

Christianity began its dialogue with Pagan (that is, Greek) philosophy in the early second century (CE, of course, as all dates from this point onward) (see Gilson 1955: 9–26); it had a breakthrough in the educated classes in the fourth century, which led to complete political take-over around mid century. Even though Pagan

---

31. The mystical significance of the "chain" is articulated with eloquence by the thirteenth-century Persian sūfī poet Jalāl al-Dīn al-Rūmī in this passage (trans. Reynold Alleyne Nicholson, quoted from Berger 1973: 72):

> I died as a mineral and became a plant,
> I died as plant and rose to animal,
> I died as animal and I was Man.
> Why should I fear? When was I less by dying?
> Yet once more I shall die as Man, to soar
> with angels blest; but even from angelhood
> I must pass on: *all except God does perish.*
> When I have sacrificed my angel-soul,
> I shall become what no mind e'er conceived.
> O let me not exist! for Non-existence
> Proclaims in organ tones: *to Him we shall return,*

which mixes the Neoplatonic concept with (italicized) quotations from the Koran.

culture constituted an indubitable threat to Faith, the breakthrough took place at the conditions of classical (Pagan) educational ideals: only if Christianity was culturally competitive would it be taken seriously by the culturally and politically decisive social strata (see for instance Laistner 1957: 44ff). We may speak of the "gentrification" of a religion and a religious community that had once primarily recruited the humble and oppressed.

The complex situation is illustrated by the life and writings of St Augustine (354 to 430)—if such things can be measured, probably the most important Christian thinker of all ages (St. Paul and Christ belonging to other categories). In his younger years he was a teacher of the Liberal Arts, from which period among other things an extensive and partially innovative work on musical theory and a highly original sketch to a treatise on semiotics (belonging under the heading *dialectic*) survive. After his baptism in 387, his voluminous writings concentrated on religious and ecclesiastical matters, many of them being concerned with education. The leading idea is that the Christian should be so polished in the Liberal Arts that he does not lose his standing among the educated; *but no more.* The emphasis on the latter point corresponds to the very strong formulations in Augustine's autobiography, the *Confessions* (X.xxxv). Here, secular curiosity, not least everything approaching scientific or purely intellectual interest, is counted as a particularly malignant variant of the concupiscence of the eye, which, one remembers, is no better than the consummated concupiscence of the flesh ("whosoever looketh on a woman to lust after her hath committed adultery with her already in his heart"—Matt. 5:28). In as far as it was at all necessary, this attitude could only open the way for a radical break with every autonomous intellectual activity beyond that which was needed for reasons of competitive power[32]—in agreement with the principle formulated in the next verse of the Gospel ("if thy right eye offend thee, pluck it out, and cast it from thee").

After the downfall of the (Western) Roman Empire, competitive power was no longer a problem. The Christian intellectual elite of the outgoing fifth and early sixth centuries tried to save as much as possible of the classical heritage (which at the onslaught of the barbarians suddenly appeared as the heritage of the Church and of that society of which the Church had become the backbone) in a situation where both knowledge of Greek and books in general were disappearing. But since this elite consisted—with only slight exaggeration—of no more than two persons, the success was limited.

---

32. It seems, however, that Augustine had a tendency to overrate what was really necessary. Few independent intellectuals of any stature come to the mind in Augustine's time, apart from Augustine himself and St. Jerome (Hieronymus), the translator of the official Latin Bible (the *Vulgate*). The latter, admittedly, tells in a letter about a dream where God reproached him of being more committed to Cicero, that is, to refined literary style, than to Christ (*Epist.* XXII, 30, quoted, for example, in Rand 1957: 106f).

The first member of the elite was Boethius (c. 480 to 525). He set himself the task to translate Aristotle, Plato, and the basic works for the quadrivial disciplines. He managed to translate a number of Aristotle's minor logical treatises and to provide them with commentaries,[33] and to translate at least large parts of Euclid's *Elements* and (perhaps) Ptolemy's *Almagest*, the culmination of ancient mathematical astronomy, together with some more elementary handbooks on arithmetic and music. The *Elements* were soon lost, with the exception of the definitions and the propositions of book I without their proofs, and the *Almagest* completely; the forgotten manuscript was rediscovered in 983 by Gerbert of Aurillac (see below, p. 60), as this scholar and future Pope tells with great enthusiasm in a letter (ed. Lattin 1961: 54); since then nobody has ever heard about it. Other works survived in monastic libraries (at first perhaps in the private libraries of affluent landowners' families), ultimately to be rediscovered, copied, and studied in the late eighth and during the ninth to tenth centuries.

The second member was Cassiodorus Senator (c. 480 to c. 550). In the 530s (at which moment he was a high-ranking official in the Ostrogoth administration) he had tried to establish a higher Christian school in Rome, where the methods of secular learning were to be used in the study of Sacred Scripture (Riché 1962: 171–177); the project had failed—immediately because war broke out, but even a similar institution in Alexandria that had inspired Cassiodorus had been shortlived. In older age he established a monastery or quasi-monastery—the Vivarium—where the copying and study of ancient texts (including first of all the Fathers of the Church) was a regular part of monastic duties (the only early monastery where that was the case, notwithstanding a widespread myth). The Vivarium, however, did not survive its founder and left few immediate traces; Cassiodorus's long-term influence was mainly secured by an *Introduction to Divine and Human Readings,* which he wrote as a not very advanced compendium for his monks.

There is a clear moral to the story of Boethius and Cassiodorus: in the West, the disappearance of much of the ancient heritage was on the whole not a consequence of ecclesiastical suppression, prohibitions, or persecution. It followed from *lack of support,* on the part of the Church as of every other body and institution. Ancient learning was *forgotten because nobody needed it* and few were able to understand even its more rudimentary levels. Ancient philosophy, science, and scholarship were forgotten because they had become superfluous by the breakdown of ancient society.

Then, after all, they were not completely forgotten, for one need subsisted, namely the need of the Church. Monks were expected to be able to read (that is,

---

33. It is worth remembering that this translation activity led him to create much of our modern philosophical terminology; whoever speaks of "terminology" and "substance" or who distinguishes "quantitative" from "qualitative" (to name but a few examples) is in linguistic debt to Boethius.

read Latin) and to understand a bit of what they read, and so were priests. Though the myth of monasteries as havens of quiet scholarship *is* a myth, monasteries did give place to some teaching and study of the Fathers of the Church and thus, indirectly, of that Pagan philosophy which had been the adversary of the Fathers and at the same time the basis on which they had had to build when explaining the Biblical account of the creation of the world (etc.); future priests were adopted into the bishop's household at an age of seven as *lectores,* and it was the task of the bishop (for centuries of the bishop in person) to teach them reading and writing. Some, though admittedly few, would go on with the Fathers and possibly with authors like Cicero as guides to good style; they might even try to pick up as many rudiments of the Liberal Arts as they could get hold of (not much, since no other sources than Patristic writings were available to any but the most lucky handful), because even these rudiments could serve to interpret the Scripture and the Fathers and to compute the day of Easter.

The system of knowledge to which the select and happy few could hope to get access was thus composed of two parts, the names of which we may borrow from Cassiodorus's *Introduction.* One is *Litera divina,* "Divine readings": Holy Scripture and the Fathers. The other is *Litera humana,* "Human readings," brief encyclopedic accounts of the basic concepts of the Liberal Arts—in Cassiodorus's own compendium concentrating on rhetoric and dialectic, in general practice restricted to grammar (including literary bits) and some rhetoric.

This, and a handful of forgotten manuscripts scattered in monastic libraries, was the scholarly legacy bequeathed by antiquity to the Latin Middle Ages. No more—but also no less.

CHAPTER 3

The Middle Ages

AN ERA OF RENASCENCES

The early Middle Ages offer little of immediate interest for the themes of modern humanities and philosophy. It could hardly be otherwise, given the meagre legacy from antiquity and the lack of alternative inspiration. But the early Middle Ages can be claimed to have been the mold in which the specific "Western" or (in the beginning rather) "Latin" interpretation of the ancient heritage was cast—as distinct from the Eastern Christian (Orthodox and other) and the Islamic interpretations, none of them less genuine nor legitimate but certainly different. Thereby the early as well as the central and high Middle Ages constitute the essential background to that idolizing return to ancient values and attitudes which became pivotal for the later rise of Humanism and the humanities.

During classical antiquity, the large majority of the population had lived in the countryside as food producers; agricultural techniques were insufficient to secure a surplus allowing more than a modest minority to be occupied in other activities. All the same, the center for all cultural innovations and for literate culture had been the cities. During the closing centuries of the ancient era, however, aristocrats had increasingly withdrawn from the city to their landed estates, or "villas". Since c. 300 CE, moreover, the social structure had begun to change. Manual workers became bound to their professional corporations and thereby became subjects of the state, which brought their actual juridical status closer to that of slaves; simultaneously, slaves and others were often settled on landed estates as *coloni,* unfree and bound to the land yet provided with their own plot—more similar to the serfs of later ages than to the chattel slaves of the early Roman Empire.

At the onset of the Middle Ages proper, this waning of slave society accelerated, and the development of feudalism began. Slaves, it is true, continued to exist and to be bound to the large estates of aristocrats and monasteries. As a rule, however, the implications of the unfree status changed. The unfree (still designated with the Latin term *servus,* whence *serf*) became a person with specific obligations and certain (though strictly limited) rights. The ideological implications of slavery, moreover, changed at least to some extent. Warrior nobility, it is true, was as contemptuous of manual work as any ancient aristocrat had been. In many periods, however, the attitude of the Church was different, as expressed in St. Benedict's Rule for the monastery in Monte Cassino (founded 529) prescribing manual work as a monastic duty. Religion itself may have played a role here, not least because participation in work could symbolize Christian humility. But as medieval monks were rarely as humble as pious historiography tends to make us believe, the absence of a professional managerial ("scribal") class outside the Church may have been an equally important factor. Monks had, willy-nilly, to participate in the management of their own estates and to take care of everything which could not be left to illiterate and probably ill-willed serfs. The Church, moreover, had to provide rulers and noblemen with staff persons in periods when territories won through war and conquest were consolidated through the establishment of administrative structures (whence comes the usage in which an English "clerk" can be an office employee or a churchman). The "scribal function," which had been culturally unproductive during classical antiquity, thus became productive once again; but compared to the Bronze Age it was now bound up with new functions (those of responsibility and not mere service for the central societal institution), and it was set in a different historical situation (the succession of antiquity, socially as well as culturally) which made it create something new[1]—not least (though not equally forceful in all epochs) the ideology that reevaluated *work* as a human and not specifically servile duty, and even a duty that called for veneration.

The early Middle Ages (reaching until c. 750) constitute a formative period, during which this reevaluation, as well as the culturally productive role of the clerico-scribal role, were only possibilities inherent for instance in St. Benedict's Rule. The same can be claimed with regard to the final feudal outcome of the transformation of the mode of production, if only we notice that two different aspects of European feudalism were contained in germ in the early structure. Firstly, there was the development away from chattel slavery and toward bondage. Secondly, the quality by which Western European feudalism is distinguished from the "proto-feudal" systems found in earlier epochs and in other parts of the world was already present in the *colonus* system and thus inherited from late antiquity:

1. One may notice the analogy with the rise of Greek philosophy, which also appeared to result from a hitherto unseen combination of social forces and structures rather than from the advent of quite new patterns.

bondage and responsibility were individual or at least familial, not a matter concerning the village community as a whole.[2] The individualism so characteristic of Renaissance and later Humanism was thus not only a combined result of the economic individualism of early capitalism and of the recapture of the literary-humanistic legacy from antiquity. It was also transmitted through the very mode of production resulting from the breakdown of ancient society—a fact not to be forgotten by the reader of these pages, which otherwise concentrate on the level of literate culture.

Returning to the latter level, which is of most immediate interest to a discussion of the humanities, we are in for a surprise. Notwithstanding the Renaissance contempt for an intermediate period seen as nothing but abandonment of true (that is, ancient) culture, and in spite of the emergence of a radically new social structure and new social values, the literate culture of the Middle Ages was—especially until the twelfth century—no less directly dependent on antiquity than the Renaissance, in particular on Roman antiquity: rather more indeed. At closer inspection of the situation this is no wonder: the collapse of the Roman Empire and Roman civilization produced no new cultural upsurge nor revival of pre-Roman, for instance Celtic, cultural patterns;[3] though conserved until long after the Roman conquest, these had finally given way to Romanization bound up with evangelization toward the very end of Christian late antiquity (cf. Brown 1971: 130f). Nor did the barbarian invaders bring much of their cultural baggage; they would rather leave the marks of their avowed inferiority behind and try to conform to the more prestigious habits of the conquered territories, in the likeness of the Ostrogothic King Theodoric of Italy, who employed Boethius and Cassiodorus as ministers and had the former executed on suspicion of ideological disloyalty. Theodoric held that

---

2. No doubt, village and similar communities existed; in most of Europe they even grew stronger during the Middle Ages, due to the introduction of the wheel plough and the ensuing development of the tilling community. What is at stake is the (predominantly) individual character of *bondage.*

3. These were best conserved in Ireland, which was Christianized without being politically submitted to the Roman Empire. Early Irish Christianity thus developed in interplay with the autochthonous Celtic elite, much less dependent upon the ancient heritage than for instance Christianity of the Gallic region. The unique character of early Christian Ireland highlights, by this contrast, the importance of the classical tradition for the form of Christianity that dominated the medieval scene in the regions once belonging to Rome.

In a later epoch, when part of the Germanic and the Nordic regions were Christianized and crystallized as *states* in a similar process (see presently), something similar to the Irish development happened in the juridico-political sphere; in other cultural domains, however, the impact of Christianity made itself felt wholesale, and—except for the vernacular literate culture of Iceland—no specific form of Christian culture arose.

"an able Goth wants to be like a Roman; only a poor Roman would like to be like a Goth" (Brown 1971: 123).[4]

This general dependence of cultural patterns upon antiquity was no hidden substructure but something of which the educated were acutely aware throughout the Middle Ages.

Firstly, the Church, the institution which more than any other (and especially more than the emerging royal power) created social coherence during the early medieval phase, expressed itself in the language of the Roman Empire—more precisely of the Western Empire—Latin. The structure of the Western Church, moreover, was framed in juridical terms, in itself a remarkable feature not shared with religious institutions in other cultures and a legacy from Rome rather than from the Old Testament.[5] As long as Roman (or Romanized) aristocratic lineages survived as landowners, finally, they occupied the upper echelons of the ecclesiastical machine and provided the monasteries with most of their monks; in this way the Church often took over the aristocrats' veneration for what they considered their specific cultural past.

Secondly, social reconstruction after the breakdown, that is, the formation of new states and systems of law, built on the foundations provided by late Antiquity, often of course mediated by Christianity. It is a recurrent pattern, from the adoption of Arianism by Ostrogoths and Visigoths to the Christianization of Denmark under Harald Blaatand and that of the Magyars under Geza and Stephan I, that warlords or chieftains trying to stabilize their command in the form of a permanent kingship would first of all enforce Christianity upon their subjects.

Thirdly, all learning built on ancient authors (including the Fathers of the Church), who were referred to as authorities. The two words *author* and *authority* (in Latin *auctor* and *auctoritas*) are of course etymologically related, the first denoting the source of a text and the second the source of power. It is characteristic of the medieval veneration for ancient learning and knowledge that no distinction between the two terms could be made.

---

4. It is characteristic of the situation and of the need for Barbarian invaders to embrace the culture of civilized society that the Barbarian rulers, when needing to symbolize their separateness in religious terms, did not do that by conserving their original tribal religion but by adopting Arianism, a variety of Christianity regarded as heretic by the Roman church. Statehood, even the statehood of Vandal, Longobardian, Ostrogothic, and Visigothic conquerors, could not be built upon Germanic religious and other cultural patterns.

5. Roman Law, like modern law, tends to be formulated as general principles. Ancient Hebrew law, like the laws of Germanic and Nordic barbarians, tends to list possible cases.

It is noteworthy that even the eleventh-century rationalization of the ecclesiastico-feudal world as consisting of "three orders" (cf. below, p. 60) may have been borrowed from the only accessible work of Plato, the *Timaeus* (cf. Dutton 1983).

Fourthly, until the twelfth century the material used for teaching was almost exclusively of ancient Roman origin: in part excerpts from Roman authors, in part Roman compendia, in part encyclopedias compiled during the Middle Ages from ancient works or from earlier medieval compilations. From the twelfth century onwards, as we shall see, more original material came in, but the stem of almost all disciplines remained ancient if no longer exclusively Roman.

The result of this acknowledged cultural dependence is the striking phenomenon of "renaissances." The Middle Ages, in fact, were marked by violent ups and downs, demographic as well as economic. In periods of social collapse and economic regression, cultural vigour and especially scholarly interests would also decline. At every upsurge, on the other hand, cultural life and interest would flourish. The striking thing is that the twentieth-century students of every such bloom speak of it as a "renaissance," in the likeness of the "rebirth" of ancient splendor in the fourteenth- through sixteenth-century "real" Renaissance. At each occasion, indeed, the dominating feature of the process was an attempt to recapture and to understand as much of the lost heritage as possible under the conditions of the day—conditions comprising intellectual prerequisites as well as the uses to which learning and other cultural activities would be put.

THE CENTRAL MIDDLE AGES (750 TO 1050)

In the early medieval period, as we saw, literate activity was concentrated in the household teaching of bishops and in the modest education offered to monks. Occasionally but not very regularly, some copying and reading of manuscripts took place in monasteries. Literary and scholarly activity, under these conditions, could hardly be expected to be anything but rudimentary and derivative, dependent exclusively on the ancient model and no vehicle for the expression of new attitudes. Such activity was not completely absent, it is true. Gregory of Tours (539 to 595) wrote a *History of the Franks.* Isidore, Visigothic Bishop of Seville (560 to 636) wrote *On the Nature of Things* and *On the Order of Creatures.* Isidore also produced the most important of all seventh-century writings, the extensive encyclopedic work *Etymologies,* containing explanations of the basic concepts of various scholarly and technical fields of knowledge (often built upon or dressed up as etymological explanations of the origin of the terms)—perhaps the most-quoted authority of the Middle Ages next to the Bible. Bede the Venerable from Northumbria (672 to 735), of whom it has been said that the scratching of his pen could be heard over the whole of Western Europe (because no other writing of significance went on), wrote an *Ecclesiastical History of the English Nation* (actually much more than a mere church history); extensive Biblical commentaries; and several works on *computus,* that is, on ecclesiastical calendar reckoning. He also translated excerpts from Isidore's *On the Nature of Things* and the Gospel of St. John into the Anglo-Saxon tongue, and he wrote an innovative treatise on metric rhythm. Valuable authors all of them, in view of the limitations imposed by the times—Gregory's and Bede's *Histories* are indeed very readable today, and Bede's presents "the basic features of scientific historiography in a way unequalled between classical times and the Renaissance" (Wrenn 1967: 63). Only Bede's computistic works, however, can be said to represent a genuine renewal, as reflected by the circumstance that they displaced everything written on the subject before and gained a position analogous to that possessed by ancient handbooks in other fields.[6] Also of possible consequence were his translations into the vernacular, in the sense that they inaugurated and may have inspired a period of Anglo-Saxon literacy whose best-known product is the *Beowulf* epos, and which the Viking invasions of the early ninth century would only bring to a temporary halt.[7]

Bede can be regarded as a portent of that first formulation of a specific medieval culture which occurred during the "Carolingian Renaissance" (which accordingly honored him with the title of *doctor modernus*). The fundament for this first bloom was a sequence of technological innovations, some of them in the military and some in the agricultural domain.[8] Most important among the changes in military technology is the introduction of the stirrup, which made it possible for a horseman to use a lance without being thrown from the saddle. From then on, heavy cavalry became the decisive armed force, irresistible to infantry in normal terrain until the advent of firearms. The change provided the drive for the juridical consolidation of emerging feudal structures: in need of armed knights and unable to support them directly, the king would distribute land with appurtenant bondsmen to noblemen against the obligation to provide armored horsemen. Agricultural innovations included the introduction of new crops (hay, protein crops) and new crop rotation systems, the application of a new harness for horses, and the invention of the wheel plough. Some of the latter innovations only entered practice gradually, but even the modest beginnings allowed demographic growth and social stabilization—first in Frankish territory, where the result was political consolidation and military expansion.

---

6. Interestingly, the earlier Irish computus tradition on which Bede draws appears to be strongly dependent on Irish pre-Christian calendar reckoning (cf. p. 53, n. 3, and McCluskey 1993: 143–147).

7. ". . . may have inspired"—but the influence is far from certain. Bede's Anglo-Saxon writings were in prose, and for more than a century all other Anglo-Saxon compositions were poetry. Anglo-Saxon prose writing only took its beginning with King Alfred of Wessex (849–899), that is, well after the "Carolingian Renaissance" (see presently). As so often, what we may be tempted to see as a starting point may rather be an early expression of a more general mood whose roots are anonymous and not connected with any single person now remembered.

8. Discussed in depth by Lynn White (1962).

The process of consolidation and expansion was inaugurated by Charles Martel, who defeated the Muslim army at Poitiers in 732, and was brought to its culmination under his grandson Charlemagne. Charlemagne took over power in the Frankish realm in 768 and died in 814 as the ruler of everything between Pamplona, Barcelona, and Rome to the south, the Channel to the northwest, Hamburg to the north, and Magdeburg and Linz to the east, with spheres of influence extending even further. Charlemagne tried to build up a centralized administration of this huge and disparate empire. One branch of his government system consisted of commissioned military leaders (*comites,* "companions [of the king]," the origin of the title *count*); the other branch of the twofold system was that of administrative control, presupposing literacy and headed by the bishops because no other body of potential administrators was at hand. For the actual working of the administrative system, a larger number of literate functionaries was required. That body was as yet nonexistent, and the only way to create it was through an organized school system.

Administrative needs were thus the motive force behind the Carolingian educational reform. The means, once again, were provided by the Church. Teaching had for centuries been an episcopal duty. Now the obligation to organize a proper school open to able free boys was imposed upon all bishops' sees (this is the origin of the "cathedral school"). Even monasteries, which had a tradition for teaching their own novices, were commanded to organize an "open" school, that is, a school whose students were not future monks.

The latter ordinance was rarely observed, and bishoprics which did not comply with the edict on schooling were more common than those overdoing the case (like the bishop of Orléans, who tried to impose general school attendance). But some cathedral schools were created, and at the imperial residence in Aachen a palace school was held for future high officials. In these places, the curriculum was taken over from (what was known about) ancient education. The Liberal Arts were considered the only possible foundation for literate education. But even if that much was known, the content of these arts was largely unknown, since few textbooks (or any texts at all) were at hand. A main result of the effort to provide for administrative needs (largely a vain effort, since the empire was split up after Charlemagne's death, and the administrative system decayed in the resulting smaller kingdoms) was thus a treasure hunt for forgotten manuscripts in monastic libraries.

Among the findings were Boethius's translations of and commentaries on Aristotle's minor logical works and a didactic poem, *The Marriage of Philology and Mercury,* by Martianus Capella (c. 365 to 440 CE) built mostly on primers of the single Liberal Arts. A curiosity is the changed status of Latin treatises on mensuration and agriculture. They had been conserved in the monasteries as handbooks for growing ancient crops; monks, as descendants of the Romanizing nobility, would eat bread rather than porridge and drink wine rather than beer (see Duby 1973: 27f). Now their rules for area computation had to fill the role of geometry, since no other texts at hand could do that (Euclid's *Elements,* anyhow, would certainly have been too highbrow). In the early ninth century, Boethius's translations on arithmetic and music were found,[9] and toward the midcentury the Irishman John Scot Erigena (c. 810 to 877), the extraordinary head of the palace school of Charles the Bald in Laon, translated several Greek Fathers of the Church and made his own attempts to reconcile Christian theology with Neoplatonism.[10]

It is remarkable that the classicizing program of the new school institution was not only felt to be necessary for the education of future officials, or at least the obvious choice. The program also aroused enthusiasm among those involved in the Aachen palace school, from Charlemagne himself (who, when not conquering Bohemia, Lombardy or the Pannonian—now Hungarian—Plain, would participate in its activities) to the students. The court in Aachen complimented itself as "Athens resurrected and united with Jerusalem", claiming thus to *stand at an even higher level than antiquity*—which we may, of course, take as an indication of the distance which separated the "Carolingian Renaissance" from real understanding of the example it had set itself.[11]

Erigena, by knowing Greek and philosophy, was an exception to the rule that all scholarship and literary activity was derived from Latin classics. Charlemagne himself was an exception of a different kind. His enthusiasm for his school and for its learning made him order that "the age-old narrative poems, barbarous enough, it is true, in which were celebrated the warlike deeds of the kings of ancient times, should be written out and so preserved."[12] As with the Frankish grammar that he also ordered to be made, these records were, unfortunately, lost in the turmoil lying ahead, and we are therefore ignorant both of the content of this early Germanic literature and of the methods and character of this early medieval example of humanistic scholarship.

Turmoil was indeed to come, undermining empire-building as well as learning. It came from Scandinavia (the Vikings) and from the Pannonian Plain, where

9. The *De musica* contains the only genuine mathematical proof known to the Latin Middle Ages until the twelfth century—quite impressive in view of the preponderant role of the quadrivial arts in education.

10. Erigena was inspired by Augustine but so much closer to real Neoplatonism that his works verged on pantheism—for which reason the only original theological and philosophical works written in Latin between 550 and 1050 were condemned by the Church. None the less, they were influential in later medieval mysticism.

11. At another occasion I have compared the self-assurance of Carolingian learning to the "cargo-cults" arising in certain Pacific islands after World War II. If you have never seen an airplane before and know nothing about its origin or construction, it is no wonder that you try whether two pieces of wood tied together may continue the delivery of canned food and flour; and if the mere ability to write is a miracle, you may easily mistake that basic tool of scholarship for Dame Philosophy herself.

12. This is told by his biographer Einhard (see presently) (trans. Thorpe 1969: 82).

Charlemagne had accomplished a "final solution" to the Avar problem,[13] only to open the way to the much fiercer Magyars. The Viking and Magyar raids caused communication inside the realm to break down. The counts made themselves de facto independent rulers. Officially, the empire was divided into three after Charlemagne's death and some initial struggle. Actually it fell apart into countless domains deprived of any law and order beyond what the local lay or ecclesiastical lord could and would enforce. Another step was taken toward the reality of feudalism, which came to be very different from the hierarchical and centralized Carolingian ideal.

In a situation where Magyar tribesmen could put the daughter of a local nobleman up for sale at the slave market in Worms,[14] there was no longer any need to train an administrative elite; nor was there peace or available wealth to uphold cathedral schools. The consequences were depicted by Walahfrid Strabo, abbot in the Benedictine monastery in Reichenau, in a preface to a *Life of Charlemagne* written by Einhard, a former student from the school in Aachen:

> Of all kings Charlemagne was the most eager in his search for wise men and in his determination to provide them with living conditions in which they could pursue knowledge in all reasonable comfort. In this way Charlemagne was able to offer to the cultureless and, I might say, almost completely unenlightened territory of the realm which God had entrusted to him, a new enthusiasm for all human knowledge. In its earlier state of barbarousness, his kingdom had hardly been touched at all by any such zeal, but now it opened its eyes to God's illumination. In our own time the thirst for knowledge is disappearing again: the light of wisdom is less and less sought after and is now becoming rare again in most men's minds.
>
> (trans. Thorpe 1969: 49f)

Cathedral schools vanished from the horizon, and so did the open schools of monasteries. In certain monasteries, the Carolingian Renaissance of learning continued as best it could (one of them being Walahfrid Strabo's Reichenau). However, the monastic reform movement of the early tenth century (the "Cluny movement") went in a different direction, toward the extension of ritual and psalm singing and toward emphasis on the worship of relics. Cluny, as it has been said, transferred the Bible from the reading desk to the altar. On the whole, the monastic environment was unable to persevere in a direction whose deeper social rationale had disappeared.

The next beginning, then, came from below. By 950 the Magyars and the Norsemen had been pacified (partly beaten, partly bought off). Administrative order was reintroduced, at first at the local (ducal and similar) level, giving rise once

13. The expression is used by another biographer of Charlemagne, Notker the Stammerer (trans. Thorpe 1969: 137). Then, as later, total extermination was meant.

14. See Bloch 1965: I, 11. There was in fact an intensive slave trade through Lorraine, conveying slaves mainly caught in Slavonic areas to Muslim Spain.

more to a need for literate officials. The spread of agricultural innovations, furthermore, provided a better economic foundation; clerks, then as ever, lived from the surplus of material production. The cathedral schools could thus regain some vigor from the middle of the tenth century onwards.

The curriculum was, once again, based upon the scheme of Liberal Arts. The starting point was the level attained in the aftermath of the Carolingian period, that is, much more adequate than what had been possible in Aachen. Still, of course, teaching was concentrated on subjects which could be understood: firstly grammar (including elementary study of excerpts from classical authors) and rhetoric (to be used, among other things, in preaching). But even some dialectic was introduced, together with the use of a newly invented or imported abacus and some geometry of the sphere to be used in astronomy, thanks not least to Gerbert of Aurillac who, before becoming the mentor of the future Emperor Otto III and eventually a Pope, was the head of the cathedral school in Rheims, and beyond doubt the leading figure of the whole movement. His teaching thus foreshadowed the flourishing of the Liberal Arts that was to unfold in the late eleventh- and the early twelfth-century school.

It is noteworthy that *no organized teaching of theology took place.* No syllabus as yet encompassed the Fathers or the Bible. Human, not Divine readings, were the aim of the new, spontaneous growth of education. The notion that learning in the pious Middle Ages was "the handmaid of theology" was started already in the Middle Ages, either as wishful thinking on the part of theologians or as a pious lie; taken over as antimedieval propaganda during the Renaissance and the Enlightenment—and once more turned into a pious lie in the era of Romanticism.

THE AGE OF THE LIBERAL ARTS

In the early eleventh century, scholars associated with the cathedral school environment (in particular the school in Chartres) formulated the political theory of their age, according to which society consisted of three orders: the (ecclesiastical) order of those who pray; the order of warriors (king and noblemen); and the order of laborers—the fact that praying also implied administration for the warring order being presupposed together with the not totally vain idea that it allowed the enforcement of some moral order on both warriors and laborers. Social reality, however, had already begun to leave this simple scheme behind in a way which was to change even the world of learning.

One factor in these changes was the relative pacification and the creation of at least local social order. Another was the steady improvement of agricultural techniques—the last great famines for centuries occurred shortly before 1050 (large-scale famine only reappeared in the early fourteenth century). The third was a gradual centralization of power in royal and, as far as the Church is concerned, papal hands.

The three factors together made possible a new beginning of trade and urban life. Towns grew up as trading centres and around craft production. These towns, like the early Greek city states, were tense with discussion and democratic claims. Unlike what had been the situation in Greece, however, the medieval towns were anything but masters of the surrounding countryside. Their democratic aspirations went no further than being allowed to govern their own affairs without interference from the feudal lord; similarly, the aspiration of single social groups (first neighborhoods, later artisans' and merchants' guilds and similar professional organizations) was autonomy. But as in Greece, the root of these democratic aspirations was the closeness of primitive-democratic experience: the structures of the urban fraternities were borrowed from kinship and village communities. As in Greece, the fraternities were composed of equals who had to find their common goals and strategies "in the middle."

Since towns would often grow up around bishops' sees, the schools were typically located within the urban environment. True, the bishop himself would on most occasions be in conflict with the urban community—he, indeed, would be the feudal lord from whose rule the town tried to free itself. But the "cathedral school" would only be loosely connected to the see. The *scholasticus,* an episcopal official, was responsible for the teaching; but other masters might teach too, in relative independence from the local ecclesiastical power (masters lived from students' fees, and were not paid from the incomes of the see). The town was thus a sounding board for the discussion in the school, and the school was a resonator for the discussions and claims of the town. The chronicler Hermann of Tournai tells us (in 1090) that the squares of his city were filled by curious crowds when Master Odo discussed philosophical questions with his students, and that "the citizens left their various employments so that you might think them fully attached to philosophy" (Werner 1976: 57, 93 n.358). Even the late eleventh-century pamphlet war between the pope and the emperor (the "Investiture Conflict," not only a pamphlet war) will have reached this environment, as suggested by a favorite argument used on both sides: that the reasoning of the other part was so poor that it was "heard everywhere in the streets and in the market-places and are gossiped over by the women in the weavers' shops."[15]

Such claims may not have been wholly untrue. Since the power structure against which the urban environment revolted was ruled by an alliance between the warring and the praying orders, and since the obvious language in which to express moral protest was religious, urban discussion and urban political claims also gave rise to a specific urban piety, which was both socially critical and potentially heretical. In the middle of the eleventh century, the first serious theological discussions in

15. This formulation (quoted from Robinson 1978: 8) is due to Manegold von Lautenbach, a supporter of the Papal side; but the elitist attitude was shared by both parties.

the Latin world since antiquity took place in this environment, concerned in particular with the nature of the Eucharist. In spite of repeated official condemnations and forced abjurations of unorthodox positions, these spread "to the whole of France, Germany and Italy," as the orthodox side would complain. Both because the argument was philosophical[16] and because the display of sacred relics was an inadequate answer to arguments, ignorant priests (certainly the majority) became a serious problem to the Church, which was one reason the Papacy backed the cathedral school movement.

A final effect of the new social situation was an awakening of interest in *astrology*, the first brief treatises on which had been translated from the Arabic in the outgoing tenth and the early eleventh century. Astrology, indeed, was a natural explanation, accessible in principle to everybody and not only to those with priestly patent on divine knowledge. Astrology thus entered Latin culture for reasons similar to those which had engendered early Greek natural philosophy. Like early natural philosophy, the astrological endeavour was humanist though not concerned with humanities, in the sense that it pointed to the possibility for human beings to reach true knowledge on their own without being dependent either on divine grace or on the grace of authorities.[17] This attitude is distinguished throughout by the epithet *humanist*, whereas the adjective *humanistic* is used when "the humanities" are involved; as a noun, however, "a humanist" will be somebody engaged in the humanities. (Renaissance Humanism, when we get so far, will be capitalized both as a noun and as an adjective, see p. 83, n. 5).

Pacification; the growth of agricultural output and of administration and urban culture; the emergence of genuine political life: together these constituted the economical and ideological background to a new, ardent interest in learning, which in the twelfth century was understood as interest in the Arts. But this time the interest was in the complete cycle of Liberal Arts, and with certain authors even in "mechanical arts" (the despised "productive knowledge" of antiquity). The prospect of future employment in ecclesiastical and lay administration made it attractive for gifted young people to attend the cathedral schools. The "twelfth-century renaissance" of the Arts was thus carried by the first environment of professional intellectuals since antiquity (first in the Latin West, that is).

---

16. It had to do with the problem of *nominalism* versus *realism:* are general concepts mere *names* invented by us to cover a plurality of objects, or do they possess *real* existence, for example as Platonic ideas? In the first case something which is obviously *bread* can hardly be *flesh* according to some higher point of view.

17. Of course astrology came to depend heavily on those authors who were termed and regarded as *authorities,* first of all on Ptolemy's *Almagest.* But these were still *human* authorities, depending themselves on *human* observation and reason and not on Holy Writ.

The scholarly culture created by these intellectuals was primarily dialectical and not rhetorical. The ideal was no longer the speaker ("the priest") but the critical peer able to produce arguments and only to be defeated by better arguments. The background in the urban environment of fraternities is obvious. But it was also humanist, as already stated in connection with the emergence of astrological interest: explanations should be accessible to human reason as presented, for example, by natural philosophy, and not have recourse the God's hidden wisdom. A nice example of this is provided by Peter the Venerable (certainly no radical in matters of theology) in his mid-twelfth-century *Summary of the Whole Heresy of the Diabolic Sect of the Saracens.* In the Koran he finds the rhetorical question "Do you not see that the birds in heaven are not sustained otherwise than by God," to which he answers (forgetting that almost the same naturalist objection could be raised against Matt. 10:29, "one [sparrow] shall not fall on the ground without your Father"): "See the simplicity of the madman who thinks that flying birds are supported not by air but by the miraculous power of God. But (as we know) fishes are supported by water and birds by air, according to an original decree and ordering of God, and not (as he thinks) by a special and invisible miracle." (Southern 1953: 40) In a world where the unfree was defined as somebody "who did not know today what he is going to do tomorrow" (because he lived according to the will of a master who might decide otherwise,[18] the free man being understood as one who knew because nobody had the authority to change his decision once it was made according to law), to procure natural or other explanations accessible to human reason thus amounted to obtaining predictability—and hence, in other words, to make man a free citizen of this world.

The title of a famous theological treatise from 1099 is telling in this respect: Anselm of Canterbury's *Cur Deus homo,* "Why God Became Human." This theme is of course a central Christian dogma, and it had been involved in most of the theological struggles of antiquity. But it had been largely displaced during the early and central Middle Ages, as it is obvious from the iconography of Christ: triumphant and ruling, not suffering on the cross. Both the familiar gothic suffering Christ and Anselm's resurrection of a forgotten theological theme are thus religious reflections of the new humanism. So is Anselm's formulation of a proof of God's existence. Anselm was a sincere believer who had no doubt in the matter; before his times, no medieval Christian would have come upon the idea that God's existence should be proved (we remember the almost complete absence of anything related to proofs even in the mathematical texts known to the central Middle Ages). But in the 1080s, the intellectual environment in a monastic school (admittedly a famous one, that of Bec in Normandy, where Anselm was abbot) was such that even God's existence could be measured by human reason.

---

18. The formulation is due to the English thirteenth-century jurist Henry de Bracton (trans. Southern 1953: 105).

Two specific twelfth-century schools and three scholars should be highlighted. First there is the so-called Chartres group, inspired by Bernard of Chartres, leader of the Chartres cathedral school from 1119 to c. 1126 (but not all members actually taught in Chartres). Bernard in known to have formulated the idea of progress in science in the aphorism that we are like "dwarfs, perched on the shoulders of giants" and therefore able to "see more and farther than our predecessors, not because we have keener vision or greater height, but because we are lifted up and borne aloft on their gigantic stature."[19] It is symptomatic of the existence of this famous group and school that the Liberal Arts are prominent in the decoration of the Chartres cathedral. The group engaged particularly in natural philosophy; one surviving work describes the creation as a natural process and argues in that connection that "it is indeed not to be believed literally that God took a rib out of the first man's side."[20] The inspiration was Plato's cosmological *Timaeus,* the most untypical of Plato's works, and the only one to be at hand in (incomplete) Latin translation; Epicurean atomism as transmitted through the Roman Epicurean poet Lucretius and through the Arabs; and the doctrine of the four elements. Aristotle's works on natural philosophy were as yet unavailable, and even the indirect presentation of his doctrines within Arabic astrological treatises had to wait a bit.

The other school to be mentioned is the open school of the Saint-Victor monastery in Paris (that it was "open", we remember, means that the students were not future monks). Its first head was one Hugh (c. 1096 to 1141), a deeply believing mystic and yet a rationalist engaged in the search for knowledge and in practical life. In 1125 he wrote the *Didascalicon,* a general introduction to studies, covering both the seven Liberal Arts and seven mechanical arts (ranging from theatre performance to trade and textile production) and the sacred readings: the Bible, the Fathers, and ecclesiastical history. During the treatment of the last subject it comes to his mind that one might question its utility. The answer is that

> Some things are to be known for their own sakes, but others, although for their own sakes they do not seem worthy of our labor, nevertheless, because without them the former class of things cannot be known with complete clarity, must by no means be carelessly skipped. Learn everything; you will see afterwards that nothing is superfluous. A skimpy knowledge is not a pleasing thing.
>
> (VI.iii, trans. J. Taylor 1961: 137)

---

19. Thus told by John of Salisbury in his *Metalogicon* (III, 4, from 1159, trans. McGarry 1971: 167). The phrase became so famous that the Danish nobleman Sven Aggesøn (ed. Gertz 1967: 29f) could identify Bernard merely as "the great master" and turn his maxim around jestingly in the late twelfth century, supposing both to be familiar to the educated elite in Denmark.

20. This rather daring deviation from literal belief is formulated in Guillaume de Conches' *Philosophia mundi* I,xxiii (translated from Gregory 1975: 196).

The examples given in the context tell that "everything" is really *everything,* or at least is not restricted to Sacred Scriptures: they deal with his own experiments in geometry, arithmetic, and acoustics and with his observations of the stars (that is, with the complete quadrivium). All were results of the curiosity of his boyhood; but though his knowledge of these fields is now modest, they are still of value.

The name of the third scholar is not associated with any specific school, even though he was driven out from several schools and contributed strongly to make Paris the paramount city of schools. It was Pierre Abaelard (1079 to 1142). He was a famous and eminent teacher, known as the master of dialectic and as the creator of "the scholastic method," and he can be claimed to have opened the way toward the modern notion of "dialectic" (new knowledge or structures engendered from contradiction). This was done in his *Sic et non* ("Yes and No"), where apparently contradictory opinions of the Bible, the Fathers, ecclesiastical councils and other authoritative authors are set forth on 158 questions regarding Christian faith and ethics, without solution of the dilemmas but with methodical advice on what to do and a general exhortation to ask critical questions as the only way to truth:

> By raising questions we begin to enquire, and by enquiring we attain the truth, and as Truth[21] has in fact said, "Seek, and ye shall find; knock, and it shall be opened unto you." He demonstrated this to us by His own moral example when he was found at the age of 12 "sitting in the midst of the doctors both hearing them and asking them questions." He who is the Light itself, the full and perfect wisdom of God, desired by His questioning to give his disciples an example before He became a model for teachers in His preaching.
>
> (trans. Piltz 1981: 82)

Bernard of Chartres as well as Hugh and Abaelard had in the main built their intellectual innovations on the Latin material handed down through the early and central Middle Ages, in combination with their own new approach to the material. During their mature years, however, a new phenomenon appeared—the "wave of translations." What happened can be seen from a well-informed fourteenth-century biography of the most prolific of all translators, Gerard of Cremona: he was "educated from the cradle in the bosom of philosophy," that is, in traditional Latin Liberal Arts; however, dissatisfied with the limits of Latin studies, he "set out for Toledo" to get hold of the *Almagest.* Having arrived, he stayed there translating the Arabic treasures "until the end of life" (Boncompagni 1851: 387ff). Another, anonymous scholar pursuing medical studies in Salerno heard that a Greek copy of the *Almagest* had arrived in Palermo; accordingly he left for Sicily, started preparing

21. That is, Christ. The Scriptural passage "I am truth" was very popular from the late eleventh century onwards. Once, as the Emperor Henry IV had used a reference to the customs of his ancestors as an argument in the Investiture Conflict, the Pope replied that "The Lord did not say 'I am custom', but the Lord said 'I am the Truth'" (Ullman 1970: 131f).

himself by translating some minor works from the Greek, and finally translated Ptolemy's *Great Composition,* as it was called in Greek (Haskins 1924: 159–162). Through the heroic efforts of these and other scholars, Latin learning gained access to most of the works known only by name and fame from Cassiodorus, Martianus Capella, and Isidore, and to a good deal more: Euclid's *Elements,* Ptolemy's *Almagest* and astrological works, Galen's medical treatises, Arabic algebra and "Hindu calculation" (with what we call "Arabic numerals"), and—not least—a fairly complete Aristotle, including his large epistemological works, the *Metaphysics,* and the books on natural philosophy and natural history. Together with these works mostly rooted in antiquity came a large collection of Arabic works serving as explanation and commentary.

At first, few scholars could do much with anything but the most elementary part of this huge meal. The translators did not in general select works to be translated from specific importance. They rushed at whatever important came within their reach and could hardly have done otherwise: the choice may not have been too varied in a place like Toledo after the Christian reconquest; few if any translators, furthermore, had received an education which permitted them to fathom the depths of the texts they translated. Importance, thus, was general, and was determined from appurtenance or affinity to the disciplines of ancient learning. The whole process reminds much of the worship of relics so current in the religious sphere, and it can legitimately be taken as a secular expression of the same general attitude. To borrow the Carolingian expression, the Middle Ages were held in the combined spell of Athens and Jerusalem. Even the twelfth-century Renaissance—whose background was social renewal and a nonderivative intellectual revolution—ended in the main, we may say, by merely shifting the emphasis from Jerusalem toward Athens, and by combining Athens (that is, Greek natural philosophy and mathematics) with Rome (Latin grammar, rhetoric, and the Latin Fathers), which until c. 1100 had been the real perspective on ancient learning.

THE RISE OF UNIVERSITIES

In the end of the twelfth century, the enthusiasm for knowledge thus found its expression as enthusiasm for the reconquered fundament of ancient learning. Whereas conservative theologians in the beginning of the century had condemned Gothic cathedrals and Abaelardian dialectic as expressions of human vanity and arrogance, those of the outgoing century aimed their spear at new enemies, complaining that many Christians (and even monks and canons) endangered their salvation by studying

> poetical figments, philosophical opinions, the [grammatical] rules of Priscian, the Laws of Justinian ["Roman Law"], the [medical] doctrine of Galen, the speeches of the rhetors, the ambiguities of Aristotle, the theorems of Euclid, and the conjectures of Ptolemy. Indeed, the so-called

> Liberal Arts are valuable for sharpening the genius and for understanding the Scriptures; but together with the Philosopher [that is, Aristotle] they are to be saluted only from the doorstep.
>
> (Étienne de Tournais, trans. from Grabmann 1941: 61)

Many Christians, indeed, would rather risk their salvation than stay at the doorstep. The flow of students to the schools, in particular the most famous schools, continued to grow. So did the number of masters, living from fees paid by their students (or, at times, from ecclesiastical incomes from elsewhere) and only marginally submitted to the authority of the chancellor of the episcopal see. Not only professional intellectuals, they were also in practice free intellectuals, as were the Sophists—a rare situation in history (nominally, it is true, being a scholar implied membership in the Ecclesiastical Order).

The most famous schools were those of Paris, Oxford, and Bologna.[22] Those of Paris and Oxford had grown out of the traditional cathedral school system with its emphasis on the Liberal Arts, whereas those of Bologna were originally law schools. In all three cases, the name *universitas* was used from around 1200. The term is simply Latin for "guild," and in Paris and Oxford the name denoted the guild-like organization which students and masters formed together in order to protect their common interests and security. In Bologna, where the teachers were regular citizens of the city and only the students came from abroad, the university was originally the student union.

> The specific character of the schools of Bologna had to do with the particularities of northern Italy. Here, the commercial revolution had begun earlier and developed further than anywhere else, and the towns had gained not only a limited autonomy but practical independence from feudal overlords. Cities like Florence, Bologna, and Siena were effectively independent commercial city republics governed by the more wealthy guilds, and the cradle of commercial capitalism. They had a much more urgent need for people trained in Roman Law than for clerks (however much the clerks of northern universities were in fact going to be administrators), and the teachers of this subject did not have the same reason for conflict with local authorities as those of Paris and Oxford.

The early history of the universities shows that scholars might well need guild protection. It also shows that this protection could be effectual. The main weapons were strikes and emigration. Students, in fact, brought their money from home. If they left a city like Paris, where they may have made up some ten percent of the population (Cobban 1975: 79), the commercial life of the city was severely hit. This was often realized by authorities, who therefore protected the scholars and gave way

22. The medical school in Salerno was perhaps equally famous, and that of Montpellier almost so; but they belong to a particular category, and I disregard them in the present connection.

to many of their claims. In other cases they did not, with the result that scholars left and settled elsewhere. Cambridge is probably the result of an early emigration from Oxford. Padua got its university modelled on Bologna when students left the latter city in 1222. Both Angers and Toulouse owed their universities to an emigration from Paris in 1229. Others could be mentioned, most of them short-lived.

Toulouse, it is true, was not a spontaneous settlement of Parisian scholars. It was, instead, established by the Dominican Order, which took advantage of the occasion when Paris was deserted. It is thus a representative of a third type of university: those founded as universities by some authority. Others belonging to that category are Naples, founded by the Emperor Frederick II in 1224, Palencia (founded by Alfonso VIII in 1212–1214), and the Papal University founded in 1244–1245.

It may seen paradoxical that authorities should found organizations meant to protect against authorities. In fact they did not. Soon after 1200 the term *universitas* came to mean something more and something different. This is a process which can best be followed at Paris, the model of most later universities (Vienna and later German universities were modelled directly on Paris, Copenhagen on Cologne, and so on) and even a model which Italian universities gradually came to approach.

Around 1200, the orientation of Parisian scholars agrees fairly well with the description quoted from a conservative theologian above. Most consequential were the "philosophical opinions" and the "ambiguities of Aristotle"—phrases that refer to interest in the metaphysics and the natural philosophy of Aristotle and to the growing fervour for dialectic (at the cost of rhetoric and grammar, the central disciplines of ancient Roman humanities and of the Liberal Arts until c. 1100). These interests led a number of scholars into nonreligious humanism and into what seems to have been a pantheist heresy, giving rise in 1210 to a process and several executions at the stake. A synod of local bishops then banned lectures on Aristotle's natural philosophy, which may indeed have been part of the inspiration.[23] This and other conflicts made the university appeal repeatedly to the Pope, who accepted the role as protector of the university while at the same time imposing adequate regulations in a number of decrees (decrees that are in fact our main sources for curricula and university organization).

A university hence became a body with a specific set of privileges, especially concerning the right of the masters to confer the *license to teach* at all similar institutions (an institution with this privilege was called a *studium generale*), and certain obligations. In Paris (and to a greater or lesser extent elsewhere) the studies were

23. See Kurdziałek 1976. The problem may have to do with the doctrines about the soul as set forth in *De anima* and elsewhere in Aristotle's "books about nature." If the soul is in some way the *form* of the body, that is, its organization as a functioning living being (*De anima* 412^a17–412^b9) then, firstly, the immortality of the soul seems dubious (CUPHOOD has no existence separate from the cup); secondly, the privilege of the human soul as compared to that of other animals turns out to be only relative.

organized in a sequence of faculties. You started studies at the age of 14 at the "Arts Faculty," where for some six years you pursued "basic studies"; the first years were spent on the Liberal Arts in general, the last increasingly on dialectic and natural philosophy. After having received your license in the Arts you might continue studies at one of the "lucrative faculties" (Canon Law or Medicine) while teaching yourself as a master at the Arts Faculty. Studies at the Theological Faculty (also "lucrative") might follow and be supported by teaching of Medicine or Canon Law. Most students left university after finishing the Arts studies, or even earlier.

ARISTOTELIANISM

Étienne de Tournais had considered Euclid, Ptolemy, and Aristotle to be equally dangerous. In 1210, however, only Aristotle's natural philosophy was condemned, which we may take as an indication that the study of Euclid and Ptolemy might perhaps endanger the salvation of individuals but was not likely to disturb the worldly tranquillity of authorities.

There were good reasons for this. Then as now, only a minority would find it attractive to invest more labor in mathematical studies than required by the syllabus (and Ptolemy's astronomy is no less demanding in this respect than Euclid).[24] Aristotle, on the other hand, offered much more direct and all-encompassing insight into the workings of the world through his natural philosophy and his metaphysics. Both because of the way they were formulated and because his teachings constituted a relatively coherent *system,* Aristotle's works also corresponded better to the deeply dialectical mood of twelfth- and thirteenth-century learning than any potential competitor.[25]

---

24. Even though the study of astronomy and astrology had originally been rooted in enlightenment aspirations, most scholars would get lost in technicalities long before they got sight of this goal, whereas those who mastered the technicalities would mostly become astrological technicians. Only insignificantly few would reach a level where they might experience astrology as "without doubt the most faithful herald of the immortal God who, interpreting his secrets, displays the Law according to which the Almighty resolved that the Heavens be made, on which he sprinkled the starry fires, testimonials of the Future" and be convinced that "this angelical doctrine makes us no less kindred of God than we are separated from the beasts by the other arts" (as formulated by the fifteenth-century astronomer Regiomontanus; translated from Schmeidler 1972: 52); technicians working from simple handbooks and schemes had no reason not to see *the Church* as "the most faithful herald of the immortal God," as they and everybody else were supposed to.

25. This formulation presupposes that we count as variants and not as competitors the interpretations of Aristotle which the Islamic philosophers Avicenna (ibn Sīnā, 980–1037) and Averroës (ibn Rušd, 1126–1198) had produced. These, indeed, were *systems* to a higher degree than the original. As a consequence, Aristotelianism was first received in the form of Avicennism, and later as Averroism.

Early thirteenth-century university scholarship was thus drawn irresistibly toward Aristotelianism, in spite of (in some cases no doubt because of) its inherent challenges to Christian doctrines—for instance that it excluded that the world could have a beginning and an end, thus denying the Creation as well as the Day of Judgement.

The prohibition of 1210 was repeated by a papal representative in 1215 and extended to the *Metaphysics*.[26] In 1231, the Pope repeated the prohibition once again, ordering at the same occasion that those who had trespassed should be absolved—a double indication that the ban had not been very effective. In 1231, moreover, a committee was ordered to prepare an inoffensive edition of Aristotle's books on nature:

> . . . since, as we have learned, the books on nature which were prohibited at Paris . . . are said to contain both useful and useless matter, lest the useful be vitiated by the useless, we command . . . that, examining the same books as it is convenient subtly and prudently, you entirely exclude what you shall find there erroneous or likely to give scandal or offense to readers, so that, what are suspect being removed, the rest may be studied without delay and without offense.
>
> (trans. Thorndike 1944: 40)

The chairman of the committee died within the year; probably for this reason, nothing came out of the initiative (Grabmann 1941: 108). In the 1230s, however, the situation became untenable for the conservatives, as even their own theological treatises were fully permeated by Aristotelian metaphysical concepts. It was clearly demonstrated not only to us but also to contemporaries that the Aristotelian system was necessary. The university environment could not do without the intellectual coherence offered by Aristotle and by no other system.

At the same time, the triumph of Aristotle was a symptom that university learning was becoming specialized, and that its close interaction with general currents was on the wane. Only within the professional environment of university masters could a climate of dialogue and controversy be regulated by the straitjacket of

---

26. Once again for good reasons, we may add. One of the books (Λ) contains the young Aristotle's "theology," still rather Platonic (reinterpreting the Platonic GOOD as the divine Unmoved Mover) and certainly in need of heavy reinterpretation if it was to be brought into agreement with Christian theology (cf. Randall 1960: 108f)—not least because the Unmoved Mover was unmoved to a degree that it would never interfere with a world whose movements it caused by *being loved or desired,* and therefore could not be identified with divine Providence.

Modern scholars turn out to have greater difficulties than their medieval predecessors with the notion of causing simply by being desired. For those who have never fallen in love with a person who did not (yet) care, the example from 2 Sam. 11:2–4 may be useful: Batsheba only needed to be there, to be seen, and to be desired in order to make King David move violently.

scholarly dialectic, and nowhere else could the quest for intellectual coherence and system become paramount.

Already during the conflicts of the early thirteenth century, the university environment was thus preparing its eventual integration into the midcentury synthesis or compromise, to which we shall return; no wonder, since the majority of students and masters were, after all, preparing for future employment within the secular or the ecclesiastical establishment.

Other social groups had gone the opposite way. Already during the second half of the twelfth century, that specific urban piety which was mentioned above (p. 61) developed into authentically heretical movements which, in particular from the early thirteenth century onwards, were submitted to large-scale persecution. The verdict of 1210 is a modest instance, and the crusade against the Cathars in Southern France the most horrifying example.

The origin of the mendicant orders is to be sought in this context. In 1208, St. Francis of Assisi had begun preaching apostolic humility and poverty, thus pursuing the same road as some of the late twelfth-century heretics; in 1209, however, his group of followers was recognized by the Pope as a regular monastic order (the Franciscans or Friars Minor), with ensuing regulations on doctrine and activity. In 1215, St. Dominic received a similar approval of the "Order of Preachers" (better known as "Dominicans"), who were to "fight heresy by means of sword as well as fire as well as tongue." For tactical, not for ideal reasons, even the Dominicans were to live in apostolic poverty; but they were also founded as a learned order, and from the beginning almost half of St. Dominic's followers were sent to Paris "in order to study."[27] Although that had never been the aim of St. Francis, the Franciscans developed in the same direction, and soon both orders received as recruits many university scholars who would rather pursue study than turn to trite clerical work.

Two eminent Dominican friars who were active in Paris also accomplished what the committee of 1231 had been unable to. Around 1250, Albertus Magnus (1193 or possibly 1206/07 to 1280) wrote a large commentary to Aristotle's *Physica,* the first of the "books on nature," telling in the preface that he did so "in order to satisfy the brothers of our Order, who now for several years have asked us to compose a book on

27. The approval of Dominic's group was unusually prompt—the Church was not too fond of the mushrooming of new monastic orders. There are good reasons to believe that the swift acceptance of the group as an official order was due precisely to the prospect of improving the intellectual level of the clergy. In spite of the expansion of schools during the twelfth century, ignorant priests had remained a problem to the Church (cf. Mandonnet 1914).

The original aim of Dominican studies was theology, and as late as 1228 it was ordered in the statutes of the Order that Dominican students "should not study the books of the pagans and the philosophers, even though they may inspect them when appropriate. They should not learn the secular sciences, nor the so-called liberal arts, unless some master of the Order or the General Chapter should dispose otherwise" (see Denifle 1885).

physics, in which they might have the perfect science of Nature, and which would enable them to understand Aristotle's books."[28] After this beginning, he continued through most of the Aristotelian corpus, and even supplemented it with books on missing subjects (one of these, *On minerals,* was used as a practical handbook for centuries).

From 1250 onwards, St. Thomas Aquinas (1225 to 1274), also a Dominican, built up a complete philosophical system in agreement with Christian faith but on Aristotelian foundations in partially Neoplatonic interpretation—the "Thomist synthesis," which was no less a systematic interpretation of Aristotle than those of Avicenna and Averroës, and which managed to reconcile both the Unmoved Mover (see p. 70, n. 26) and the understanding of the soul as a form (see p. 68, n. 23) with Christian doctrines.[29]

A famous dictum is often used to sum up the essence of the Thomist doctrine: "Grace does not abolish nature but brings it to perfection." The implication is that (Aristotelian) natural philosophy is considered valid and is allowed to explain as much as it can; "Grace" (that is, divine intervention and explanation) only enters as a complement and where natural explanation fails. Similarly, the principles of "natural law"—those principles which can be derived from Aristotle's political philosophy as understood at the time—are accepted as valid in any society, Christian as well as non-Christian; revealed truth (the teachings of the Bible) can only specify and complement them, but cannot abolish them.

The dictum was not a mere philosophical principle. It was also a rationalization of the division of the university into faculties and of the autonomy of the Arts and Medical Faculties (and, where it existed, of the faculty of secular law); in these, the study of natural philosophy and of presumed natural law should be allowed without constant policing on the part of the theologians.

Another feature of the Thomist system had similar implications: its emphasis on the Aristotelian division of knowledge into separate domains, each derived from

28. The quotation asks for several terminological commentaries. Firstly, *physics* (*physica*) does not mean what it means to us; it is simply the ancient Greek term for "nature," the inherent properties of things and beings. *Science* (*scientia*) designates any field of knowledge which is organized according to the precepts set forth by Aristotle.

In view of the recent regulations of Dominican studies (cf. p. 71, n. 27), this role not only of a single Dominican scholar but of his fellow-brothers in general is quite striking.

29. In the beginning, Thomas's theological doctrines were met with some resistance, not least on the part of Franciscan theologians, who would rather stick to Augustine's more directly Neoplatonic teachings (but Dominicans and others were also involved). After having been declared the official doctrine of the Dominican Order, however, the system was adopted in the 1330s as the official philosophical stance of the Church as a whole (remaining so until the Vatican II Council in the 1960s). From this moment on (and only then) is it legitimate to speak about a full Aristotelization of the Catholic doctrine.

its own specific set of principles or axioms. Once again, this agrees with the compartmentalization of university knowledge into the domains of separate faculties, each governed by its own principles and not supposed to encroach upon the territories of the others. Obviously, one exception to the general principle of mutual autonomy should be remembered: the teachings of other faculties were ultimately to be completed by (and hence also to be compatible with) "Grace," that is, theology and its basis in revealed truth.

A document from the Arts Faculty in Paris from 1255 (trans. Thorndike 1944: 64f) tells the curriculum for the coming academic year in terms that presuppose the content to be already familiar but the lectures often too superficial in view of the difficulty of the texts. Apparently only the mature level is concerned, and the list seems not to be complete. With this proviso, Aristotle overshadows everything else, being accompanied in the document only by a few Boethian commentaries to his logic and some texts on grammar.

Aristotelianism had thus won the day. It would be mistaken, however, to see the outcome of the process as a victory for the radical thinkers of 1210. What won the day was an Aristotelianism that had been moulded by the "repressive tolerance" of the Albertine-Thomist synthesis (to use a term which was coined to describe an analogous process in our own century), and the environment in which it won was no longer a major threat to social and intellectual stability. In 1210, in connection with the ban on Aristotle's natural philosophy and the condemnation of the heretical priests, the diffusion of theological treatises translated or written in the vernacular had been forbidden; mid-thirteenth-century university annals offer no similar evidence of interaction between scholarship and lay religiosity.

THE COMPROMISE

The transformation of Aristotelianism exemplifies a general trend of the mid-to-late-thirteenth century toward balance or compromise.

Balance was a general social phenomenon. For a while—that is, as long as moderate economic expansion continued—open fighting between Papacy, secular rulers, nobility, and commercial towns had declined or ceased; large-scale revolts in towns and in the countryside were phenomena belonging to the past (and, as it turned out, the near future).

Within the university, the masters of arts had become a semi-autonomous but also an isolated professional group. This is appropriately demonstrated by one of the condemnations of supposedly heretical scholars that did take place.

The scholar in question is Boethius de Dacia (*fl.* c. 1275), who was accused of being an "Averroist," that is, a proponent of an Aristotelian system which had not gone through the Thomist domestication. In a treatise *On the Eternity of the World*, he distinguishes, on one hand, "the truth of [Aristotelian] philosophy," which asserts this eternity, and on the other, "the truth of Christian Faith, which is absolute truth," which denies it. The style of the work, not least the use of emphasis and jokes, leaves no doubt in me that Boethius is sincere in equating the truth of Faith with genuine truth.[30] The truth of philosophy—thus the solution to the apparent dilemma of the "double truth"—was only *established* as a consequence of the creation of the physical world, and it will be abolished at the last judgement; between these two limits, it is obvious, neither beginning nor end of the world can take place. As a master at the Arts Faculty, that is, as *a philosopher* (no longer, we observe, a teacher of the liberal arts), none the less, Boethius explains it to be his duty to pursue *the truth of philosophy.*

The underlying inclination toward mysticism goes against Thomas's belief that reason and faith can be harmonically combined, and points forward towards certain late thirteenth- and early fourteenth-century philosophers, such as Meister Eckehart, Duns Scotus, and William of Ockham—not to mention later Protestant currents. The proclamation of an autonomous sphere of knowledge which the philosopher should pursue, however, is in line with the Thomist program even if it goes beyond the limit which Thomas found acceptable. When, in 1277, Boethius's theses were condemned by the Paris bishop, it is also characteristic that some of Thomas's opinions were included (Thomas had died in 1274); ultraconservatives saw no decisive difference.

That Thomas's as well as Boethius's stance is to be explained with reference to the sociology of the university institution, and not solely from the impetus of Aristotelian epistemology, is indicated by the failure of attempts to secure autonomy for domains *within* the complete range of subjects covered by the masters of arts.[31] The domain which achieved epistemological autonomy was thus not defined by epistemological criteria, that is, by shared methods or by the subject-matter to be dealt with; it was demarcated by a purely social institution. Autonomous knowing could be accorded to people who ran an autonomous institution and who knew to do this without disturbing the compromise which this institution had made with stronger powers.

The effects of the professionalization of university teaching thus merged with those of the violent suppression of heretic movements (the Cathar crusade!) and of the primitive-democratic tendencies of towns for which ecclesiastical and royal

---

30. One example: who denies that the dead will be resurrected is a heretic. But he who tries to prove it by means of reason is a fool! (ed. Sajo 1964: 51). Thus speaks a genuine believer who appreciates the use of reason but feels that the mystical experience of his faith goes deeper.

31. One such attempt was made for mathematics by Jordanus de Nemore, one of the two best European mathematicians of the thirteenth century—in my opinion clearly the more original of the two. Even his closest followers either did not understand his aim or did not bother (cf. Høyrup 1994: 195–197).

authorities were responsible: the connection between universitarian and popular politico-religious discourse became tenuous and mostly nonexistent.

It is true that certain fourteenth-century scholarly conflicts on the border between philosophy and theology were connected with important political conflicts—but mostly conflicts between the Papacy and secular royal powers. The only scholarly conflict with heavy impact on fourteenth-century popular heretical movements (which were important, since the social compromise did not outlive the thirteenth century by many decades) was located within the Franciscan Order. It concerned a group within the order (the *spirituales*) which refused its development away from absolute poverty toward participation in the scholarly world and in the "Scholastic compromise."[32] It is thus merely a pseudo-exception which, when inspected more closely, confirms the rule that the scholastic compromise implied an interruption of the connection between popular and scholarly discourse. Only toward the very end of the fourteenth century was this dialogue to be revived sporadically in the wake of John Wycliff's activity, with consequence only in the Hussite movement and the early Reformation of the fifteenth and early sixteenth centuries.

One particular development should be mentioned which contributed to severing the ties between scholarly and popular discourse. The masters of arts were gradually loosing their position as free intellectuals. This changed the character of universities as seen from within and eventually undermined the autonomy of the "artists". One reason for this change of condition is that specific chairs were established, often at colleges supported by an endowment. Originally, *La Sorbonne* in Paris was one of these, endowed by the theologian Robert Sorbon and meant to shelter students of theology; the College system of Oxford and Cambridge also has its origin here.

Another reason, for a while more important, is that an increasing number of teachers were Dominicans or Franciscans. These were primarily members of their order and not of the university understood as a guild. They would therefore not be solidary during strikes, and could be suspected of being more loyal to Church and "Grace" than to their fellow masters and to "Nature."[33] Initially, the friars were therefore met with strong resistance by other masters. In the end, however, they had to be accepted, among other things because universities needed the Papacy as an

32. That we should rather speak of a "Scholastic" than of a merely "Thomist" compromise follows from the interpretation of the main tenets of Thomas's philosophy as expressions of broader movements in the culture and institutional framework within which Thomas and other university scholars worked.

33. Certain decrees censuring the behavior of scholars belonging to the mendicant orders tell us that some of them were actually more interested in "Nature" than in "Grace." As already hinted at, entry into a friars' learned order could be a way for scholars to remain scholars instead of leaving the intellectual environment of the university.

ally against local authorities—in Paris, where the conflict was strongest, probably also because Thomas Aquinas the Dominican was found by the students to be a far more interesting teacher than his secular competitors.

THE FOURTEENTH CENTURY

As mentioned in passing, the multilevel balance of the mid-thirteenth century did not last long. When seen in the context of demography and economic history, it can in fact be understood as the brief interlude between the crisis of growth of the early High Middle Ages and the crisis of decline of the late Middle Ages.

The cultural bloom of the late eleventh and the twelfth centuries had grown out of demographic expansion based on improved agricultural methods and of the rise of towns and commercial economy which it made possible. By the outgoing thirteenth century, the new techniques and increased population pressure had exhausted the soil in many areas, and population growth stopped. To make things worse, the climate began deteriorating. In the 1320s, protracted warfare between England and France set in; this "Hundred Years War" went on with interruptions until the 1450s. It was followed by bankruptcies among the largest North Italian bankers, who had invested in quick victories. Worst of all was probably the plague, which swept over Europe in the end of the 1340s and cut the badly fed population on the average by some thirty-five percent. In many of the commercial towns, violent rebellions and civil war broke out between the mercantile patricians and poor artisans and workers.

The population decline in the countryside created a shortage of manpower, thus leading to a reduction of the value of landed property. Attempts to increase the exploitation of feudal peasants only resulted in rebellions, which at least in England and France were far more successful than those of the working population of the towns. The Church, the largest landowner of all, was impoverished; enforced political submission of the Papacy to the French king led to conflicts with other secular rulers and in the end to the Schism of 1378–1417, where two (for a short while three) rival popes existed.

These political and economical turmoils affected the universities and university learning in several ways. First of all, recruitment changed, becoming increasingly dominated by the upper social levels; gifted peasant's sons, quite visible during the thirteenth century, became rare in universities.

The reduction of ecclesiastical income from landed property after the plague also affected the universities directly, since the Church (and local churches) had financed much of what went on in universities: students going to the higher faculties and possible teaching the Arts on the same occasion, as well as colleges. Both the intellectual level and the social status of university activity was lowered in this process; a reform edict from Paris from 1366 shows this quite clearly in its attempt to repair some of the damages. It also proclaimed that the "scholars hearing their lectures in the [Arts] faculty sit on the ground before their masters, not on seats or benches raised above the ground, as was the custom when the studies of the said faculty were more flourishing, so that occasion of pride may be removed from the young." (trans. Thorndike 1944: 246)

In the end, this development killed much of the intellectual creativity of the university environment. Yet during the decades of incipient crisis, that is, until the effects of the plague attained their full impact, certain developments took place which are not only interesting in themselves but also illustrative of the interaction between institutional environment and style of thought—and thus generally informative about the humanities as a social enterprise.

Some of the philosophical developments and some of the conflicts between different philosophical schools were superficially mixed up with the political conflicts of the day. Political partisanship and conflicts, however, were not the driving forces behind the increasingly sophisticated approach to dialectic and natural philosophy. Instead, a highly original approach to the problems of language and meaning and an unprecedented attempt at mathematization of natural philosophy were introduced. They were certainly not meant as investigations of practical discourse nor of real nature; instead we must see them as *testing and display of the key professional tools and disciplines of the masters of arts:* logic and natural philosophy. Structurally, this is similar to what was done by the Fara scribes around 2500 BCE and to processes taking place in many environments populated by professional intellectuals. The outstanding sophistication of the fourteenth-century developments and their isolation from common sense and from all everyday concerns were only possible, however, because the masters of arts as a group were highly specialized and professionalized, and because their professional activity (as long as they remained masters of arts) was itself disconnected from everyday intellectual practice (be it administration, juridical practice, secretarial functions for rulers, or preaching).[34]

Contemporaries were quite aware that something new was produced. They spoke of a *via moderna* in philosophy, a opposed to the *via antiqua*. The latter term covered not only (and not so much) ancient philosophy in general as the kind of Aristotelianism that had established itself during the thirteenth century—not least as embodied by Albert and Thomas. Like Aristotle, the *via antiqua* was "moderately realist," that is, it held that "universals" ("Cuphood," etc.) are real but only exist by partaking in individuals (the Dog as a species is no free invention but the shared "form" of all individual dogs).

---

34. This does not imply that no participant in the movement was engaged in such functions, which is certainly not the case. What is important is that *the environment which defined* what was of interest and produced the norms governing philosophical activity was disconnected from worldly affairs.

The *via moderna,* on the contrary, was nominalist and proto-positivist. "The Dog" is nothing but "a puff of the voice," to quote a favorite expression, and much effort was invested in exploring the relation of language and logic to reality.[35]

In a certain sense, the *via moderna* was thus built on Aristotelian concepts, and it investigated problems arising within Aristotelian logic and Aristotelian natural philosophy. But it did not feel obliged to take these concepts as Aristotle or the commentators of the *via antiqua* had interpreted them. The backbone of the mathematization of natural philosophy, for instance, was the idea that the Aristotelian *qualities* could be varied continuously in numerical degree.[36] In spite of its Aristotelian fundament, the approach of the *via moderna,* and even its way to discover problems, was hence quite new.

So new in fact, and so different from anything which had come before, that many aspects of fourteenth-century philosophy were not understood during the Early Modern period but only on the background of twentieth-century semantical theory and abstract algebra—that is, when seen in the perspective of disciplines which themselves are products of highly specialized and professionalized academic environments.

Some broad features of the development from c. 1150 to c. 1400 can be summed up as follows:

— Scholasticism, which literally means nothing but the *learning of (medieval) schools* from 1050 onwards, ripened into the particular style of the "mature" medieval university. This has come to be the normal interpretation of the word,

---

35. Without pursuing their particular ideas and doctrines, we may list some of the foremost representatives of the *via moderna:* William of Ockham (c. 1285–1349), Jean Buridan (b. 1285, d. after 1358), Richard Swineshead (*fl.* c. 1340–1355) and Nicole Oresme (c. 1320–1382).

36. For instance, cold, heat, moisture, and dryness—the qualities which were bound up with the doctrine of the four elements and with medical doctrines. To those who have grown up with thermometers and hygrometers, numerical gradation of these qualities is a matter of course, but according to traditional Aristotelians they might well "admit of variation by degree," as stated by Aristotle (*Categories* 10^b26), but it would be as meaningless to ascribe numbers to the degrees of cold as to the degrees of justice or health. It is remarkable that the main idea behind later mathematizations (be it of physics, biology, or linguistics), namely that the numerical values to be ascribed to a quality should correspond directly or indirectly to something which can be *measured,* was totally absent from the fourteenth-century "quantification of qualities." The concepts of which it made use were in the style of "twice as cold" or "three times as healthy," which we will probably find just as absurd as did the proponents of the *via antiqua* (even though, admittedly, we have got accustomed to the equally absurd numerical marks given in school). The quantifiers, on their part, did not claim that their ascription of numbers had any relation to reality; they were probing the capacity of their tools "according to hypothesis," as they would tell.

often colored by the negative attitude of Renaissance polemicists toward the style.

— Whereas the cognitive interest of twelfth- and early thirteenth-century university learning was often emancipatory (whence the enormous enthusiasm for the new learning), that of the late thirteenth- and fourteenth-century university was rather legitimizing, an ideological support for status interest and status consciousness.

— Through the reconquest of ancient *philosophy* (as opposed to the remains of polished Roman *culture*), the twelfth and early thirteenth centuries had reached that "Athens" which medieval scholars had only dreamt and spoken of until then. Truly, this Athens still clung to the *texts* of antiquity, using the Abaelardian (so-called "Scholastic") method to make them agree; but through the sophisticated innovations in semantics and logic and through the quantification of qualities, university scholars had even produced something *new*, starting from, but not really restricted to, Aristotle.

— On the other hand, a scholarly culture had been created which seemed increasingly irrelevant even to educated people outside the university sphere. Only the astrological counter-current, which, admittedly, grew quite strong at the universities of the later fourteenth century, seemed to carry a relevant message. Fourteenth-century Scholasticism can, on the whole, be seen as a brilliant but late intellectual afterglow of a general social compromise between conflicting forces which had since long ceased to be tenable.

THE POST-MEDIEVAL UNIVERSITY

As this anachronistic orientation combined with the effects of impoverishment in the late fourteenth century, a genuine intellectual decay set in. Already in the outgoing fourteenth century, university learning is no longer adequately described as oriented toward sophisticated logic, semantics, and (bookish or hypothetical) natural philosophy. It was oriented toward the sophistication of the earlier fourteenth century, that is, toward what had been created and canonized before 1350; we may speak of "Aristotelianism by inertia." New works were still written, but mainly compendia introducing students to the original thinking of the early century. Very broadly speaking, the decay process accelerated after 1400. It is characteristic that the "new" books (that is, books not written during classical antiquity) that were printed in the university towns around 1480 would mostly either be compendia written a century before or original treatises written between 1200 and 1350! Exceptions exist, but they are rare.

Grosso modo, universities had become fossilized, uncritically dogmatic schools for administrators, physicians, lawyers, and priests—and most of them retained that character until 1800 or later. This does not imply that nothing new entered university learning for 400 years. But curricular novelties entering a university during these centuries would mostly be a hundred years old or more, except in cases where it was the result of a reform guaranteed and enforced by higher authorities. To mention but one example, Isaac Newton's infinitesimal calculus (created in the late seventeenth century) only entered the curriculum of his own university (Cambridge) during the 1820s. In contrast, Thierry of Chartres had used books for his teaching in Paris in 1145 that had been translated no earlier than 1140 in Toledo (this was before the invention of printing!).

In brief, universities had become enclaves isolated from the real life of their period—from what went on in the humanities, in natural science, and in general culture. No wonder that university scholarship and university education were ridiculed and parodied in Thomas More's *Utopia,* François Rabelais' *Gargantua et Pantagruel,* Jean-Baptiste Molière's *Le malade imaginaire,* and Ludvig Holberg's *Erasmus Montanus.*

CHAPTER 4

The Renaissance

RENAISSANCE AND HUMANISM

The basic feature of that "real life" to which universities only reacted passively and torpidly was a thorough (though of course not quite sudden) transformation of economic and societal structures. With local stops and starts, towns and commercial activities continued their growth to the point where the commercial capital of towns became the main determinant of economic life: agricultural production under more or less feudal conditions was still the major component of the economy, but even agricultural production was to a large extent made for the market and not for local consumption.

In Italy—the cradle of the Renaissance movement—many of the commercial towns had been independent city republics ruled by the burghers (the members of artisans' and trading guilds) or by the commercial patriciate (the merchants' and bankers' guilds alone) at least since the twelfth century. From the late fourteenth century onwards, the prevailing tendency was a constitutional change toward some kind of monarchic rule, or toward republics ruled by a nobility emerging from the commercial patriciate but increasingly burying its wealth in landed property. In Northern Europe, where towns had never gained more than autonomy (most developed in Flanders and the German Hanse), the growth of mercantile capitalism was linked with an increasing growth of statal power at the cost of feudal "freedoms"—most markedly in Tudor England and the Netherlands.

Culturally, this development was reflected in growing self-consciousness outside the ecclesiastical and universitarian spheres. In Italy, on which we shall concentrate first, the new culture flourished most conspicuously in the vicinity of the new princely courts—not least in Rome, the Papacy being the most wealthy and the most powerful of the courts and behaving quite in the manner of a secular court.[1]

1. In the middle of the fourteenth century, before courts and courtly culture developed in the city states, several important participants in the new cultural movement were connected to

It may seem a paradox that the cultural expression of the new age was most vivid in an environment which in some respects was retrograde; after an expansion of mercantile capitalism during the thirteenth and fourteenth centuries, the Italian city states were moving toward what has been called a "re-feudalization." The paradox is only apparent, however, unless one has a very mechanical view of the connection between socioeconomic structure and cultural expression. The new aristocracy of Italy was certainly as eager to gain honor as the feudal knights of twelfth-century France. But the conditions on which honor could be gained were different. Precisely because the ground had been prepared, and because of the connection to a still vigorous commercial and urban life,[2] honor came to be based on an aristocratic transformation and accentuation of those cultural values which had developed and established themselves during the fourteenth century.

Central to these values are ideas and practices covered by the terms *Renaissance* and *Humanism.* Both the Renaissance and the Humanist movement originate in mid-fourteenth-century Italy, Francesco Petrarca (1304 to 1374) and Giovanni Boccaccio (1313 to 1375) being pivotal figures; and both spread to the rest of Western and Central Europe over the following two centuries. "*The* Renaissance," it is true, was not a current expression during the period itself. But metaphors of rebirth were current, and more widespread than they had been during the Middle Ages. The cultural movement of the fourteenth to sixteenth centuries is also spoken of as a renascence with much better reason than the various medieval revivals. These are, indeed, best described as attempts to restore what had been lost (vain attempts until the twelfth century). That renascence of antiquity which took place during the Renaissance epoch, on the other hand, was really a new birth of ancient material to (new and different) life. The "Renaissance renascence" can thus be interpreted as a *reassimilation of forgotten aspects of ancient culture* to that new development which had started in the late eleventh century, and whose first products had been the Gothic cathedrals, the "twelfth-century renaissance," Scholastic culture, and quite a few other things not mentioned in these pages.[3]

Avignon, at that time the abode of the pontifical court. One among them was even outstanding: Petrarca.

2. Investigating the birthplaces of 600 writers and artists born between 1340 and 1520 and constituting the Italian "cultural elite," Peter Burke (1972: 36, cf. 293–302) finds that 60 percent come from large and middle-sized towns, where only 13 percent of the population was found.

3. As Marie-Dominique Chenu, an illustrious Dominican scholar, once said at a congress on medieval philosophy, after a session discussing the *artes* in the twelfth century: it is a pity that time has not allowed us to discuss the *ars amoris,* which, after all, also had an appreciable success in our epoch.

That the feeling of closeness to the ancients—as comrades in arms rather than as "authorities" in medieval style—might penetrate even the private life of Renaissance intellectuals is illustrated by a famous letter written by the politician, political philosopher, and historian Niccolò Machiavelli (1469 to 1527) in 1513.[4] Having played the wrong card in Florentine politics, he had retreated to a small estate; in the letter he tells how he spends the day. Two passages are significant:

> . . . Having left my wood, I go to a fountain, and from there to my aviary. I bring a book, either Dante or Petrarca, or one of those minor poets, like Tibul, Ovid, and the like: I read about their amorous passions and their loves; I remember my own, meditations which I relish for a while. . . .
>
> When evening has come, I go home, and enter my cabinet; and already at the threshold, I take off my everyday clothes, covered by dirt and mud, and dress in robes suited for the royal or pontifical court. Thus, decently costumed, I enter the ancient courts of the men of antiquity where, gently welcomed by them, I nourish myself by that food which is truly mine, and for which I was born. I have no shame to speak with them, and ask for the motives of their actions, and they, thanks to their humanity, answer me. For four hours I experience no trouble whatsoever, I forget all my distress, I no longer fear poverty, death does not frighten me: I consign myself entirely to them. And since Dante says that there is no knowledge if one does not retain what he has understood—I have taken down from these conversations what I found essential, and I have composed a booklet. . . .

—namely the treatise *The Prince,* which was condemned publicly and studied eagerly in private by countless statesmen in subsequent centuries.

Even the term *Humanism,*[5] like *the Renaissance,* is anachronistic, though in a different manner. Strictly speaking, it refers to the Renaissance concept *studia humanitatis. Studia humanitatis,* however, was no general study of the humanities, of human beings as participants in a culture, nor *a fortiori* an expression of the idea that the human being (thus neither God nor tradition nor Nature) is the center or measure of all things. *Studia humanitatis* was in the main a study of what Cassiodorus had termed *litera humana,* more precisely the study of the subjects which were central to *good style:* (Latin) grammar, rhetoric, poetry, history, moral philosophy. A *Humanist* was somebody engaged in or teaching the *studia humanitatis.*

---

4. Letter to Francesco Vettori, 10 December 1513, (ed. Bonfantini 1991: 94–98).

5. In order to minimize ambiguities, I shall capitalize *Humanism* (and its derivatives) when the term refers to the Renaissance phenomenon; *humanism,* when not capitalized, refers to an attitude which can be found in all ages—cf. p. 62. Protagoras (above, p. 34) was a humanist, but no Humanist.

The immediate and practical reason for cultivating these exercises of good style was their utility (a key word in texts from the age). Humanists were secretaries (literally: somebody initiated in the secrets of the master), counsellors or chancellors to patricians, high prelates, princes, and city republics; their tasks would be to write the official correspondence of their employers, the history of their family or of the city itself, and what else needed to be taken down in good style.[6] They were also teachers training others to perform these tasks.

But even though the immediate purpose of the human studies was utilitarian, their implications and impact went much further—the "humanity" of the ancients referred to by Machiavelli certainly does not refer to their training in the *studia humanitatis:* Humanist culture was moulded upon the literate culture of the Roman upper class and hence also regarded as the symbol and the guarantee of personal and especially civic qualities—utility, indeed, was always meant as civic utility. As a consequence, sons (and a few daughters) of families belonging to the upper social echelons, and even sons of princes, were educated by Humanists, or sent to their schools in order to learn to speak and write Ciceronian Latin as if they had been native speakers. The age was one of individualism—both within the broader sphere of the urban patriciates, where economic structures were reflected in ideology, and among the princes themselves, whose world was unstable enough to require specific personal distinctions from anybody who was to gain or conserve power. Even though these distinctions had little to do directly with Humanist culture, their abstract expression as *individualism* provided the connection. Even princely self-esteem (and esteem on the part of others, which would certainly have political importance) could not build on the mere possession of princely social status: status had to be combined with qualities belonging to the prince *as a man*—a man like others, but a better man than others.

To master the *study of humanity* was thus, automatically, to be *a better man.* The reason that the term Humanism could become and stay popular (and the reason that it coincides with the name given by Old Babylonian scribes to *their* specific qualities) was this inherent ambiguity.[7]

---

6. This professional role of the Humanists is reflected in the scientific method they used when involved in textual criticism, from Petrarca onwards. Trained in the techniques of exposing fraudulent juridical documents, in particular forged donations of privileges or "freedoms"—and, one may safely assume, familiar with the techniques of forging documents that might escape detection—they made use of the same techniques when proving (to mention the most famous examples) that a privilege allegedly granted by the Roman Emperor Constantine to the Pope in the fourth century was in fact expressed in the language of the seventh or eighth century (Lorenzo Valla, 1440), and that a body of writings attributed to the semidivine and fully legendary Hermes Trismegistus (supposed to prove that ancient mystics had had access to fundamental Christian teachings already at Moses's time) was written in the typical Greek of late antiquity (Isaac Casaubon, 1614).

7. In fact, the ambiguity was triple: in Cicero's Latin, *humanitas* would often mean "kindness" or "gentleness" (this is also the usage in Machiavelli's letter). Though neither Renaissance

In a technical sense, Humanism was nothing but a return to the ideals of pre-Abaelardian literate culture: Latin grammar *cum* literature, and Latin rhetoric. As it had happened at every cultural revival, the Humanists took their material from the ancient heritage. But the two undertakings are separated by a leap in quality which makes this technical comparison highly misleading. The study of literary fragments in traditional grammar had aimed at familiarizing students with sentences and grammatical structures. The study of Latin literature in the schools of the Humanists aimed at knowing them intimately enough to use them in allusions and for producing the right connotations when you expressed yourself, and hence also at understanding *their* allusions and connotations. Whereas traditional grammar had (mostly, and grossly speaking) used its literary fragments as a phrase book for tourists ("Can you tell me way to the barber's shop?" "What is the price of a cup of coffee?" etc.),[8] the aim of the Humanists automatically forced them to read the historians as history, the tragic authors as tragedy, the poets as poetry—precisely as Machiavelli did: "I read about their amorous passions and their loves; I remember my own."[9] Since one of the ever-recurrent presuppositions of ancient Latin literature was the importance of Greek letters, they would take up the study of Greek literature to the extent that it became possible, thanks not least to the assistance of Byzantine scholars.

The new approach to antiquity served as a pretext for emancipation from the fossilized rationality of late Scholasticism—better, perhaps, as a pretext for disregarding it as irrelevant. *True Reason* was the reason of antiquity (which implied that Seneca's and Cicero's Latin moral meditations replaced Aristotle's ethics in Thomistic interpretation—no philosophical progress but probably more adequate for practical life in the ruling strata of the city states). *Good Latin* was the Latin of Cicero, and not the crude and supposedly degenerate Latin of the Middle Ages. *True Christianity* was the Christianity of St Augustine, and not the Thomistic synthesis between Christian theology and Aristotelianism. *True logic* was that of the ancients, and not the sophisticated semantics of the via moderna of the fourteenth-century university.

How far even the best Humanist minds had moved away from the thought and discussions of the *via moderna* is illustrated by a satirical passage in Thomas More's *Utopia* from 1516. It is told that

---

princes nor Renaissance Humanists were characterized by particularly kind manners, this could only enhance the appreciation of the term.

8. Some twelfth to thirteenth-century medieval scholars, it is true, would engage in ancient literature as literature, as revealed by Étienne de Tournais's complaint (see p. 66). But rare were those who, like the Danish historian Saxo Grammaticus (celebrated for that by Erasmus of Rotterdam), were able to approach and—we will therefore guess—really to appreciate the style of the ancient writers.

9. "Automatically" indeed in the long run, but not immediately, nor without contradictions. Cf. below, p. 91f, on pedantry and detail-thrashing in much Humanist teaching.

> Of all those philosophers whose names are famous in the part of the world known to us, the reputation of not even a single one had reached [the Utopians] before our arrival. Yet in music, dialectic, arithmetic, and geometry they have made almost the same discoveries as those predecessors of ours in the classical world. But while they measure up to the ancients in almost all other subjects, still they are far from being a match for the inventions of our modern logicians. In fact, they have discovered not even a single one of those very ingeniously devised rules about restrictions, amplifications, and suppositions which our own children everywhere learn in the *Small Logicals.* In addition, so far are they from ability to speculate on second intentions that not one of them could see even man himself as a so-called universal—though he was, as you know, colossal and greater than any giant, as well as pointed out by us with our finger.
>
> (ed., trans. Surtz & Hexter 1965: 159)

His reverence for Plato and Aristotle notwithstanding, More is obviously a nominalist by inclination: *men* exist, but MAN, the universal or "second intention" of the word *man*, does not. One might therefore have expected sympathy with fourteenth-century nominalism and its more recent heirs at the universities (that which "our children everywhere learn").[10] Instead, More is so much disgusted by the pedantry and technicalities of the discussions that he rejects the current wholesale.

THE WIDER CONTEXT

Humanism was only part of, and a specific expression of, a broader movement, even though it was certainly the expression that was most intimately connected with the new aristocratic rule. This is already obvious from the courtly function of the Humanists as advisors and secretaries - guilds might well employ a painter to decorate their guild-house or to paint a picture to be donated to a church, but would have no use for a Humanist—and from the class of young people who were educated by the Humanists. The connection between Humanism and the highest social strata is born out by the pattern of recruitment: few of the Humanists were of lowly social origin, while such a parentage is the main rule for artists.[11]

---

10. It is true that these heirs were no longer nominalists; but they kept a sophisticated terminological and conceptual tradition (more or less) alive that allowed formulation of the problem.

11. In his investigation of the origin, social status, and activity of 600 members of the "cultural elite" of Renaissance Italy, Peter Burke (1972: 39, cf. 41 and 66f) finds that the "known fathers of [320] painters/sculptors/architects include 96 artisans/shopkeepers, compared to 40 nobles/merchants/professionals. The fathers of [231] writers/scientists/humanists include 95 nobles/merchants/professionals, compared to 7 artisans/shopkeepers." As pointed out by Burke, this information is hardly representative, since a lowly origin is more likely to have

To dissect a broad cultural movement into constituent parts is always somewhat misleading, both because no list can be exhaustive, and because the cuts of the dissecting knife create the illusion that the resulting sharp boundaries are inherent in the movement itself. None the less, such a dissection may be a necessary first step.

Beyond Humanism, the following constituent parts or aspects of the Renaissance movement may be particularly important.

Firstly, the writing of poetry and other literature in the vernacular. Early writers in Italian are Dante Alighieri (1265 to 1321), Petrarca, and Boccaccio. All of them also wrote in Latin. Petrarca and Boccaccio are counted among the founding fathers of learned Latin Humanism; Dante, who by one generation is too early in date to belong to the Humanist movement itself, was held in high honors by the Humanists, some of whom (for example Marsilio Ficino, whom we shall meet repeatedly below) also took care to translate his Latin works into Italian. It is hence obvious that the Humanist movement, in spite of its veneration of Latin literature and ancient culture, was not isolated from that creation of a vernacular literate culture which is one of the best-known aspects of the Renaissance movement. Since much of the courtly service of Humanists had to be performed in the vernacular, this alliance is hardly astonishing.

Secondly, the renewal of the visual arts (painting and sculpture). These arts, whose practitioners had been regarded rather lowly in antiquity, now associated themselves with architecture, becoming thereby legitimate in the eyes of many Humanists.[12] Obviously, this distortion of ancient value judgements by people who were convinced to adhere to the standards of antiquity must have had specific reasons: the importance of these arts in the life of court and town, and as further expressions of lay human self-consciousness.

The latter statement calls for a commentary. The vast majority of paintings still dealt with themes from the Bible or the lives of the Saints—maybe 95 percent in the 1420s, and 80 percent in the 1530s, according to Peter Burke (1972: 27f, cf. 118); in this respect there is nothing particularly lay about the visual arts. But the

---

been hidden by the son—but then, according to the statistics, more likely to be hidden by a Humanist than by an artist, which confirms the conclusion that Humanists were more closely associated with the upper crust than artists. The fact that artists were trained as apprentices (who were boarded in return for their work) and Humanists in school and university (which were not free, and where no money was earned) also tended to close the latter path for talents with insufficient economic background.

12. True, the statistics quoted in n. 11 show that the upper classes would still give higher value to a literate career, and many members of the social elites would repeat the ancient view that artists *qua* manual workers were to be looked down upon. But this argument, far-fetched as it may seem to us (as Peter Burke [1972: 70] ironically observes, fighting with a sword was as much manual labor as cutting marble), was often set forth in a way that suggest it to be less obvious than it had been in Aristotle's time, and it never went undisputed.

uses to which paintings were put, the way themes were dealt with, and the claims on background formulated by those who ordered the paintings show that neither painters nor all customers were moved exclusively by pious feelings.[13]

Already from Petrarca onwards, the biography and the autobiography came into favour—not least the biography of the artist or other creative personality. As asserted by the sculptor and goldsmith Benvenuto Cellini (1500 to 1571) in the very first sentence of his autobiography (ed. Camesasca 1985: 81), "Men of all conditions who have made something excellent, or something that really appears excellent, should, if truthful and honest, write their life with their own hand." The position of creative characters, indeed, is much more prominent than in the ancient biographical collections—what a Plutarch tells about Archimedes occurs as a digression in his biography of the Roman general Marcellus, one of whose soldiers killed the genius. This demonstrates, firstly, a new interest taken in the individual personality: biography of generals and statesmen may be made from interest in military and political history, but the biography of creative personalities in general (soldiers, soldier-intellectuals and artists on a par) must have their personality[14] as their focus. Secondly, it shows that the artist and the Humanist were regarded as personalities *par excellence:* at first by themselves, since they were the ones who wrote the biographies and autobiographies; but since they had a public, also by this larger public. The writing of biographies of elite intellectuals is thus quite as much an expression

13. Some customers certainly were predominantly moved by piety or by a wish to demonstrate piety. The Renaissance movement was, precisely, *a movement,* and neither a culture shared equally by all members of society, nor the product and mind-set of a precisely definable social group. For many customers, paying a painter (at best, if you could afford it, a famous one) for a holy picture will simply have been the recognized and obvious way to express religious devotion, given the general context of Renaissance culture.

14. Their personality, as manifested in their creations and their public activity, but not their inner psychical life. Considerations like those made by John Donne in a sermon in 1626, that

> I throw myself down in my chamber, and I call in and invite God and his angels thither, and when they are there I neglect God and his angels for the noise of a fly, for the rattling of a coach, for the whining of a door. I talk on, in the same posture of praying, eyes lifted up, knees bowed down, as though I prayed to God; and if God or his angels should ask me when I thought last of God in that prayer, I cannot tell. . . . A memory of yesterday's pleasure, a fear of tomorrow's dangers, a straw under my knee, a noise in mine ear, a light in mine eye, an anything, a nothing, a fancy, a chimera in my brain, troubles me in my prayer.
>
> (ed. Craik & Craik 1986: 178)

are not too far from what Augustine wrote in the first pages of his *Confessions* 1200 years earlier—but they are unthinkable in anything written between Petrarca and Machiavelli.

The rise of interest precisely in *creativity* is discussed in Kemp 1977.

of worship of the universal genius in the style of a Leonardo da Vinci or a Michelangelo as a mere consequence of veneration for or love of art and writing.

The interest in biographies and autobiographies is not likely to amaze a modern audience, which is often at least as curious about the life and loves of the artist as about his works. But the Renaissance interest is a strong contrast to what we encounter in medieval intellectual culture, where even important personalities are often poorly known. Much of our information about the life of Thomas Aquinas has only come down to us thanks to the process conducted prior to his beatification. Jordanus of Nemore, the major thirteenth-century mathematician mentioned above (p. 74, n. 31), is known exclusively from his works, even though these works are referred to repeatedly in the writings of contemporaries who appear to have known him personally. No single trace of his personal life has survived; we are ignorant of his birth, of his death, and of his nationality, and only the distribution of manuscripts allows us to conclude that he must have taught in Paris.

Another manifestation of the same kind of individualism is the importance of the private letter as a means of expression, reminiscent of Old Babylonian and ancient Roman literate habits. Still another is the entry of religious devotion into the private sphere: a large part of the market for smaller paintings and for mass products like woodcuts was constituted by private people who wanted to have Virgin Mary (or some saint) at home (perhaps in every room) as a focus for worship rather than as a piece of decoration.

It may seem strange that Neoplatonism became a dominant philosophy, given its hierarchical top-down structure, which holds the "Great Chain of Being" to channel divine influence and power through all orders of existence. One aspect of the explanation may be the importance of courtly culture—the late fifteenth-century Medici court in Florence was a center for Neoplatonism, and courts have a natural bent for seeing the world in a top-down perspective. But other factors like the alliance of Neoplatonism with the occult current (see presently) since late antiquity will certainly also have played a role. This is clearly demonstrated by a curious reinterpretation of the doctrine formulated for example by Marsilio Ficino (1433 to 1499), the most important Neoplatonist of the Renaissance and active precisely at the Medici court. According to Ficino, *Man* is no longer a subordinate unit in the Chain: he is the central, active mediator, binding together the upper and the lower orders and acting upwards as well as downwards—in a way the human being becomes more important than the divine One itself. Notwithstanding that emphasis is shifted from single personalities to that universal—MAN—which the Utopians were unable to see, Ficino's interpretation establishes harmony between Neoplatonism, humanism (*not* capitalized), and Humanism, giving thus Protagoras a kind of revenge over Plato. Although the observation builds on a pun and not on genuinely shared meanings, we may notice that the particular regard for *the universal genius* can be understood as regard for the most obvious representation of *the universal, MAN.*

Even though most of the participants in the Renaissance movement were sincere, and many of them even deeply religious Christians, the total movement can thus be seen legitimately as a lay movement. Firstly, it was not subordinated to the church in its function as a religious body; secondly, it tended to see existence and even religious themes in the light of practical, civic life.

In this connection it should once more be remembered that the Papacy often functioned more as a lay court than as a religious center. Intellectuals who worked for the Pope or for other high ecclesiastical officers were thus integrated in a courtly rather than in the ecclesiastico-religious sphere—we may remember Machiavelli's "robes suited for the royal or pontifical court." Even though in one sense the Renaissance and the Reformation are phenomena growing from the same soil, the Reformation (in particular the Lutheran variant) was also a reaction to the all too visible transformation of the Papacy into a lay princely court. It will be remembered that the spark which set fire to the Reformation conflagration was the commercialization of indulgences, which was meant to finance ostentatious building activities in Rome.

Individualism, laicism, human self-consciousness, and "realist" art are aspects of the Renaissance which have often given the impression that the Renaissance is the first phase of the Enlightenment, following upon the obscurity of the Middle Ages. During the last three to four decades, however, other aspects of the Renaissance have come to the fore (indeed a natural consequence of the better understanding of the High Middle Ages as anything but intellectually dark): antirationalism, mysticism, and alchemical, astrological and other "occult" undertakings, that is, undertakings which were to be kept "hidden" (*occultus*) to the unworthy eyes and ignorant minds of the uninitiated multitude. These aspects of the Renaissance do not contradict those that were discussed until this point and which constitute the traditional picture of the period. Instead, the "darker vision of the Renaissance," as it has been called, demonstrates that the received "bright" interpretation is superficial. Ficino (to name but one instance), the eminent Humanist who translated Dante, the Neoplatonic philosopher who made *Man* the key figure of the Great Chain of Being, was a firm believer in astrology and magic. Ficino translated the writings of Hermes Trismegistus (cf. p. 84, n. 6), which were taken to contain the summit of occult teachings—indeed to such a point that Renaissance occultism is often spoken of as "Hermeticism."[15] Ficino,

---

15. It seems likely that Hermes's success as the exclusive embodiment of occultism is due to the need of those (Humanists and would-be-Humanists) who wanted to furnish their occultist sympathies with ancient legitimization. Without the appeal to Hermes and a few pseudo-Aristotelian treatises, occultism would have been a too unmistakably philo-Arabic affair, and thus to be denounced by true Humanists.

If this interpretation is valid, Hermes fulfils a function for Renaissance occultism which is strictly parallel to Archimedes's function with respect to Renaissance mathematics—cf. below, p. 97.

and many others with him, demonstrate that the Renaissance is not the indisputable victory of Reason over either Faith or obscurantism; it was just as much a way to dissociate oneself from a late Scholastic rationality which had proven irrelevant, and thus a necessary step toward the establishment of a *better rationality*—"better" not in the abstract or in any absolute sense but in relation to the actual historical situation and actors.[16]

HUMANIST SCHOLARSHIP, PEDANTRY, AND THE HUMANITIES

Shared etymology notwithstanding, the Humanist movement should not be confounded with "scholarly practicing of the humanities." This much should be clear from the above. It should also be clear, however, that the two are connected. It can even be argued that the origin of the modern humanities as a separate yet internally coherent enterprise can be traced back to the Renaissance Humanists.

At the outset, the connection between Humanism and the humanities concerns literary studies—more precisely, classical literary studies. Even though prominent Humanist teachers held the aim of their teaching to be the production of better leading citizens, the path believed to lead to this aim, that the pupils should "learn to speak and write Ciceronian Latin as if they had been native speakers" in order to follow both the meaning and the connotations of the ancient texts, passed through immense thickets of mostly very pedantic studies of the details of the ancient literary heritage and of the contexts to which its terms referred. One mid-fifteenth-century example of "pedantry raised to the second power," namely a commentary not to an ancient text but to a single line from another commentary—an explanatory text which had been written by the pioneer of Humanist teaching Guarino Guarini (1374 to 1460), and which tells in this line that the Crab is both an animal living in the water and a celestial constellation—"goes on for more than a page. [Ludovico da Puppio] lifts a complete list of the signs of the Zodiac from

16. The importance of this distinction between "absolute" and "local" rationality is highlighted by the attitudes to witch hunting. Jean Bodin (1530–1596), a trained Humanist and lawyer, a father of comparative and historical legal studies, and one of the grandfathers of modern political sociology (thus certainly as modern a mind as could be found), suggested that those who refused to believe in sorcery should be burnt along with those who practiced it. In contrast, a Spanish inquisitor, trained in the scholastic tradition of Canon Law, managed to analyze a giant witchcraft epidemic in Basque country as a psychological mass panic, putting thereby an end to witch burning in Spain in 1613. In general, the secular judges of sixteenth and seventeenth-century France, generally taught in the Humanist tradition, were much more severe than contemporary Italian inquisitors, who were priests and certainly closer to the tradition of medieval rationality (Febvre 1948: 12f; Parker 1980: 23; Henningsen 1980; Monter 1980).

Servius, with the months they were held to rule and the spheres of the planets that were assigned to their control. Only then does he pass on to Guarino's original level of simple lexical distinctions, and even so he finds it necessary to amplify and to explicate Guarino's already very simple latin. . . ." (Grafton & Jardine 1986: 13f)

One may wonder why anybody would pay for this kind of education. Grafton and Jardine (p. 23f) point to three reasons. Firstly, Humanist education "was modish; it was in vogue with the elite." Secondly (not wholly unconnected to this fashion), the skill "to speak *extempore* on any subject in classical Latin, the ability to compose formal letters to order in the classical idiom, the ability to teach exactly as [one] had been taught, were all valuable assets" in fifteenth-century Italy, whether he was to "serve as an ambassador, or secretary to a government department, or could become an advocate, a priest, or a professor of the *studia humanitatis* in his turn."[17] Thirdly, this kind of schooling

> fostered the sort of personality traits that any Renaissance ruler found attractive: above all, obedience and docility. Much of the time in . . . classroom was spent . . . passively absorbing information, accumulating and classifying predigested and processed material. Active participation, like the formal disputation . . . which had figured prominently in medieval training, played a comparatively small part in the programme; hence the insignificant place of dialectic or "active thinking" in the course. The consequences of this were much as they had been in late antiquity, or as they would be in the seventeenth and eighteenth centuries: students became accustomed to taking their orders and direction from an authority whose guiding principles were never revealed, much less questioned.[18] . . . Fluent and docile young noblemen were a commodity of which the oligarchs and tyrants of late fifteenth-century Italy could not fail to approve.
>
> (Grafton & Jardine 1986: 24)

Pedantry is not to be mistaken for scholarship—in so far as its essence is to be repetitive and opposed to original thinking, it comes close to being the opposite.

---

17. In this connection we should remember that this was an age without dictionaries and encyclopedias of the kind we know. Everybody who needed them had to internalize the dictionary and the encyclopedia from the commentaries to texts.

18. We may also remember the above analysis of the "double-bind" effect of Old Babylonian scribal schooling (cf. above, p. 22). Those who had gone through the Humanist school would certainly be no less sure of themselves as a special and higher class of people than the scribes had been.

Physical punishment, that other way to inculcate docility, also appears to have been as common in the Renaissance as in the Babylonian school. According to the repeated warnings of Erasmus (1466–1536) against transforming the school into a torture chamber, it was often administered without prior offense and with the sole but explicit purpose of teaching students humility (see for instance Thompson et al. (eds.) 1978: III, 40; IV, 326–331).

Yet pedants, if they do not find the material at hand which they need, may be forced into making original work themselves. They may constitute a public willing to appropriate and pay for what has been produced by other, more original minds. And some of those who pass through their hands may learn with sufficient facility to be able to go on with their own creative work on the stable foundations that pedantry at its best can provide.

All of this happened within and to the Humanist movement. Machiavelli's letter to Vettori leaves no doubt that his familiarity with ancient letters had left pedantry far behind and permitted him to draw on them both for personal consolation and as primary material for his formulation of political theory. Lorenzo Valla (1407 to 1457), whose denunciation of the Constantine Donation was mentioned above (p. 84, n. 6), went so far as to criticize Priscian, the most respected ancient grammarian.[19] He also developed a whole program (in part transmitted in teaching and public lectures, in part in writing) which approached and emulated ancient elite culture as a culturally informative whole and not its details alone; which reinstated dialectic (though Platonic rather than scholastic) in Humanist education as a precondition for creative understanding; and which thus really participates in the *renascence* movement.

Willingly or not, by upholding Humanist teaching as a modish trend, even the pedants also contributed to creating a need for new texts and (after the invention of printing) a market for text editions with scholarly commentaries. Of particular significance was the import, spread, and printing of Greek texts. These had been presupposed as obvious background knowledge by the classical Latin authors, and any reading of the latter which did not share that background was therefore bound to be inferior. On the other hand, getting to the point where one understood the Greek texts adequately was a challenge calling for more systematic and analytical thinking than the mere continuation (be it expanded) of the Latin tradition in the liberal arts of grammar and rhetoric. This held on the level of textual and grammatical understanding, but also—since many of the Latin texts that were available had once been written as popularizations or simplified versions for the use of less well-read fellow citizens—on the level of substance. Getting behind the Latin texts thus contributed to making the Renaissance reconquer that metropolis of ancient thinking of which Latin culture was never more than a periphery. Such reconquest had been attempted with much consequence already in the twelfth century, it is true. At that moment, however, utterly few translators had come in touch with the Greek texts

---

19. See Chomarat 1988. The point of the critique is that Priscian tries to fit Latin into the categories of Greek grammar; Valla recognizes that the structures of the two languages are different. Being more interested in language as a carrier of culture and discourse than as the container of literary relics, he even claimed Latin to be the more meritorious of the two languages.

themselves (and not many more with the Arabic translations), and the use of the Latin versions had largely been absorbed in the Scholastic synthesis. This may be the main reason that the disciplines which built on Greek texts were dismissed by the early Humanists; another reason, which may rather have been a pretext, was the non-Ciceronian language into which the sacred texts had been translated.[20] In any case, Petrarca and his contemporaries had done their best to reduce legitimate learning to what could be learned through the ancient Latin authors.[21]

Another source for a new and more refined understanding of language was the production of literature in the vernacular, if not by the pedant members of the Humanist current then all the more by its creative participants. From proper experience they discovered the difference between a language that had been expanded and polished through extensive literary and scholarly use and a language that had not gone through this process. Latin was clearly felt to be better suited than the vernacular for literary purposes, as expressed by Dante and accepted by most fourteenth- and fifteenth-century Humanists, since

> that discourse which is so arranged that it exhibits human thought clearly is meritorious when doing this, and most meritorious is that discourse which does it most clearly; wherefore, inasmuch as Latin makes clear many things conceived in the mind which the vernacular cannot, as those know who possess both kinds of discourse, the merits of the former is greater than that of the vernacular.
>
> (*Convivio,* 1,5,80, trans. from Borzi et al (eds) 1993: 888)

—if only, as Dante continues his Italian composition, writing certain works in Latin instead of the vernacular would not have been as useless as gold and pearls buried in the ground.

---

20. That the language may first of all have been a pretext is suggested by the fate of the translation of Archimedes from the Greek made by the Neoplatonist and Dominican friar William of Moerbeke in 1269. It was printed by Niccolò Tartaglia in 1543, who seems not to have believed it impossible to make the translation pass for his own work (Clagett 1970: 229). The printing history of Euclid's *Elements* tells a similar tale: the version which was first printed (in 1482) had been made by Campanus of Novara around 1260. A Latin translation directly from the Greek was published in 1505 by Bartolomeo Zamberti, but for decades all new editions would either follow Campanus or give the two versions in parallel—Campanus's being apparently understood as mathematically better and probably easier as far as language is concerned (personally I subscribe to both judgements), while Zamberti's text, seemingly sounder in principle, may have been felt to be unnecessarily tortuous (that it was made from an inferior manuscript is a later discovery).

The moral in both cases seems to be that once the subject was regarded as interesting, objections to the language became secondary.

21. Certainly not because they did not want to read the Greek authors: they merely did not possess the texts, or—as a wretched Petrarca realized when he finally managed to get hold of a Greek Plato—they could not read them.

However, the Humanists did not stop at such value judgements. Some of them continued to write some of their works in Italian, the choice of language depending on genre and intended use or public. Since others did not agree, or would have chosen differently in specific cases, the debate about the relative merits of the two languages continued, and the self-defense of those who wrote in the vernacular forced them to make up their minds about *the reasons* for the difference, and led to understanding of the process by which vernaculars are transformed by being tools for literate discourse. In this way, certain writers came to approach semantics through the mutual dependence of linguistic form and content. Leone Battista Alberti (1404 to 1472), a prolific author in both languages and particularly known for his *Ten Books on Architecture* and as the co-author (with Brunelleschi the architect and painter) of perspective theory, engaged deliberately in the process of adapting the Italian tongue to its new uses. Lorenzo Valla summed up the new insights in words which may not astonish us but are far from trivial when compared to his fourteenth-century precursors or to ancient theories (cf. p. 45, n. 30 and p. 55, on the Stoic understanding of language and on Isidore's *Etymologies*):

> Indeed, even if utterances are produced naturally, their meanings come from the institutions of men. Still, even these utterances men contrive by will as they impose names on perceived things. . . . Unless perhaps we prefer to give credit for this to God who divided the languages of men at the Tower of Babel. However, Adam too adapted words to things, and afterwards everywhere men devised other words. Wherefore noun, verb, and the other parts of speech per se are so many sounds but have multiple meanings through the institutions of men.
>
> (trans. Gravelle 1988: 376)

Language was not the only field where the Humanists tried to connect insight in the historical process with the attempt to shape the future. History itself was another. As in the case of language, the perspective was restricted to antiquity plus the present time—the medieval interlude was rarely looked at.

History was understood to be more than the mere writing of annals. Source criticism was not the strong point of Renaissance historiography, apart from the unmasking of forgeries; nor was the understanding of historical change. The fundamental idea was similar to what came to be called *uniformitarianism* in nineteenth-century geology: the processes which went on in the past are of the same kind as those which take place in this very moment (cf. below, p. 153, n. 7). For many Humanists, from Petrarca onwards, this reduced history to a reserve of everlasting moral lessons. Others asserted that history was not simply "past events or even the recollection thereof but rather . . . their accurate description according to an order which was topical and chronological. History was concerned above all with causes, dealing as it did with motives, acts and consequences. History's interest in vicarious experience gave it a common ground with oratory, but it was distinct because of its method and its 'verisimilitude'." (D. R. Kelley, paraphrasing George of Trebizond, *in* Schmitt et al, eds, 1988: 749) This was the approach which, when combined with the uniformitarian presupposition that the reasons for and consequences of Moses's actions would be no different from those of a Roman emperor or a Renaissance prince or city state tyrant, permitted Machiavelli to use his conversations with the ancients not as a mere reservoir of lessons but as primary material for a comparative treatise on political strategies.

As the Humanist movement spread beyond Italy and produced so-called "northern" (in particular French, Dutch, German, and English) Humanism, some of the beneficiaries would rather use its prestige for propaganda purposes or fit its insights into preconceived schemes. French early to mid-sixteenth-century lawyer Humanism—a strong movement—would prove that everything valid in this world (language, knowledge, art) was originally produced by the Gallic forefathers of the French who had taught the Hebrews and the Greeks;[22] Lutheran theologians insisted on understanding history in terms of Augustine's four world monarchies (Kelley 1988: 750); and so forth. But the spread of Humanism beyond its native Italian ground, where the leftovers of antiquity could be found everywhere though in ruins and half buried in the soil, to countries where antiquity was only to be traced in libraries, also accelerated the further formation and shaping of humanistic scholarly disciplines, at first along the lines which were already described above.

A "SCIENTIFIC RENAISSANCE"?

To the received picture of the Renaissance belongs, together with writing in the vernacular, "realist" art, and worship of the universal genius, also the idea of the "scientific Renaissance": the Renaissance is seen as the era in which Copernicus told the Sun to stand still in the center and the Earth to move, when Galileo broke the spell of Aristotelian physics, when Harvey discovered the heart to be a pump, and when Descartes invented analytical geometry. It was the era where Europe discovered and described (and often conquered) regions of the world of whose existence the ancients had never heard, or only heard through fabulous rumors.

This is true, excepting details. But it is no less true that if we are to locate it with regard to the "two cultures" of our own times, the fourteenth and fifteenth-century Renaissance movement was mainly humanistic, concerned with rhetoric, letter-writing, literature, history, visual arts, and mathematics (including astronomy). Natural science beyond mathematics and astronomy was only represented by occult interest in "the secrets of Nature," and by a fervor for "natural magic" applying this occult knowledge. Technology was often regarded rather highly, in part because it was understood as part of architecture (a highly respected component of

---

22. An impressing array of citations will be found in Cifoletti 1996.

ancient culture) and in part because of its *public or civic utility* (better, as a result of both explanations in combination).

Regarding the absence of a non-occult "natural science" it is important to remember that at the time this could only mean *natural philosophy*, which would identify it with the Aristotelian philosophy of universities. Astronomy was bound up with astrology, and even though the two together continued the medieval tradition, they were not tightly bound up with the Scholastic tradition but rather—when not a mere tool for prediction—an instance of interest in the secrets of nature (cf. the quotation from Regiomontanus the astronomer-Humanist on p. 69, n. 24).

The understanding of mathematics as a Humanist subject has several explanations. Firstly, mathematics had its root in antiquity, and in particular in the person of Archimedes. Archimedes was mentioned by many Latin authors as an eminent servant to his king and country; further (abstractly) as the most subtle among geometers and, indeed, the most ingenious of minds. There were thus good reasons that already Petrarca wrote several biographical notices about him, even though he did not know as much as the title or the topic of any of the mathematical works. In this way, Petrarca and other Humanists paved the way for a legitimization of abstract mathematics as the activity of the supreme genius.

Next, from the 1430s onward, mathematics came to be applied in the theory of perspective, and thus to be connected to both architecture and painting.

Thirdly, mathematics was centrally concerned with harmony and proportion, and mathematical harmony and proportion had been taken already by classical authors as symbols for social harmony and for the just character of the unequal distribution of social power and wealth (cf. above, p. 31)—metaphorically seen, mathematics was a way to moral and political philosophy.

On all three accounts, mathematics was hence legitimized by close connections to central themes of early Renaissance culture (supplementary, less directly connected reasons for legitimization could be mentioned).

But precisely the same reasons made mathematics a Humanist subject.[23] The interest of certain Humanist scholars in mathematics should therefore not be mistaken for scientific interest in nature, and it is actually only from around 1500 that we can speak of the beginnings of a *scientific Renaissance* through a transformation of Humanist thought.

Several symptoms of this transformation can be traced. In the late fifteenth and early sixteenth century, Leonardo da Vinci (1452 to 1519) filled his notebooks

23. We may find this classification unfamiliar—but ours is not necessarily better founded. There is, indeed, no particular reason to count mathematics as one of the sciences that investigate the physical world, apart from the historical accident that *advanced* mathematics was first used *as a tool* in the physical sciences, and only later and with less predictive success in economics and certain other social and human sciences.

with anatomical, botanical, and other studies and with sketched inventions and mechanical inquiries. Some of these served the naturalist precision of his paintings or were correlated with his activity as an architect and a military engineer; but many point further, toward theoretical scrutiny of the mechanism of vision and toward theoretical investigation of mechanical principles. In the same decades, printed editions of ancient works on natural history and botany first facilitated "material textual criticism," that is, comparison of the text and illustrations as handed down and possibly distorted through a complex manuscript tradition with real plants and animals; but soon they also kindled interest in local fauna and flora, with the result that the insufficiency of the ancient books was discovered (in part they were simply erroneous or distorted beyond repair, in part they described Mediterranean species which differed from those found in Switzerland and Germany). The discovery of new worlds beyond the oceans—with vegetation, animals and populations never heard of—proved an even more powerful challenge to the books (cf. Grafton 1992).

Botany, in the form of herbals, had a traditional function in medicine. In the early sixteenth century, however, a new medical doctrine based on alchemy and not on herbs was introduced by Paracelsus, alias Theophrastus Bombastus von Hohenheim (1493 to 1541). Its roots are not in Humanism, but rather in the philo-Arabic, non-Hermeticist occult tradition (cf. p. 90, n. 15). None the less, "iatro-chemistry" (medical chemistry) was a great success in early sixteenth-century Humanism; it became a matter of teaching for physicians (which means that it was no longer *occult*), and led to appreciable progress in chemical knowledge—less, perhaps, in actual cures for the sick, apart from the treatment of syphilis by mercury.

The appearance of several roughly contemporary symptoms pointing in the same direction suggests that they are precisely *symptoms,* and that the real cause of the transformation is to be found at deeper levels; one may suggest that a world which was transformed technologically and socially at an accelerating pace could no longer be served by a merely literary intellectual culture. New elites (and groups with elite ambitions) might pay the necessary lip-service to Humanist culture in order to gain recognition for their professions, and Humanists might become aware that "civic utility" had come to encompass more than just literary service to the Prince supported by architecture and military techniques.

This is just what the Humanist Georgius Agricola (1494 to 1555) argues in Book I of his *De re metallica,* which he had begun writing in 1530. This is one of several famous works which, while demonstrating the Humanist-legitimate character of technology in general, integrated the description of actual procedures with as much scientific insight as could be produced.

Agricola was as good a Humanist as any, but he was also a physician and competent enough to earn a fortune in the mining business; he can thus legitimately be taken as a representative of the *new* elites. The no less famous Petrus Ramus (or

Pierre de la Ramée, 1515 to 1572) embodies the traditional elite trying to widen the perspective of Humanist studies.[24]

What Ramus attempted was nothing less than a complete reconstruction of all knowledge under the aegis of Humanism reconstructed as a universal "method." Aristotelian logic—much too stiff to be adequate (cf. p. 118, n. 15)—was to be replaced by the "natural" logic of ordinary discourse: but this discourse turns out to be that of the ancient rhetors and poets. Aristotle's natural philosophy—again concerned with matters that are much too abstract to Ramus's pedagogical taste—should give way to actual nature: but again, actual nature was to be found in the Elder Pliny's anecdotic *Natural History* and in Vergil's *Georgica,* a didactic poem describing countryside life. Ramus is much in favor of the knowledge of ordinary practitioners—but what he wants to find with them is embedded university knowledge; that autonomous knowledge which technical practitioners really possessed not only does not interest him, its very existence is refused.[25] Geometry should be useful, not theoretical; none the less, geometry is understood as Euclid's *Elements,* merely expurgated of the proofs and provided instead with references to the practical applications of the theorems. All in all, there is no real interest in empirical natural knowledge (and certainly none in theoretical knowledge, which is emphatically denounced), nor in the useful knowledge actually possessed by the technical professions; but great efforts are spent in order to uphold the supremacy of Humanist studies. Though it may sound like a parallel to Agricola's advocacy of technical knowledge (and is often understood as such by modern interpreters), close scrutiny of its hidden premises shows Ramus's call for utility to be a disguised repetition of the similarly worded claims of Petrarca and other early Humanists—not significantly broader than theirs but dressed up as if it were.

---

24. It is disturbing to discover that precisely Ramus epitomizes the attempt of the old cultural elite to conserve its preeminence: although of noble ancestry, his grandfather had been a charcoal burner, and his father a farmer; Ramus himself had to fight his way into and through school and university, which may explain much of his arrogant self-assurance, as well as the bitterness with which he attacks his adversaries.

25. Ramus's *Scholae mathematicae* (1569) offer striking exemplifications of this blindness (which in all probability is intentional and fraudulent—Ramus knows the traditions whose existence he refuses well enough to borrow wholesale from them). Thus, not only Columbus's nautical mathematics but also typesetting and the "mechanics of bombardment" are claimed to owe their invention to a diffusion of mathematical knowledge in Germany sparked off by the transfer of the astronomer and mathematician Heinrich von Hessen from Paris to the University of Vienna in the 1380s (p. 64f). Italy, on the other hand, puzzles Ramus: he recognizes that practical mathematics flourishes here more than anywhere else; but universities have few chairs in mathematics (p. 107). What Ramus pretends *not* to know is that practitioners' mathematics had been taught systematically in Italian towns since the later thirteenth century, but in nonacademic "abacus schools" (Fanfani 1951; Franci 1988).

Works like *De re metallica* were not known from antiquity, apart from Vitruvius's work on architecture and Columella's handbook for landowners, *De re rustica,* the title of which Agricola echoes intentionally. Along with this new genre, the 1540s produced a sequence of major works which better fitted the traditional genres but surpassed the best works from antiquity: Nicolaus Copernicus's *De revolutionibus orbium coelestium,* Andreas Vesalius's anatomy (*De humani corporis fabrica*), and Girolamo Cardano's algebraic *Ars magna* are usually considered to mark the watershed.

All three works were printed, Copernicus and Cardano by the same scholar-printer in Nürnberg in 1543 and 1545 respectively, Vesalius in Basel in 1543. Thereby they rapidly gained influence outside universities, though their public was certainly dominated by university graduates (see Drake 1970). As they were soon followed by other works, confidence arose that better knowledge of nature than what had been inherited from antiquity could be established—and, moreover, that the *belles lettres* and the classical tradition did not constitute the apex of possible knowledge. The formation of this conviction, and not the mere production and printing of major books, constitutes the real transition to the "scientific Renaissance."[26]

It may be difficult to trace the emergence and stabilization of a conviction, but a *terminus ante quem* can be pointed out. At the beginning of the seventeenth century, a number of ideologues for the new science drew the consequences of the discoveries, establishing that new rationality for which the Renaissance had paved the way—not least Francis Bacon (1561 to 1626), Galileo Galilei (1564 to 1642), and René Descartes (1596 to 1650). All three, though each in his own way, were strongly critical of both traditional natural philosophy and of Renaissance Humanism; all three—each one, again, arguing and putting the accent in his own way—emphasized the necessity of making new observations and experiments; and whereas Bacon was an exception on this account, the others accentuated the need to have observation and experiment guided by *new* theories, which as far as possible should be structured mathematically.

Since the "scientific Renaissance" starts so late, it is customary to count the three heroes as Renaissance figures. In those fields where Petrarca and Giotto count as founding fathers, nobody would include the seventeenth century in the Renaissance. This may therefore be a convenient point to examine the attitude to the purpose of different kinds of knowledge as it had developed over the Renaissance.

Natural philosophy was still *theoretical* in outlook and *practical* or *emancipatory* in its destruction of "idols" (Bacon's term for general classes of fallacy and mistake), as it had been in the pre-Socratic era. But the *technical perspective* was already rising above the horizon, with Bacon as its main prophet and Galileo as a practitioner. For

26. This new mood is a suggestive but delayed parallel to the changing gauge by which artistic quality was assessed. Already in the fifteenth century, *fantasia* had replaced imitation as the gist of the visual artist's work (cf. Kemp 1977).

this reason, natural philosophy no longer restricted its questions to how nature would behave when developing without impediments, as it had done in the Aristotelian tradition (cf. p. 41, n. 25). For technical uses it was also of interest how nature would behave when adequate situations were produced (for instance that a piece of iron suspended in a thin beam as a pendulum would move so regularly that it might be used in a clock); keeping in mind the harsh criminal legislation of the times, we might formulate the new view in the maxim that "nature, like a criminal, is likely to reveal more of her secrets when submitted to torture than when speaking freely on her own."

Humanism, on the other hand, which had started out as *technical knowledge* (letter-writing, rhetoric), tended to loose this character. The *belles lettres* were no longer a necessary model for effective political action; they became something *beautiful,* entertaining, edifying, educating, or a subject for investigation. Literature *as art* and *humanities as scholarship* tended to diverge, after having belonged together during the Renaissance. The humanities as scholarship were further accentuated by the rise of textual criticism and by new, more critical trends in the writing of history.

A final aspect of the "scientific Renaissance"—in part an outcome, in part a parallel, in part even a precondition—is the dismissal of magical and Hermetic thought: certainly neither instantaneously nor by everybody, nor however by scientists alone.

An illuminating example is provided by Johannes Kepler's discussion of the feasibility of astrological prediction in the first (and indeed major) part of an astrological calendar for the year 1602. We know, thus Kepler, that the Sun influences what goes on on the Earth, through the heat generated by its rays. The influence of the Moon is visible in the tides, and the phenomena of spring tide and neap tide demonstrate that even the conjunction and opposition of Sun and Moon have consequences. This should not provoke wonder in a universe governed by geometrical harmony (this was Kepler's conviction), and we may suspect that even the "aspects" of other planets (conjunction, opposition, the "harmonious" angular distances 120, 90, and 60 degrees) influence our existence, albeit to a lesser degree because of the weakness of the rays of these planets. Aspectual astrology is founded on physically real phenomena and therefore not to be dismissed *a priori,* although its influence can only be of a universal nature: it may affect the weather, or the movement of humors in humans (and thus human health) in general, but it cannot govern the fate of an individual directly. The Zodiac, on the other hand, is a purely human convention, coming—like the meanings of language according to Lorenzo Valla, we may add—"from the institutions of men." Whether planets stand in one or the other of these artificial divisions of the heavens cannot influence what goes on in physical reality, neither in general terms nor concerning the individual.[27]

---

27. The very first passage deserves to be quoted literally, both because it suggests even greater skepticism and on account of its ironical tone (trans. Field 1984: 232):

The heart of the argument is the distinction between human symbolization and external reality. Symbols and names are arbitrary; they do not reveal the essence of things but only what we have chosen to put into them. They can only influence by being understood, that is, planets and other entities deprived of mind are outside their range. Kepler's rejection of Zodiacal astrology thus builds on the same fundament as the rejection of doctrine of "signatures" by most seventeenth-century physicians; according to this theory, introduced by Paracelsus and widely accepted in sixteenth-century medicine, the shapes of leaves or roots of plants were signs telling their medical utility—that is, they were symbols written by Nature and to be read by men.[28]

Valla's stance regarding the nature of language, as representative of the late Renaissance view, is thus no superficial analogy but an important substructure for the new thinking. Unexpectedly, one of the roots of the disentanglement of natural science from magical thought thus drew nourishment from the scholarly transformation of the *studia humanitatis,* where the relevant insights had been reached already in the fifteenth century.

Another root soaked in the development of the natural sciences themselves. The emphasis on *new* observations, independent of ancient books and beliefs and

---

> The public expects a mathematician to write annual Prognostications. Since at the approach of this year, 1602 from the birth of Christ our Savior, I decided not to pander to the public's craving for marvels but rather to do my duty as a philosopher, namely to limit the scope of such Prognostications, I shall begin with the safest assertion of all: that this year the crop of prognostications will be abundant, since, as the crowd's craving for marvels increases, each day will bring an increase in the number of authors.
>
> Some of what these pamphlets will say will turn out to be true, but most of it time and experience will expose as empty and worthless. The latter part will be forgotten while the former will be carefully entered in people's memories, as is usual with the crowd.

—or, as it was expressed in 1530 by the Florentine historian Francesco Guicciardini, as sober-minded a historian as his friend Machiavelli (*Ricordi* 57) (trans. from Pasquini 1975: 78):

> How lucky are the astrologers compared to the rest of mankind! They, telling one truth together with a hundred lies, gain trust in such a way that even their lies are believed; others, telling together with many truths a single lie, are no longer believed even when honest. This is a consequence of the curiosity of men who, craving to know the future and having no other way, are inclined to run after those who promise to tell.

Though not the last to reject astrology, nor were astronomers and natural philosophers the first.

28. See, for instance, Eamon 1994: 214 and Thorndike 1923: V–VI, *passim.*

using mathematics not as an emblem of qualitative insight but as a framework within which quantitative measurement could be correlated with theory, tended to eradicate beliefs whose only basis was literary.[29] In a rather acrimonious counterattack on an opponent, Galileo summoned him to distinguish between science and literary works like the *Iliad* or *Orlando furioso,* in which "the least important thing is whether what is written in them is true."[30] Wholly outside the sphere of science, natural as well as human, but clearly in the Galilean vein, is the answer given by a Roman prelate in 1676 when sorcery was blamed for the sickness of the Emperor's wife: "In Rome we do not really believe in such spells" (Parker 1980: 24). A symptom of a spreading conviction that "science" had disproved the possibility of magic is Louis XIV's decree from 1682 that sorcerers were to be treated as charlatans (Monter 1980: 35).

But the ousting of magic is certainly not to be explained exhaustively by the development of humanistic scholarship and the new critically inquisitive attitude of natural scientists alone. The reasons are complex and to be found on many levels. The new organization of religious life after the Reformation and the Counter-Reformation will have played a role (but for a while a highly ambiguous one—nowhere was witch hunting as frenetic as in Lutheran territory). So will events like Casaubon's exposition of the Hermetic writings as late ancient forgeries (p. 84, n. 6), even though their importance should not be overestimated: grand beliefs—be they in fundamentalist religion, in progress or its futility, in magic, or in the general potentialities of science—do not depend on specific arguments as much as on general patterns of experience which decide whether arguments count and what is to count as an argument. Despite their immense respect for St Augustine, neither the High and Late Middle Ages nor the Renaissance had bothered about Augustine's irrefutable argument against the possibility of astrological medical prognostication.[31]

---

29. This does not mean that traditional beliefs were always rejected a priori by scientists. In 1664–65, the Royal Society of London undertook a large-scale investigation of the effects of May-dew on organic change (A. B. H. Taylor 1994). But precisely this episode shows that traditional beliefs were controlled, not accepted on the face value of tradition or written authority.

30. *Il saggiatore,* trans. Stillman Drake in Drake & O'Malley (eds.) 1960: 183.

31. *De civitate Dei,* V.2–3 (ed. McCracken et al. 1957: II, 138–146). Briefly, the argument runs as follows: Often twins have quite different medical histories. Of course, as the astrologers defend themselves, the twins are not born at precisely the same moment; but then, Augustine counters, changes in the heavenly configuration that are too small to be observed may change everything. The stars may still influence our health; but we shall never be able to learn how. The similarities which are often found between the medical fates of twins are rather to be explained from natural causes (in the tradition of Hippocrates and early natural philosophy): the condition of the parents at the moment of birth; similar nourishment; similar climate and situation; similar kinds of exercise.

Cardano, not only a physician and an outstanding mathematician but also a fervent astrologer, would not have been impressed by Kepler's argument: according to his philosophy the whole cosmos was alive—and thus, he might have objected, participating in the same universe of symbolic meanings as human beings.

The most important strain in the intertwined causality behind the decline of magic is probably the technological successes of the outgoing Renaissance. It has been held that all magical thought is abortive technology. The thesis is certainly an undue simplification which omits important aspects like the symbolic self-expression through the enactment of magic rituals.[32] Similarly, the high medieval and Renaissance interest in natural magic, *qua* its participation in naturalism in general, was certainly a cultural expression of humanist attitudes. But it *also* expressed technological aspirations; moreover, the crux of the humanist message was precisely this technological optimism—at times spilling over as epistemological optimism, as when Regiomontanus held astrology to provide us with insights into "the Law according to which the Almighty resolved that the Heavens be made" (cf. p. 69, n. 24). Likewise, as it was argued (p. 90, n. 15), the specific Renaissance interest in Hermes was largely derivative, as a way to dress up a broader interest in occult knowledge—not least natural magic.

All technologies fail occasionally, which is normally not taken as a reason to dismiss technology in general. As long as no serious competitor was at hand, natural and related magic could therefore survive. Some of the recipes worked, and in the absence of theory this made just the impression which in modern parlance is spoken of as "pure magic" (whoever has spilled a glass of red wine on a white carpet and seen the vacuum cleaner remove all traces together with a pound of salt the next day will know the feeling). But in the course of the Renaissance age, "genuine" technology developed at an accelerating pace, often based on systematic experimentation. This did not disprove the magical alternative—technology by definition does not prove, and the promises of natural magic often concerned what could *not* be done by ordinary techniques: yet it created a general experience of what could and what could not be done; it showed that old books were no reliable guides in this field—what they promised to do often turned out to be impossible, while much could be done of which they had never dreamt; it taught that results were not obtained by spells and other symbols but by physical intervention, often combined with measurement and

---

32. Less theoretically expressed: incising the name of your secret love in the school desk is a way to confirm to yourself your passion, and remains so whether or not you believe it to be also a means to conquer the chosen one.

The identification of magic as pseudo-technology is customarily ascribed to the anthropologist Bronislaw Malinowski. As so often, the original presentation of the idea is less simplistic, and involves the emotional function of magic as a central component—see B. Malinowski 1955/1925, chapter V.

calculation; and it demonstrated unambiguously that nature could effectively be treated as a mindless object, irrespective of Cardano's and similar views.

Experiences like these allowed the transfer of Valla's understanding of the character of language, suggested Kepler's dismissal of zodiacal astrology as physically impossible, and gave rise to the new epistemological optimism of Bacon, Galileo, and Descartes. While Renaissance *scholarship* was certainly much "darker" than often assumed in the literature until a few decades ago, Renaissance *technology* thus legitimizes the traditional "bright vision."

CHAPTER 5

THE EARLY MODERN EPOCH AND CLASSICISM

A SHIFTING CENTER OF GRAVITY

Only one of the three main prophets of the new science (namely Galileo) was an Italian, whereas one was French and one English. This is symptomatic of a general displacement of the European economical, political and intellectual centre of gravity and of a new balance of power in these and other domains.

Part of the background is the discovery of America and the development of new trade routes and of the whole American market as well as the market for American products. At first, it is true, the American boom benefitted to Spain.[1] Spain, however, was unable to take advantage of the opportunity and adapt its socioeconomic structure to the new situation. The successful completion of the Spanish *Reconquista* (the conquest of the Islamic territories begun in the eleventh century and brought to an end in 1492) and the inflow of American gold were the fundament for an impressing cultural efflorescence (*El siglo de oro,* c. 1550 to c. 1650, represented in literature by names like Cervantes, Lope de Vega, Góngora, and Calderón). But its high points lay in *art* (literature as well as painting) rather than in renewals of scholarship or worldview, for which reason I shall not describe it systematically but only return to it in connection with a broader discussion of the Baroque. Even its fruitful tendency not to observe the "classical" norms as established in the Renaissance was rather a matter of practice derived from the aims of art than of theoretically argued principle—a description which also holds for the broader current of Baroque art and literature (cf. below, pp. 122ff).

1. Portugal gained similar though smaller advantages through systematic piracy in the Indian Ocean (dressed up in later historiography as "trade on India").

Instead of falling to Spain, the long-term benefits of the Atlantic trade were eventually reaped by England, the French Atlantic cities, and the Dutch provinces which freed themselves from Spain in the seventeenth century. (The German area was ravaged in the Thirty Years' War, 1618–1648, and lost most of its vigor and probably the larger part of its population.) This is one of the reasons that the economic centre of Europe moved to the northwest. Another reason, which is also the basic reason that the northwestern countries could appropriate the gains from the Atlantic trade, is the transformation of the sociopolitical structure of the countries themselves: in England and France, centuries of intermittent internal and mutual warfare had weakened the feudal nobility, and in (uneasy and sometimes unstable) alliance with the mercantile bourgeoisie, the royal power constructed a more centralized state. In the late sixteenth to early seventeenth century, both countries slid toward absolutism. An even more outspoken alliance between the semimonarchic Republic and the upper bourgeoisie resulted from the Dutch emancipation.

Economically, an equilibrium between a predominantly feudal mode of production and a global structure geared to mercantile capitalism emerged. Within the framework of historical materialism (but not respecting the simple categories of the textbook version) one might say that the feudal mode of production lost its hegemonic role within the social formation and was replaced in this position by mercantile capitalist relations. Feudal relations became subordinate much in the same way that eighteenth- and nineteenth-century plantation slavery was subordinated to the world market and to European capitalism. The balance of forces was of course different in France, England, the Netherlands, Basel, and the German more or less autonomous city republics; moreover, it varied over time in each of these places, as the balance between the different centers was itself subject to temporal change.

In spite of this new hegemony, and even though the Atlantic trade "represented the future" (that is, carried features which in later capitalist socioeconomic structures would become even more conspicuous), the fundament for European economy was still agriculture. Until the eighteenth century, it was therefore France (whose population was much larger than that of England, not to mention the Netherlands) that took the economic and political lead. For the same reason, the French court (to which much of the surplus took its way) and other institutions associated with the French royal power became the focus for cultural development. The characteristic innovations of the seventeenth century were thus linked to the appearance of full-fledged absolutism and to French courtly culture.

COURTLY CULTURE AND CLASSICISM

In many ways, Versailles (and everything it stands for) is a parallel to San Pietro (and everything that building stands for). The courtly culture of France was thus a continuation of the courtly cultures of Renaissance Italy. But French society of the later seventeenth century as a whole was very much different from the society of Renaissance Italy, the former being centralized in one state with one cultural focus (certainly with important exceptions in provincial and popular culture) and the latter pluralist, with many political centres and many cultural foci. As hotbeds for culture, art, and worldview, the two societies thus produced quite different crops.

This can be illustrated by the changing concept of *academies.* The term first turns up in the Italian Renaissance, borrowed of course from Plato's school but rather understood in the beginning in the likeness of Cicero's villa *Tusculum,* as a locus of cultured leisure. The Academy was the place where the prince or patrician met with his Humanist, artist, and philosopher friends to be one of them for a while. When meeting in Academy, then, the "friends" would take their seats not according to rank but in order of arrival.

In the sixteenth and seventeenth centuries, princes (in particular the Medicis in Florence) would found specialized academies, such as the *Accademia del disegno* ("Academy of drawing") and the *Accademia del cimento* ("Academy of Experiment"), as abodes of research and mutual inspiration. It was the secretary of the latter institution who explained in 1664 that the Medici Prince Leopold when participating in the meetings "likes to act as an Academician, and not as a Prince. He is content to play the second role only on occasions when there is a question of expense, generously supplying the needs of the Academy." (Lorenzo Magalotti, quoted from Middleton 1971: 56f)

Similar groups of literary or scientific peers, only without a princely protector, appeared in France around 1610–1630.[2] They were not allowed to stay private, however; instead, they were given the status of official institutions by Cardinal de Richelieu (the architect of French absolutism) and his successors, with financial support from the state but also with specified responsibilities and statutes: *L'Académie,* and *L'Académie des sciences.*

Among the tasks of the French Academy was (and is) to make a dictionary, that is, to decide about what was correct language. This obsession with *rules* is characteristic of the whole French Classicism. The beginning of genuine humanistic scholarship that had come out of the late Renaissance was absorbed into prescriptive poetics and aesthetics. The favorite form of analysis of a literary or other artistic product is an aesthetic judgement which follows the pattern that "this poem/painting/building *is good because it observes rules A, B and C,* but it *is not supreme because it fails to agree with rule D.*"

There are thus fair reasons to regard the cultural domain as the sphere where absolutism was best realized (though at the cost of regarding the popular genres as not worthy of notice even if still going strong). It is true that Louis XIV claimed to

---

2. Further references in Hahn 1971: 5–7.

be the state[3] and thus to rule independently of all custom and precedent; none the less, the actual working of the political and administrative apparatus was a patchwork of new rules superimposed upon but not fully suppressing old customs and "freedoms," and themselves developing into insuppressible privileges. *Only in art* could François Rabelais and Pierre de Ronsard be declared to be simply bad taste, as done for instance by Nicolas Boileau in 1674 in *L'Art poétique* (Chant II, ed. Sainte-Beuve 1868: 202) when he first sets out the rules for idyllic poetry[4] and afterwards condemns first the poet who does not follow them (the passage is omitted here) and next the one who follows them with insufficient elegance:

Telle qu'une bergère, au plus beau jour de fête
De superbes rubis ne charge point sa tête,
Et, sans mêler à l'or l'éclat des diamants,
Cueille en un champ voisin ses plus beaux ornements:
Telle, aimable en son air, mais humble en son style,
Doit éclater sans pompe une élégante idylle
Son tour simple et naïf n'a rien de fastueux,
Et n'aime point l'orgueil d'un vers présomptueux.
Il faut que sa douceur flatte, chatouille, éveille
Et jamais de grands mots n'épouvante l'oreille.
[. . .]
Au contraire [Ronsard], abject en son langage,
Fait parler ses bergers comme on parle au village.
Ses vers plats et grossiers, dépouillés d'agrément,
Toujours baisent la terre, et rampent tristement:
On dirait que Ronsard, sur ses pipeaux rustiques,
Vient encor fredonner ses idylles gothiques,
Et changer, sans respect de l'oreille et du son,
Lycidas en Pierrot, et Philis en Toinon.[5]

Seen from our point of view, it seems at least bizarre that an idyll can only be elegant and thereby acceptable if its characters carry Greek names. Being brought

3. Never in these words, it seems; but the apocryphal saying corresponds well to his actual strivings.

4. Even though certain ideals—thus simplicity and elegance of language—held widely, different genres indeed obeyed each their own rules.

5. In my translation, and despoiled of versification

As a shepherdess, on the most beautiful festive day,
does not weigh down her head with arrogant rubies,
and, instead of mixing the gold with the shining of diamonds,
picks the most beautiful adornments in a neighboring field:
Thus, lovely in look but humble in style,
should shine without ostentation an elegant idyll.
Its simple and naïve goings have nothing ornate,

up in the late aftermath of Romanticism, we also tend spontaneously to find it more than bizarre that accordance with pre-established rules should be the main gauge of artistic quality (this is why readers will probably have accepted without objection nor further thought the above remark about the "fruitful tendency" of Spanish *siglo-de-oro* art not to respect the classical norms). As a matter of fact, however, rules did not prevent a number of artists from making magnificent work—among those who were close to Boileau we may mention Molière and Jean Racine. An important part of the explanation is that rules were not really pre-established but to a large extent abstracted from the actual art of the period[6]—more so indeed than was realized at the time, when it was believed that these rules expounded the real canon of ancient art.

In reality, and as always, not only the form but also the content of the art of the epoch expressed its own outlook and explored its own dilemmas. Racine's tragedies were concerned with individual psychology rather than with fate; more than once, Molière's comedies came close enough to urgent political and moral conflicts to bring him into acute trouble. Though the phenomenon of rules was an expression of court dominance, the actual content of art (and thus even the actual rules derived from it) reflected the overall experience and societal situation of at least the literate classes, with all their tensions.

Different authors and artists would certainly orient themselves differently within the field of tension, depending both on their personality and on their public. Racine, when the dilemma became too severe, stopped writing tragedies and became

---

> and do not love the haughtiness of pretentious verse.
> Its sweetness should please, tickle, awaken,
> and never scare the ear with excessive words.
> . . .
> To the contrary, [Ronsard], using gross language,
> makes his shepherds speak as they speak in the villages.
> His base and vulgar verse, deprived of attraction,
> always kiss the earth, and crawl lamentably:
> one would say that Ronsard even hums
> his barbarian idylls to his boorish shawm,
> changing, with no respect for the ear and the timbre,
> Lycidas into Pierrot, and Philis into Toinon.

Boileau and his generation were certainly not the last to be more interested in developing schoolmastering literary criticism than poetics or (with a modern term) literary theory or insight. In his very introductory note to *L'Art poétique,* Charles-Augustin Sainte-Beuve, the recognized leading authority in French literary criticism of his days, explains that "this poem is admirable because . . . ."

6. Another factor is evidently that the post-Romantic folklore identification of artistic creativity with contempt for all rules is no less superficial than the infatuation with rules.

a courtier. More than others, Molière makes us aware that artistic innovation was rooted in the burgeoning *bourgeois public sphere* (see below, p. 127) while being supervised by the state.

Independently of the question whether the preoccupation with rules and taste kills or fosters artistic creativity, this obsession had one important consequence for the understanding of the role of art and culture: measured by explicit standards derived *de facto* from the artistic products of the later seventeenth century, ancient works could not help but fail. It was, put sharply, impossible for Sophocles to be a better Racine than Racine. After a half-century of skirmishes, the "battle between the ancients and the Moderns" broke out definitively around 1690: Homer was full of implausibilities and outright errors, Terence and Seneca were crude compared to Molière and Racine, Montaigne's essays were in better style than the Younger Pliny's letters. After fifty years where literary culture and scholarship had detached itself from the trend inaugurated in the natural sciences by Bacon, Galileo, and Descartes, it joined the ranks. It did so, indeed, not only because of the rules for literary taste but also because "the century of Louis XIV" was superior to antiquity on *all* accounts. Charles Perrault, who launched the onslaught in a poem entitled precisely *Le siècle de Louis le Grand* and read before the Academy in 1687, exalted the newly invented microscope and the telescope and the worlds they had discovered as part of the argument (Perrault 1688: I, 9).

The outgoing seventeenth century is thus the time which finally broke the spell of antiquity in literate culture, by creating another *via moderna* better fit for the actual world than that of the fourteenth century. The counterattack formulated in a translation of the *Iliad* in 1711 (Aldridge 1968: 78), that whatever reproaches were made against Homer could be made with equal right against the Old Testament, proved more dangerous to established religion than to modern culture, and can be taken as an expression of enlightenment *malgré lui,* and thus as a harbinger of the process described in the following chapter.

FROM SCIENTIFIC TO PHILOSOPHICAL REVOLUTION

Of more direct importance for this impending development, however, were the repercussions of the "scientific revolution" *in philosophy,* that is, the way philosophy understood the "scientific Renaissance" and drew its own general consequences concerning the acquisition and nature of knowledge and about human life.

As was mentioned above (p. 100), the early seventeenth century had produced the ideology of a new science; this, however, could only be done convincingly because the same epoch had produced astonishing new scientific *results,* replacing the tradition of astronomy and natural philosophy, which had developed undisturbed (when at all developing) since antiquity, with something quite new, which was furthermore understood to be quite new. For the sake of brevity we may restrict ourselves to the following high points:

— Kepler's *New Astronomy,* from 1609, which, firstly, had dismissed those perfect heavenly circles which still formed the fundament of Copernicus's heliocentrism, and had replaced them by ellipses; and which, secondly, had abolished the distinction between Heaven and Earth, arguing that the same physics was valid above and below the sphere of the moon (which, as it followed, was no longer a crystal sphere but a mere elliptic orbit in space; this common physics, not the ellipses, was what in Kepler's eyes made his astronomy "new").

— Galileo's derivation of the laws of free fall and ballistics (etc.—published 1638) by means of mathematics combined with experiment, which brought an end to Aristotle's physics and relegated 300 years of critical but non-experimental discussion of its problems to the archives of the history of science and philosophy.

— William Harvey's discovery of the circulation of the blood (1628), which had no less cataclysmic effects on the faithful repetition of Galen's classical doctrines with minor addenda and corrections.

— The inventions of the microscope and the telescope, which had opened worlds whose mere existence had never been imagined—from the sperm cell to the mountains of the Moon and the moons of Jupiter.

— And finally, as the culmination, Newton's *Philosophiae naturalis principia mathematica* (1687), which replaced Kepler's purely empirical laws (the elliptic orbits, the relation between period and distance from the Sun, etc.) and his qualitative physical speculations by four simple laws and precise mathematical calculation.

In the first place, the justifications which participants in the movement gave for their methods and the arguments they set forth in defense of their right to disregard tradition developed into a new philosophy of knowledge. Bacon, Galileo, and Descartes were already mentioned as ideologues of the new science, and this may be the place to tell a bit more about them.

Bacon was the one of them who was least important as a participant in the scientific movement itself; as stated in one biography, "his standing as a scientist . . . is low" (Lea 1979: 564b). Much of his fame in the later seventeenth century (which does not do full justice to his actual opinions) rests on his emphasis on experience and induction: concerning a specific quality like heat (the only case he treats), many experiments should be made, and it should be observed when heat is present (for instance, in sunlight) and when it is absent under otherwise similar circumstances (for instance, from moonlight); only in this way would one be able to find the "simple natures" which determine phenomena. Like the "experiments" of Renaissance natural magic and alchemy, Bacon's were meant to be qualitative and unencumbered by precise measurement and mathematics.[7]

7. As it has been said repeatedly, Bacon did not recognize important science when he encountered it. He rejected Copernicus's heliocentric astronomy as uninteresting, together

In this respect, Galileo's stance was quite different. In his opinion, the Book of Nature is widely open to our eyes, but it can only be read by the one who knows the language in which it is written: the language of mathematics. Like Bacon he would perform experiments (his notebooks reveal more of them than his published works do), but his experiments would (sometimes) involve careful measurement, undertaken as they were in order to test mathematically formulated hypotheses.

The cornerstone of Descartes' philosophy was a radical rationalism, according to which one should start from self-evident truths alone; but Descartes combined the rationalist principle with application of mathematics—his *Discours de la méthode* was indeed published in 1637 as a common introduction to his analytical geometry, to a work on optics making ample use of geometry, and to a third treatise containing an equally geometric analysis of the rainbow. Descartes also made a bold compromise with experimentation and empirical investigation, leaving to them to decide at such points where metaphysics derived from self-evident principles was mute or ambiguous (most points, of course).[8]

Evidently, these three philosopher-scientists did not advance as a closed phalanx; in fact, if reduced to one catchword each (induction / mathematization / self-evident truths), they present us with no overlap at all. It would be mistaken, however, merely to see them as complementary, as insisting on different aspects of the scientific process. The underlying themes of their writings are, indeed, rather similar: the rejection of tradition; the importance of precise observation and experiment—the whole current they inspired would speak of itself as "experimental philosophy"; the use of critical and analytical reason; and (except for Bacon) the prominence of mathematics as a tool and a language. Still another theme which is also shared though formulated differently is a *mechanistic* view of nature as some kind of sophisticated clockwork or a piece of pneumatic machinery.[9]

---

with Napier's invention of the logarithms, the eminent importance of which for navigation (etc.) no real practitioner could have overlooked in early seventeenth-century England

But his philosophy provided an underpinning for another facet of seventeenth-century science, which possesses less present-day prestige but which was quite important in its time and for the eventual creation of more prestigious breakthroughs: the fact-finding and fact-ordering activity in fields where no adequate theory was as yet possible—as it went on for instance in the botanical gardens where the flora of all known parts of the world was cultivated and classified.

8. Cf. also Gewirtz 1941.

9. Bacon was an atomist, seeing everything as composed of small material particles in motion; even the "spirit" providing the active forces of the system consisted of a subtle fluid somehow similar to air. Although his actual theories are not formulated so as to depend on such considerations, Galileo suggests in his *Letter to the Grand Duchess Cristina* that the planetary system may in some way be driven by the rotation of the Sun (trans. Drake 1957: 213f). Descartes, in part V of the *Discours de la méthode* (ed. Alquié 1963: I, 621ff), describes the function of the heart more or less like that of a steam engine, whose heat makes

All these themes gained broad influence in seventeenth-century philosophy. Most bizarre for modern eyes is probably the triumph of the *geometric method:* metaphysics as well as ethics and theology were set out in the style of Euclid's Elements and Archimedes's statics, with definitions, axioms, theorems, and proofs. Descartes had done so on one occasion (even though he presented his *Géométrie* in "non-geometric" essay style), and he was followed by Baruch Spinoza (1632–1677) and others. Blaise Pascal (1623–1662), who only employed the geometrical method strictly in his mathematical writings, none the less declared it in general to be the most perfect of humanly possible methods (*De l'esprit géométrique et de l'art de persuader,* ed. Chevalier 1954: 576f).

More durable was the success of *empiricism:* all knowledge comes from empirical observation of the phenomena of the real world—or, in a classical formulation due to Thomas Aquinas (*In De Trinitate* I,3), "nothing is in the mind which was not first in the senses." The empiricist program was formulated by John Locke (1632–1704), in continuation of Bacon and under inspiration from the achievements of Robert Boyle and "such masters as the great Huygenius and the incomparable Mr. Newton" (*An Essay Concerning Human Understanding,* ed. Fraser 1959: 14). Strictly speaking, empiricism had also been the basis of Aristotle's philosophy. According to Aristotle, however, experience was to lead to a finite set of pre-existent, immutable, and exhaustive principles (cf. above, p. 42). Already for this reason, Locke's versions constituted a radical innovation (Bacon's "simple natures" were closer to Aristotle). Moreover, Aristotelians of later ages (and especially those of the fossilized university) tended to pay only lip service to the principle of experience, which made Bacon's and Locke's contributions to "experimental philosophy" even more innovative.

Not only methods and epistemology were borrowed from the new science and its spokesmen. Even the mechanistic view was taken over as a general principle, mostly in the radical atomistic variant (cf. p. 45, n. 30). Atomism had been known to the Middle Ages and the Renaissance in part from Lucretius and in part through Aristotle. In the early seventeenth century it was broadly adopted because of its agreement with the mechanistic view. It was still suspected of being atheistic, but Christian versions were produced in France as well as England.

What makes atomism and related views important in our context is primarily the use that was made of it in explanations of human nature and human society. On one hand, mechanism was applied directly: to Descartes and in particular to his followers, human beings as well as animals were machines who differed from other automata in complexity but not in principle—the only difference between man and

---

the blood expand into the arteries (whereas Harvey had seen the heart as a pump); explains the movements of the planets as the movement of vortices in a liquid; and interprets light as small particles, whose speed of rotation determines their color.

animal being the presence of a "ghost" in the human machine, that is, a soul.[10] On the other hand, atomism served as a metaphor and a model: as seen by Thomas Hobbes, society is composed of social atoms (translated into Latinized English: *individuals*) who, in the state of nature, are as indifferent to each other as atheist Epicurean atoms; only reasoned self-interest may make them realize that all will have better chances of survival if they surrender their individual liberty by a contract to a strong ruler who may force some order upon everybody and prevent them from cutting each other's throat.[11] According to Locke, who accepted Descartes'

---

10. There is some doubt as to whether Descartes himself shared this radical stance to the full; in a letter from 1646 (ed. Alquié 1963: III, 693ff) he admits the possibility that some kind of thought be connected to the organs of animals, similar to the kind of subconscious control of our limbs that allows us to walk without reflecting upon how we do it. But even the animals that seem most intelligent are not likely to possess an *immortal* soul, he argues: if some animals do so, all should possess it—but this seems implausible in oysters and sponges.

Whatever the shades of the reasoning, however, we recognize the dichotomy that made Kepler reject zodiacal astrology. It is not the universe as whole which is an ordered cosmos kept together by a universal spirit or by God's intentions, nor the universe itself that is alive and ultimately governed by spirit and meaning. The universe is a heap of mechanical devices, of which some have the privilege to be provided with a ghost capable of conscious thought (and perhaps salvation).

11. The fundament of the argument is thus the concept of 'human rights' belonging naturally to each individual—first of all the right to protect oneself. As explained in *Leviathan,* Chapter 14 (ed. MacPherson 1968: 189–192):

> And because the condition of Man, (as has been declared in the precedent Chapter) is a condition of Warre of every one against every one; in which case every one is governed by his own Reason; and there is nothing he can make use of, that may not be a help unto him, in preserving his life against his enemyes; It followeth, that in such a condition, every man has a Right to every thing; even to one anothers body. And therefore, as long as this naturall Right to every thing endureth, there can be no security to any man, (how strong or wise soever he be,) of living out the time, which Nature ordinarily alloweth men to live. And consequently it is a precept, or generall rule of Reason, *That every man, ought to endeavour Peace, as farre as he has hope of obtaining it; and when he cannot obtain it, that he may seek, and use, all helps, and advantages of Warre. . . . .*
>
> From this Fundamental Law of Nature, by which men are commanded to endeavour Peace, is derived this second Law; *That a man be willing, when others are so too, as farre-forth, as for Peace, and defence of himselfe he shall think it necessary, to lay down this right to all things; and be contented with so much liberty against other men, as he would allow other men against himselfe.*

But certain rights cannot be given up by this social contract:

mind-body dualism, complex thought was built from simple ideas resulting from experience, in the way physical objects were composed of atoms (a piece of ice in the hand will produce the "perfectly distinct" simple ideas of coldness and hardness, and the concept of ice will thus be composed from these ideas).[12] Mental association, moreover, was explained as ideas that were "strongly joined."[13]

Vaguely linked to the ideology of the scientific revolution, if in most cases only through the willingness to reject received opinions and through the application of critical and analytical thought, were the varying doctrines of natural law, as represented by Hobbes, Hugo Grotius (1583–1645), Samuel von Pufendorf (1632–1694),

---

> Whensoever a man Transferreth his Right, or Renounceth it; it is either in consideration of some Right reciprocally transferred to himselfe; or for some other good he hopeth for thereby. For it is a voluntary act: and of the voluntary act of every man, the object is some *Good to himselfe.* And therefore there be some Rights, which no man can be understood by any words, or other signes, to have abandoned, or transferred. As first a man cannot lay down the right of resisting them, that assault him by force, to take away his life; because he cannot be understood to ayme thereby, at any Good to himselfe.

No human individual, thus the first presupposition, can be claimed to give up his rights in exchange for an advantage falling to other individuals or to the community. And no fear of damnation or hope of eternal felicity remains, since no advantage is possible after our physical death. Individuals are really *atoms,* and Hobbes's social atomism is no less atheist than the Epicurean version.

The idea of rights inherent in the very fact of being human had been under way for at least a century, and was clearly expressed by Shylock in William Shakespeare's *Merchant of Venice* (III, scene 1, ed. Alexander 1951: 237):

> I am a Jew. Hath not a Jew eyes? Hath not a Jew hands, organs, dimensions, senses, affections, passions, fed with the same food, hurt with the same weapons, subject to the same diseases, healed by the same means, warmed and cooled by the same summer, as a Christian is? If you prick us, do we not bleed? If you tickle us, do we not laugh? If you poison us, do we not die? And if you wrong us, shall we not revenge? If we are like you in the rest, we will resemble you in that.

None the less, Hobbes's thought, with its coupling of universal rights to radical individualism and to the conviction that "you've only one life to live," was certainly innovative.

12. *An Essay Concerning Human understanding* II.ii.1–2 (ed. Fraser 1959: 144f).

13. Locke only used this extension of the atomistic model to explain *faulty* thinking; but contemporary and subsequent thinkers would use attraction and repulsion between ideas to explain the functioning of the mind in general (for those of them who considered even the soul a kind of subtle matter, the atomism of ideas was no mere analogue, as by necessity it had been to Locke)—see Vartanian 1973: 136ff.

and Locke. Admittedly, as in the case of empiricism, the concept of 'natural law' was not new—it had been one of the basic tenets of the political theory of Thomas Aquinas and other Scholastics. But Thomas's Nature was Aristotelian, and Thomist natural law tended to be both theological and Aristotelian and had been conceived as an answer to the problems posed by the thirteenth-century balance between the Church, autonomous bodies, and feudal rulers. Seventeenth-century natural law was thus something new. Through Hobbes, as we have seen, it was essentially related to the mechanical world picture and independent of religious doctrines; the problems which it answered were those posed by the interaction between the new nation states and by the internal political structure of these, and the individualism on which it was based corresponded to the emerging social structure of capitalism, however much it borrowed ideas and terminology from mechanism.

Also indirectly linked to the scientific revolution, but more directly to the persons and philosophies of Descartes and Pascal, was a novel approach to the understanding of language, the "general grammar" of the Port-Royal school.[14] From one point of view, it is true, it simply revived the semantic approach of Stoic and scholastic grammatical theory, assuming that *logic* was the basis of language and that the function of language was to express thought. But inherent in the philosophies deriving from the scientific revolution was a new conceptualization of logic. To Aristotle and all subsequent logical theory until the end of the Renaissance, logic had (roughly speaking) been considered to deal with classes of objects and attributes,[15] for which reason the basic category of grammatical theory had always been the word-class understood in isolation[16] (whence to be defined from its inflection, the verb, for instance, from its possession of tense). "General grammar" instead saw the sentence and the correspond-

---

14. Rieux, et al. (eds.) 1975 is a translation of its founding achievement, the *Grammaire générale et raisonnée,* first published in 1660. Its main theoretician, the theologian and logician Antoine Arnauld (1612–1694), was a disciple of Descartes and a friend and associate of Pascal.

15. The standard exemplary syllogism may elucidate how: "All men are mortal; Socrates is a man. Thus . . . ." The major premise was read as a statement that the class of "men" has the common attribute "mortality"; the second as a statement that the single individual "Socrates" belongs to the class of men. Putting such conditions on the phrases participating in a logical argument avoided fallacies of the kind "Red is a color; my coat is red; thus my coat is a color"; but it often made it impossible to relate the reasoning of normal or scientific discourse to the theory of logic, as a result of which both Bacon and Descartes had rejected it as useless. Even Ramus had made the point, we remember (see p. 99), but without having anything to propose that might clarify the matter.

16. "How many are the parts of speech? Eight. Which? Noun, pronoun, verb, adverb, participle, conjunction, preposition, interjection"—thus begins Donatus's fourth-century *Ars minor* (ed. Holtz 1981: 585), which was the most-read grammar textbook during the Middle Ages and the Renaissance (and well beyond), and which became the model for most early vernacular grammars.

ing judgement as the elementary building stone of language and started its analysis from there, in agreement with an equally new understanding of logic[17] as concerned with the actions of the human spirit in "conceiving, judging, reasoning, and ordering" (ed. Roubinet 1964: B v), all resulting in or dealing with sentences. The outcome was not only a more adequate understanding of syntax but also a delimitation of word-classes where inflection became secondary and meaning regulated by syntax primary.

SCHOLARLY AND THEORETICAL ACTIVITY

Classicism and the "philosophical revolution" are the most conspicuously innovative contributions of the seventeenth century to our present picture of the humanities; but the traditional branches of humanistic scholarship continued their development, and new theoretical insights were obtained.

Much of what went on in humanistic scholarship perpetuated the Humanist traditions of the later Renaissance, at times for better and at times for worse. The victory of the moderns over the ancients was a French victory; it happened late in the century; and it was far from complete. Classical literature (including Hebrew letters, not least in Lutheran countries where only the Greek and Hebrew Bible and not the Latin translation was assumed to be God's original word) still dominated scholarly studies, and philological progress made its way rather automatically, due to the continuation of hard and systematic work within an unbroken tradition; but scholarly progress was often submerged in a morass of pedantry.

The institutional focus of this tradition could no longer be private Humanist schools; they had died long ago. Already during the Italian Renaissance, however, certain universities had employed illustrious Humanists as teachers (not least new universities created by princes—thus Ferrara, created and protected by the House of Este). As the products and habits of Renaissance Humanism crept into the general mentality of the cultured classes, (often but not necessarily pedantic) Humanism established an alliance with postmedieval dialectic and disputation and thus found a dwelling. As the Jesuit Order, established as part of the "Counter-Reformation" (that mid-sixteenth century movement within the Catholic Church which sought to redress the abuses which had provoked the Reformation rebellion and to uproot all heretical movements and tendencies), established its own high-quality schools for the sons of the elite, even these schools came to harbor high-quality Humanist scholarship aimed at teaching.[18] The better Lutheran universities, being the best

---

17. Published in 1662 in *La Logique ou l'art de penser* ("Logic of Port-Royal"), anonymous but written in fact by the same Antoine Arnauld together with Pierre Nicole (1625–1695), another friend of Pascal's.

18. The Jesuits also attempted to take over control of universities, emulating the Dominicans of the thirteenth century. They succeeded in many parts of Catholic Europe, but with less consequence than in the thirteenth century because of the diminished social role of universities.

available institutions in their segment of Europe, fulfilled a similar function both socially and as far as scholarship was concerned—not least Wittenberg, where Martin Luther's close associate Philip Melanchton had taught, and where Tycho Brahe and Hamlet had studied, the latter according to Shakespeare. Everywhere, the soil from which the succession of Renaissance Humanism grew was watered with religious orthodoxy and service to those in power.

A different kind of continuity with Humanism was present in a particular approach to the study of language. As mentioned above (p. 96), a strong current in later sixteenth-century French lawyer Humanism had done its best to prove that all valuable arts, including writing, had been taught by the ancient Gauls to the Greeks, the Egyptians, and the Chaldeans. A similar current had existed in the Netherlands, and in 1569 one Goropius Becanus had published his proof that Dutch-Flemish had been the language of Paradise and the source of all other languages, built on highly imaginative etymological constructions.[19] A number of seventeenth-century scholars took up the problem of etymological links between languages while rejecting many of Goropius Becanus's more fanciful ideas. At the same time they took over the idea that Persian, Greek, Latin, Celtic, Slavonic, and Germanic languages had a common origin in Scythian—at times with the purpose of using this to prove that precisely *their* language represented the common source most faithfully. In spite of this latter aim, ridiculed ever since, the insights which they provided made it difficult to hold that Latin and Greek possessed special letters of nobility, or that Hebrew was Adam's language, all other languages being created in the Babylonian confusion. Just as in the victory of the moderns over the ancients, though less intentionally and with more modest effects, they contributed to preparing the Enlightenment and to undermine religious dogmatism.[20]

The effort to connect the vernaculars and, on the part of some scholars, to prove their historical superiority,[21] was a consequence of the general higher status of these

---

19. This connection may be illustrated by an abbreviation of the initial steps of G. J. Metcalf's summary (1974: 243) of Goropius Becanus's 70-page attempt to find the real meaning of the tribal name *saxon:* it must be connected to Latin *saga* (sooth-sayer); but this comes from Dutch *segunen* (to bless), a derivative from *sagun.* The latter is composed from *sac* and *gun. Sac,* on its part, means the opposite of the (apparently freely invented) *cas* (supposed to mean "diminish"), which has the same letters in opposite order; and so forth.

20. This could still be done with reference to Biblical history. Gottfried Wilhelm von Leibniz (1646–1716), in his studies of language a direct heir of the current under discussion and discoverer of the Finno-Ugric language family, was able to connect the Celto-Scythian (now Indo–European) and the Aramaic (now Semitic) language families with two of Noah's sons. However orthodox this may seem, it annihilates the Biblical explanation of the multitude of languages.

21. Goropius Becanus's way to prove this was not the only possibility. "General grammar," seeing logic as prior to language and common to all human minds, also deprived the classical

languages and thus, ultimately, a result of the gradual spread of general schooling to social strata who had no use for Latin (not to mention Greek and Hebrew). It is no accident that the process had started around 1300 in Dante's Florence (cf. p. 94), where half of all children were taught elementary reading and writing, and relatively many boys continued their education in the merchants' "abbacus school." The Reformation brought the educational tidal wave to countries it had not touched before, enforcing at the same time increasing literacy and that welding of vernaculars into literate languages which Italian Humanists had undertaken in the previous centuries.[22] The Renaissance writers in the vernaculars had exerted themselves to increase the *copia* of the vernaculars, that is, the abundance and differentiation of their vocabularies and stylistic possibilities. This had also been a deliberate policy of Ronsard and his circle in the sixteenth century. In the second half of the seventeenth century the process had gone so far in many countries and produced stylistic norms and appropriate terminologies for so many genres that it needed not be continued.[23] Boileau's attack on Ronsard, and the whole classicist attempt to simplify language and style, demonstrates once again that a watershed had been reached and left behind.

Already Machiavelli had used history as a fundament for political theory. Equally in the early sixteenth century, Thomas More had been inspired by Amerigo Vespucci's account of the customs of American Indians from 1504 when writing the *Utopia.* The later sixteenth century, as well as the seventeenth, continued this incipient development of anthropology and of some shared parent of political philosophy, political sociology, and philosophy and comparative studies of law. Part of this (represented not least by Hobbes, Locke, and Grotius) was formulated within the framework of natural law and was dealt with above. Early anthropology, on its part, soon developed from utopian-critical reflection into a practical tool for the Christian mission and

---

languages of their privileged status (and in fact Arnauld and Lancelot argue without distinction from all languages they know). Other French scholars, similarly interested in the logic of grammar, close to classicism and inspired by the Cartesian tradition, would distinguish "between (1) an ancient type of language, with free word order and frequent inversions, and (2) a modern type in which the fixed arrangement of words gave a true picture of the natural sequence of thoughts or ideas" (Le Laboureur, *Avantages de la langue française sur la langue latine,* written in 1669 and summarized in Diderichsen 1974: 287).

22. Comparison of, for example, Albrecht Dürer's early sixteenth-century letters with written German from the later part of the same century illustrates the immense impact of Luther's translation of the Bible. Though produced in a country with much stronger traditions for vernacular writing (from Chaucer to Shakespeare, and even as far back as Bede and Alfred!), King James Version (1611) also had a strong influence on literate English. In many social strata, the Bible and the Psalter were the only books a family would possess.

23. The integration of new fields of knowledge into vernacular culture might still call for systematic creation and perfection of terminologies. As a rule, however, new fields would from now on be quite new, and thus equally new in Latin and in the vernaculars.

for colonialism and is better characterized as ethnography (*description* of unfamiliar people). Modern anthropologists still use much of the material collected by missionaries. These, indeed, have produced the only available reports regarding many nations as these looked at the moment when their interaction with European colonialists began; many of them, furthermore, were sufficiently sympathetic to those whom they described to apply a broader perspective than that of their employing organization. None the less, the tie between the mission and the European mother institutions and colonial administrations did much to determine the kind of questions that were posed and answered. By the seventeenth century, moreover, most European political philosophers (and intellectuals in general) had become so convinced of European superiority over the savages that they did not bother to draw on whatever material was available.[24] Finally, many missionaries' reports remained as manuscripts in the archives, and thus were not available. For these reasons, the influence of this early ethnography remained limited, and we shall not pursue the matter.

THE PROBLEM OF THE BAROQUE

Not infrequently, the seventeenth century as a whole is spoken of as the "Baroque Age." If this is a sensible characterization, one may wonder why the Baroque was mentioned only once in the preceding pages.

If *Baroque* meant nothing but "seventeenth-century," of course, we should not wonder; in this case the term would be empty. But the point in the characterization is different: it implies that a particular mentality —paradigmatically expressed in a certain kind of Church architecture (for instance San Pietro), a certain kind of painting (say, Caravaggio and Rubens), a certain kind of literature (say, Góngora and Andreas Gryphius)—dominated at least the artistic sensibility of the whole century. If such a dominance existed, one must expect it also to have had an impact in other domains of thought.

If. But *whether* it existed depends very much on delimitations, and at least for the purpose of the actual question, a picture with marked contrasts will be most

---

24. The formation of this attitude in the course of the sixteenth century can be followed in the iconography of the Adoration of the Magi. One of these Three Wise Men, as it is known, was supposed to be black. In the beginning of the century, he appears in the same princely apparel as the other two and with a similar retinue; but towards its end he is currently depicted as masters would imagine their slaves.

The only non-European nation which in the second half of the seventeenth century was generally looked at with respect was China, which the Jesuit missionaries hoped to convert and described with great admiration, and whose ideographic writing system was felt by many philosophers (in particular Leibniz) to be close to the "mathesis universalis," an artificial language of almost algebraic character which they hoped could avoid the ambiguities and confusion of ordinary language (another version of the belief in the "geometric method").

useful. Instead of seeing, for instance, Boileau's and Racine's classicism as just another but rather different kind of Baroque, we shall therefore regard it as a counter-current, as a reaction, and ask then what characterized the Baroque *stricto sensu.*[25]

In its origin, the Baroque was closely connected with the Counter-Reformation and with the Jesuit Order—so much so, indeed, that "Jesuit style" often serves as another name for the Baroque. Seen from this angle, art was to serve the purpose of stimulating faith. The artistic program endorsed by the Church aimed at achieving this spiritual awakening through a strong sensual and emotional appeal. Movement, tension, and contrast; monumentality and rich decoration, looking for effect rather than derived from some kind of "inner necessity" (for example, facades rising considerably higher than the church building behind—in remarkable contrast to the flying buttresses of Gothic cathedrals, which actually serve the stability of the building); dramatic uses of light; total planning of an impressive "artistic environment" where decoration, architecture, and surrounding urban space were part of a *Gesamtkunstwerk*—these were adequate means for realizing the aim, and indeed became characteristic of the Baroque proper.

But evidently the Counter-Reformation was only a starting point. The Baroque became a general court culture (although in bridled form precisely in Versailles because of interaction with the classicism of literate culture), and came to be connected both with Lutheran orthodoxy (which just as much as reformed Catholicism asked for controlled piety) as well as Pietism (no current used erotic and sexual imagery as strongly for spiritual purposes as did precisely Pietism).

It is evident, too, that the general characteristics of the Baroque style would unfold in many different ways, from one country to the other, from one genre to the other, and from one artist to his colleague. Pedro Calderón de la Barca (1601–1681), author of countless *autos sacramentales* serving precisely the diffusion of popular piety, could end his drama *El alcalde de Zalamea* by showing on the scene the dead body of the rapist-officer seated in the garrotte. The dazzling styles developed by Spanish poets such as Luis de Góngora y Argote (1561–1627) and Francisco Quevedo y Villegas (1580–1645) would make the spiritual impression depend instead on the intellect and not on such violent appeals to the passions, making use of artful metaphors that had to be reflected upon and combined in order to reveal their sense, or were only meant to be understood by the learned.

However much this Spanish (and related Italian) poetry was only aimed at a narrow public, it reveals another feature of the Baroque: the elements that it used—

---

25. This is also the conclusion of René Wellek (1973: 195a), in his discussion of the use of the term *baroque* in the history of literature: "The term *baroque* seems . . . most acceptable if we have in mind a general European movement whose conventions and literary style can be fixed narrowly, as from the last decades of the sixteenth century to the middle of the eighteenth century in a few countries."

in its convoluted poetical metaphors, its paintings, and its distortions of the orderly architectural space of the Renaissance—were predominantly of ancient or Biblical origin.[26] But they were, in a way, not taken seriously in themselves; they were decoration (*rich* decoration), or they were pretexts for something different (which was certainly taken seriously); even in religious painting, the religious motif was a pretext for light, shadow, and tension, and these, at least as much as the motif itself, were the true carriers of the emotional appeal. Greek mythology, like the elements of architecture (columns, circular arcs, etc.) and even Biblical history were reservoirs defining the limits of what could be done. But they remained primarily reservoirs, and the limits they defined were precarious and unstable: columns might be cut in spirals (as were Bernini's four bronze columns around the sepulchral chapel in San Pietro[27]), and the circular arcs might be broken. Just as much as classicism, though in almost opposite terms, the Baroque depended paradoxically on the ancient heritage and was unable to free itself however much it reinterpreted and recombined its elements (one is reminded of the relationship between fourteenth-century *via moderna* and Aristotelianism). No wonder that *echo,* treacherous and delusive repetition but none the less repetition, was a favorite metaphor and a favorite poetical technique (cf. Koch 1994).

At least the aims of the Italian and Spanish literate Baroque found expression in theoretical treatises; though their norms are radically different from those of Boileau, they may count just as legitimately as literary theory, and in several ways they even seem more modern.[28] In other fields, however, it is not easy to point to innovative scholarship reflecting directly the "Baroque mind-set."

This can be explained at several levels. One explanation—crude but worthwhile—looks for the sociocultural affiliation of the Baroque. Fundamentally, the Baroque

26. This of course is not true of seventeenth-century Dutch painting with its predominantly bourgeois public. But precisely this difference of themes and style as well as public reflects that a Rembrandt should not be counted to the same Baroque current as Rubens the Flemish court artist.

27. In the case of which, by the way, the metaphors of *reservoir* and *limits of what could be done* acquire a literal meaning: the columns were made from ancient Roman bronze pilfered from the Pantheon.

28. However, Boileau's as well as the Baroque treatises are more adequately compared to the literary and artistic manifestos of our century (those of Futurism, of Dadaism, etc.) than to authentic theory.

The "modernity" of Spanish "Góngorism" (etc.)—the dependency on metaphors and allusive meanings, the rejection of immediate comprehensibility and "imitation" of reality—is another parallel to the *via moderna.* Enlightenment and Romanticist critics regarded it as scornfully as Thomas More regarded the descendants of the *via moderna,* as empty artificiality; only twentieth-century modernists (thus Lorca and his generation of Spanish poets) would discover in seventeenth-century poetics material that could be reinterpreted so as to fit their own poetical venture.

was connected with court and church, and more specifically to their representative function: piety, awe, or higher spiritual significance were to be imparted by means of emotional appeal or allusive use of metaphors. The Baroque *Gesamtkunstwerk* was a theater scene where *Truth* was to be displayed, and accepted by the public as displayed, not to be argued in clear and emotionally neutral terms. Innovative scholarship, as we have seen (p. 109), grew out of a culture of dialogue and explicit argument, of a culture where even the Medici Prince was in principle *inter pares,* and *primus* only when the economical costs of scholarship had to be paid.

Biographies show that Baroque art and poetics were not unmediated expositions of an ecclesiastical and courtly program; many outstanding Baroque artists and theoreticians had no less trouble with the mighty of this world than Molière. But this does not change the basis for their art, and on the level of content, the same explanation can be repeated in different terms. In Baroque culture, with its emphasis on effect, "the least important thing was whether what [is asserted] is true," precisely as in the *Iliad* and the *Orlando furioso* and in a ceremony. But this attitude was precisely—as argued by Galileo—what was inadmissible in science (and, we may add, not only in natural science). That separation of reality and symbol which was essential for the new science (cf. p. 102) was not easily integrated with a culture so wholly oriented toward symbolization as the Baroque (albeit toward external symbolization chosen for the effect and not taken quite seriously).

This strength of the Baroque mind-set in relation to the production of art and its weakness as a basis for scientific scholarship is clearly illustrated if we look at such developments which in some way or other do reflect the Baroque mood.

One example was already mentioned and briefly discussed (p. 120): the etymological school of Goropius Becanus and his successors, with its fanciful but artistic rather than critical use of arguments. Another instance—almost an archetype—is offered by the seventeenth-century curio cabinets. The better of these were to become starting points for later historical, archaeological, and naturalist museums, and quite a few were created by learned and competent collectors (for instance, Athanasius Kircher, to whom we shall return, and the Danish polymath Ole Worm). But in contrast to the botanical gardens (cf. p. 113, n. 7) they did not aim at orderly fact-finding, nor were they systematically concerned with a specific field. Their aim was to exhibit the striking or amazing—the rarities of nature, of human art, and magic *pêle-mêle.* Natural magic and occult thought are near at hand—but as in the case of the Baroque use of ancient mythology no longer taken quite seriously.

In larger scale, a similar orientation is found with the polymath authors of the time, of whom the Jesuit Athanasius Kircher (1602–1680) may be taken as an outstanding representative. More than forty books of his are known, dealing with almost all scholarly fields, from Coptic grammar to the construction of telescopes and the use of burning mirrors. Even the single books, however, consist of such mixtures. As an example may serve his *Musurgia universalis* from 1650, a breathtaking work of more than 1150 folio pages. There are numerous observations regarding acoustics, harmonic theory, and musical instruments. But the framework is that of "universal music," musical harmony as the fundament of everything (so much so that it is impossible to pigeonhole the work as dealing with either "natural" or "human" science). The exposition is strongly oriented toward the domain of the marvellous and even the magical—for instance, it is discussed (Kircher 1650: II, 232f) whether the Pied Piper of Hameln could lead away first rats and next children by the natural power of music, and concluded that the piper was the Devil himself who, upon God's decision, carried away the Hameln children to Transylvania. In the discussion of "Phonocamptic magic, that is, Echo, the nature of the reflections of the voice, and its marvellous effects," fish are believed (II, p. 240) to come when called by name because the ancient Roman encyclopedic naturalist Pliny tells so, and when "hierarchical harmony, that is, the harmony of the angels distributed in nine choirs" is the topic, we encounter the mystical properties of the number 4 (II, 448f). On the whole, the many solid observations and reflections are drowned in a mass of trivialities, curiosities, marvels, and anecdotes which turn up because they fit (in the style of Baroque decoration), with little critical afterthought as to whether they are true or relevant. We are much closer to the tradition of natural magic than to Galilean or Cartesian philosophy. Kircher's is a universe where the *Iliad* and *Orlando furioso* are just as valid arguments as technical experience, textual criticism and Archimedean geometry—and his works demonstrate how Baroque thought, even when scholarly, tended to produce art and entertainment rather than science.[29] The virtual absence of the Baroque from the main part of the preceding chapter is thus no paradox.

---

29. More so, of course, in the works of a polymath than when an established scientific discipline with its stricter norms was involved, which might bridle excesses. One such example (described by Henk Bos 1993) is to be found in the development of seventeenth-century geometry. Descartes, in his analytical geometry, had produced a tool that allowed solution of geometrical problems by means of algebraic computation—an approach which was wholly different from the construction of solutions by means of ruler and compass alone, canonical since antiquity. Instead of accepting this new tool to the full, however, Descartes and his followers used the algebraic tool to transform problems so that they could be solved by means of curves—more exotic curves, like moving parabolas, but still distorted echoes of the circle and the straight line. Indubitably the outcome was mathematical *science* in the strictest sense; none the less it was a dead end, whose existence mathematicians have done all they could to forget since the moment analytical geometry was accepted to the full.

Another example is offered by Isaac Newton himself, idol of the eighteenth and nineteenth centuries because of his works on optics and on the "mathematical principles of natural philosophy." In these, Baroque influence is not to be traced. In his works on Biblical chronology, on the other hand, he was not restrained by the norms of an established discipline; accordingly, the interests if not the methods are no less expressions of a Baroque posture than Kircher's *Musurgia*—and they were ignored just as efficiently by the ashamed physicists of future centuries as the abstruse curves of the geometers by *their* successors.

CHAPTER 6

The Enlightenment

THE APPEARANCE OF THE "PUBLIC SPHERE"

The absolutist states (not only the French) had supported art and scholarship (if occasionally "as the rope supports the man in the gallows"). Classicism and the rule of taste had resulted. The *public,* however, was not composed solely of court nobility and officials, and in so far as courtiers and officials took part of the public, they did not participate solely qua courtiers and officials (that is, the way they would participate in a ceremony, where they would play their prescribed part and obey; as theatre public they would protest if they were not pleased). Late seventeenth-century art is to be seen in the context of the incipient *public sphere* of bourgeois culture.[1]

The roots of this public sphere are to be found in the neighbourhoods and guilds of the medieval town and in similar peer organizations of the Middle Ages in general. From the eleventh century onwards, towns had continually given rise to the emergence of organized groups of equals, both within the (proto-)working class, the (proto-)petty bourgeoisie, and the (proto-)bourgeoisie (cf. above, p. 61). From the Renaissance onwards, as we have seen, members of the latter group especially, and members of those courtly and intellectual circles which were closest to the mercantile patriciate, came increasingly to regard themselves as autonomous individuals, and this fact imprinted their communities.

1. The term public sphere translates Habermas's *Öffentlichkeit.* An *Öffentlichkeit* or a *public sphere* can be explained as a *forum for the formation of shared interpretations of the world* and of *shared will.* The public sphere, in other words, is the social substratum for the formation of ideology (as this elusive concept will be defined below, p. 342). Characteristic of the public sphere of bourgeois culture (the *bürgerliche Öffentlichkeit*) is that interpretation and will are founded upon discussion and argument; the "representative public sphere" of feudalism and of the absolutist state (as prominent in the appearance of the Baroque) is the *display of truth* in ceremony and ritual; it is thus no real *sphere of the public* but a scene at which the public is expected to gaze, accepting the message of authority and power.

The Academies of the Renaissance present us with a related phenomenon: an organization of select intellectual peers (considering themselves as autonomous individuals *par excellence*) in a closed network based on discussion and argument, either scholarly or around artistic products, depending on the case. Around the middle of the seventeenth century, the Masonic and similar organizations imitated this pattern of the closed organization of equals, often, indeed, the social composition of the lodges was similar to that of the "amateur academies" of the Italian Renaissance: intellectuals, intellectually alert noblemen, and wealthy bourgeois. Also related, but older and more broadly based socially, is the proliferation of heretic and dissenting religious communities from the twelfth to the eighteenth century.

The late seventeenth century, after the ultimate failure of absolutist attempts in England, brought a transformation of this pattern: *open* circles, centered for instance around coffee-houses, where any topic of general interest could be discussed—political, religious, moral, or artistic—and where anybody possessing the necessary cultural qualifications might participate. In the early eighteenth century, the salons of the modernizing, town-oriented fraction of the French nobility fulfilled a similar role. These open circles are the genuine first prototype of the *bourgeois public sphere*,[2] where *truth is to be found in the middle* but where each participant also possesses *a private life*, a sanctuary not to be revealed publicly—an idea which had not been present in Greek or primitive democracy—and where cultural level and neither mere citizenship nor social status *per se* determines who has access.

A possible explanation of the rise to prominence of this distinction between the private and the public sphere may be the increased role of the printed word. Until the middle of the seventeenth century, public spheres based on argument had mainly involved direct confrontation between participants. During the Reformation, it is true, ample use had been made of printing, but in the vernacular, that is, for the general public, mostly for producing satirical leaflets and stamps. In the later seventeenth century, periodicals had begun to turn up (including the first scholarly journals, generally connected to the scientific academies), and in the early eighteenth century they became a major vehicle for moral and other discussion, escaping every official control.[3] This anonymous medium (more anonymous because writers would often work under pseudonym, not least when clandestine printing was involved) facilitated a distinction between the public and the private.

---

2. This description (which, from the Masons onward, is grossly the one given by Habermas) is highly idealized; the actual minutes and organization of coffee-house discussions are much more fuzzy: some came to drink beer and chatter, other came to discuss *in other rooms*. The *salons* of noble houses, on their part, were certainly only open to a select class of people. Still, the idealization catches an essential structure.

3. Popkin 1991 is an illuminating examination of the journal institution (newspapers but particularly magazines) as carrier of a new form and organization of knowledge.

In this environment of printed debate a number of writers circulated, not as random participants but as main actors. Here they discussed with each other and with other intellectuals, here they found their public, and here they presented their products to the public, setting or at least formulating the agenda for the discussions.

Writers of all sorts participated. Cronique scandaleuse as well as politico-moralist journalism and essay writing found their basis in the environment, and the novel came to maturity as a literary genre (often no less moralist than the journalism and essays).

The public was as heterogeneous as the "active" participants, and its interests were directed at different parts of the literary production. Master artisans, bourgeoisie, enlightened nobility, and officials can all be found. Even though Denmark and even northern Germany were relatively peripheral in the process, Holberg's *Den politiske kandestøber,* "The Tinkerer Turned Politician," is representative, with its satirical presentation of the Bremen pewterer Herman who wastes his time discussing politics instead of taking care of his workshop.

Given the heterogeneity of both kinds of participants, no universally valid characterization of the environment can be made. Often, however, the environment was politically critical in one way or another. In France, absolutism was a main target, less often as a principle than because of its actual working, together with the irrationality of feudal "freedoms" which came increasingly to be seen as freedom from burdens such as taxation that *should* be carried by everybody. In England, the adversary was often a political system dominated by the nobility in alliance with the king, but general or specific moral decay was not forgotten. In Germany, where educated officials made up most of the public, outworn local princely and noble power was under attack.

THE ENLIGHTENMENT MOVEMENT AND ITS WORKERS

The writers involved in the higher levels of this environment (which excludes rank-and-file journalists and leaflet writers) constituted a milieu of free intellectuals, and came as such to formulate the intellectual movement which was baptized *the Enlightenment* ("les lumières," "die Aufklärung"). In particular in France, the members of the movement came to be known as "the philosophers," *les philosophes.*

Part of the intellectual background was the victory of the moderns over the ancients, in the sense that it legitimized both the presentation of new thought and the new literary genres in which the participants in the movement expressed many of their ideas. More direct was the impact of the philosophical interpretation of the scientific revolution, including empiricism, the principle of natural law, the mechanistic view, and the ideal of mathematization of fields where this made sense. The "geometric method," on the other hand, was given up together with the dream of a *mathesis universalis,* and both were indeed unsuited to the undogmatic Enlightenment endeavour (only Kant would resurrect the geometric method in weakened form); Voltaire, in his *Dictionnaire philosophique,* would speak of the "ridiculous parading of geometry in the treatment of a metaphysical and moral topic,"[4] in direct polemic with Spinoza.

Most important, but at a level which conditioned the impact of intellectual developments rather than competing with them, were unquestionably those changes which had shaken the social world: the economic modernization of Britain and the institutionalization of constitutional rule after the Glorious Revolution of 1688 ("English liberty," the contrary of "freedoms", at least as it was understood in France); these were the changes which were reflected in the new philosophical doctrines. Further, the development of French and other absolutisms into incurable routines that prevented similar changes from taking place outside Britain; and, of course, that emergence of the bourgeois public sphere which provided the basis for the movement.

There is thus little in the intellectual luggage of the Enlightenment (beyond a general confidence in the moderns and hence in the possibility of intellectual progress) which has to do with the humanities. But like the "enlightenment currents" of Ionian Greece and the early twelfth century, the movement was humanist in the sense that it presupposed and supported trust in actual, living humanity and its senses and reason as better sources of truth, knowledge, and morals than traditional or theologically sanctioned authorities.

As a first step, we may approach the Enlightenment through its carriers, those intellectuals who from our point of view embody the movement, and who in their own time formulated its themes. Rather than trying to give a general definition of the category, for which the movement is too diverse, we may list some prominent representatives and groups.

An outstanding Danish/Norwegian example is *Ludvig Holberg* (1684–1754). He worked within the framework of the nascent Danish bourgeois public sphere which he himself helped develop (notwithstanding his persiflage of the poor pewterer Herman of Bremen who will rather discuss the business of princes than mind his own—Herman and his friends are shown not to possess the culture that qualifies for participation in the public sphere); like many other Enlightenment writers, moreover, Holberg attacked not the *principle of absolutism* but its badly administered real-life versions.

Holberg's writings span widely: from satirical poems and culturally-critical comedies through politically and morally critical fiction (*Niels Klim*) to essays (the *Epistles*), history (concentrating on economy, institutions, and customs and not on the history of glorious wars), and natural law.

The main current, however, was the French Enlightenment, which in its first generation was strongly inspired by English science (Newton), philosophy (Locke),

4. Edition of 1770, translated from Naves & Benda (eds.) 1967: 539.

and politics (constitution and relative tolerance). The major individual figures were Montesquieu, Voltaire, Diderot, and Rousseau; collectively, the "materialists" and the physiocrats were also important.

Charles Louis de Secondat, baron de Montesquieu (1689–1755) wrote politico-juridico-philosophical theory (most celebrated is *L'Esprit des lois* from 1748, inspired not least by the English constitutional system) and political and moral satire (the *Lettres Persanes,* from 1721—quantitatively dominated by sweeping derision of French society and customs, but certainly strongest in its attack on the oppression of women, formally located in the harem of one of the Persian gentlemen who describe the strange manners of Paris).

> The best known aspect of *L'Esprit des lois* is the theory of the three powers (XI,vi): in any political system, there is a legislative, an executive, and a judiciary power. If all three are united in the same person or the same social body, we get tyranny, as in the Ottoman Empire and in the Republic of Venice. Most European kingdoms allow moderate liberty, because the power to judge is delegated to the citizens. The only system giving true liberty to the citizens is the one prescribed by English law (not necessarily actual English practice, as Montesquieu points out): all able-minded citizens elect representatives to the legislative power locally (the nobility is allowed a specific House, since they would be disloyal if not allowed special influence—the English experience had shown so); the King executes; and citizens are drawn by lot to judge according to the law.
>
> But the work is much richer. Even if tyranny and liberty depend on actual social institutions, human temper is determined by the environment in which people live, in particular by the climate (XIV; XVII). Good legislation (like that of the Chinese) is the one which opposes the vices generated by geography; but actual government and legislation are largely determined by the environment—too intense heat turns you into a coward prone to accept tyranny.

François Voltaire (1694–1778), certainly the leading figure of the French Enlightenment, was an apostle of Newtonianism, English empiricism and English liberty. Being himself only of moderate talent in mathematics and physical science (though on one occasion he delivered an "Essay on the Nature of Fire and its Propagation"[5] to the Academy of Science), he had his friend and mistress Madame du Châtelet (a competent physicist and mathematician) translate Newton's *Principia*—evidence for the importance he ascribed to the new natural science as foundation for that moral improvement of society which was his central interest.

---

5. That is, on the nature of heat. The essay was based on experimental work and seems to have been quite good both in the eyes of contemporaries (the Academy ordered it printed) and from a modern point of view (see Partington 1961: III, 606f).

His *bête noire* was the Catholic Church (of which he simply spoke as "L'Infâme"), or more precisely its intolerance; as a result, he himself became the scapegoat of Lutheran as well as Catholic *dévôts.* His historical writings, meant to describe and hence to further the transition from barbary to civilization, included the history of art, learning, and culture, and tried to draw laws from the data. The short novel *Candide ou l'optimisme,* one of the most widely read Enlightenment works, is a witty dismissal of all facile optimism and of any belief in automatic progress.

Denis Diderot (1713–1784) was the principal of the monumental *Encyclopédie ou Dictionnaire raisonné des Sciences, des Arts, et des Métiers* (1751–1780), which was immediately recognized as the central manifestation of the Enlightenment project and indeed its greatest publishing success. According to the title, its topics are the sciences, the arts, and the productive trades; in fact, it also treats of all the other central themes and concerns of the Enlightenment, including moral discussions, politics, theoretical science (under which also are *arts libéraux,* now to be understood as "science as culture"), and *arts mécaniques.* There is thus nothing reminding us of the twentieth-century distinction between humanities, natural sciences, and social sciences; instead *reason is one, and its purpose is human welfare*—in agreement with the roots of the Enlightenment movement in the general educated public rather than in a body of specialists (or a body representing the interests of the state). Among the contributors we find practically all the leading French Enlightenment writers, but also many less famous figures who happened to possess specialist knowledge about one or the other topic.

> Diderot also wrote novels in the English "sentimental" style, presenting among many other things a Humean, proto-behaviorist theory of knowledge in dialogue form in *Jacques le fataliste et son maître* (written c. 1773, only published posthumously in 1796). In later years he formulated a break with the idea that all sciences should emulate Newtonian mechanics (the eighteenth-century version of the "geometric" ideal for scientific reasoning) and came very close to Freudian psychological perspicacity in *Le Neveu de Rameau* (written between 1761 and 1774 and repeatedly quoted with approval by Freud).[6] Also in later years he formulated his rejection of absolutism very clearly: *just* absolute rule is not the best but the worst form of government, because it dulls the people's thirst for liberty and the sense of political responsibility; three

---

6. The essential passage—a dialogue between the narrator and "Rameau's nephew" (who shares quite a few characteristics with Jean-Jacques Rousseau, to whom we shall return)—runs as follows (ed. Benac 1962: 479); the "little savage" is the nephew's much-beloved son:

Narrator: If the little savage was left to himself, if he conserved all his imbecility and combined with the scarce reason of the infant the violence of the passions of the thirty-year old man, he would wring his father's neck and sleep with his mother.

Nephew: That proves the need for a good education; and who disputes that?

> consecutive generations of just rulers may transform any people into a horde of obedient slaves.[7]

Jean-Jacques Rousseau's (1712–1778) main immediate impact was in the theory of education (*Émile,* from 1762, where he developed ideas first presented by Locke); but he also wrote on the foundations of social life (and was even asked to propose constitutions for Corsica and Poland). In this domain he went well beyond Voltaire's denial of automatic progress, emphasizing that the shift from barbary to civilization had meant progress in "intelligence," not in reason, and that intelligence could be used—*and had been used*—to suppress, not to liberate, man.

> In *The Social Contract* (like *Émile* from 1762) Rousseau asks how society *could* be organized in a way that was better than the savage state (given that the existing state of affairs was definitely worse) and answers that society had to be constructed around the *General Will* of the citizens. This General Will, he explains, is not the sum (we might say, the average) of discordant particular wills (that is, a compromise) but the will to constitute society for everybody—the explicit will, we may say, to *find and accept* that middle which is bearable for everybody. Answering in advance to latter-day ultraliberalists who deny society the right to interfere with private property (for instance through taxation), he points out that property can only exist through society: without society and the common will to define mutual rights, only possession can exist, which so to speak only lasts until you fall asleep; the inspiration from Hobbes is undeniable and undenied.
>
> Since individuals become citizens by alienating their individual liberty, becoming instead participants in the shared freedom of everybody, the General Will may intervene wherever it finds it necessary. Rousseau emphasizes that this right of the collectivity should be exercised only when really necessary for general welfare; none the less, the possibility to read into Rousseau's work the idea of intolerant and monolithic (totalitarian) democracy was demonstrated during the French Revolution by Robespierre and others. Other readings influenced other phases in the Revolution as well as sundry later constitutions.

The *materialists*—Julien Offroy de La Mettrie (1709–1751), Claude Adrien Helvétius (1715–1771), and Baron d'Holbach (1723–1789)—were openly atheist, whereas Voltaire and many of the other *philosophes* were deists, professing belief in a highest benevolent intelligence who had created the world.[8] They accepted Descartes'

---

7. *Réfutation suivie de l'ouvrage d'Helvétius intitulé l'Homme,* ed. (Vernières 1964: 619f).

8. The deist (whom Voltaire calls a theist) is, according to the article "Théiste" in Voltaire's *Dictionnaire philosophique* (translated from Naves & Benda (eds.) 1967: 399)

> firmly persuaded of the existence of a supreme Being equally good and powerful, who has created all extended, growing, feeling, and cogitating

view of man as an automaton but discarded the mind-body dualism, dispensing completely with "the ghost in the machine" and insisting instead that matter itself is spiritualized and potentially intelligent. La Mettrie's central work carries the title *L'Homme machine,* and its organic-deterministic view of the human mind contributed to opening the way to psychiatric *treatment:* if the mind is not free and responsible for itself, the physician may try to alleviate its pains by changing the conditions on which it functions. La Mettrie as well as Holbach used the machine-man model as the foundation for a morality based on human pleasure and contentment,[9] rejecting thus the Dostoyevskian maxim that "if God does not exist, then everything is permitted." Inherent both in their atheism and in their particular variant of the machine view of human nature was the refusal of the doctrine of original sin (which, in the Enlightenment view, was an all-too-convenient excuse for ecclesiastical rule): the core of human nature is not evil but malleability; whether a human being turns out virtuous or wicked depends on education. In these respects—the denial of original sin and the ensuing inherent wickedness of man, plus the belief in the power of education—Rousseau agreed. His much-ridiculed belief in the "natural" goodness of man means nothing but this (cf. p. 132, n. 6).

The Physiocratic school of economic thought (François Quesnay, Anne-Robert-Jacques Turgot, and others, active between c. 1750 and c. 1775) rejected mercantilism and emphasized agricultural production as the real source of social wealth. This break was no mere shift of theoretical explanatory principles but another illustration of the new aims of Enlightenment thought. Mercantilism, indeed, had not been a *theory of societal wealth* but a *technique to create military power.* According to mercantilist thought, the state was to favor exports, minimize imports, and thus build up reserves of bullion—not as a magical token of wealth but as the necessary means to pay soldiers and a navy. *Mercantilism,* the main economic doctrine of the seventeenth

---

> beings; who perpetuates their species, who punishes crimes without cruelty, and compensates benignly virtuous actions.
>
> The theist does not know how God punishes, how he favors, how he pardons; he is indeed not so audacious that he flatters himself to know how God acts; but he knows that God acts, and that he is just.

9. In this respect they follow the lead of Montesquieu. In the latter's discussion of the influence of the climate on human temper, he speaks explicitly about men as "machines" which, in the South, "delicate, weak, but sensitive, render themselves to love which, in a seraglio, is born and calmed incessantly"; or which, in the North, "healthy and robust, but heavy, finds its pleasure in everything which may move the spirits: hunting, travelling, war, wine" (*L'Esprit des lois,* XIV,ii). No doubt that Montesquieu's machines can really feel; no more doubt that they are really *machines,* according to the discussion that precedes.

In the same vein, Diderot (ed. Vernières 1964: 564) applauds Hobbes for changing Descartes' "I think, thus I am" into the observation that "I feel, I think, I judge, hence a lump of matter organized as I am can feel, think and judge."

century, had thus been meant as a tool for statal power.[10] Physiocratism, to the contrary, investigated the conditions for general human welfare within society, in good agreement with the general orientation of the Enlightenment movement.

Another significant current was the *Scottish Enlightenment,* represented among others by Hume and Smith.

David Hume (1711–1776) is important because of his radical continuation of Locke's empiricism—so radical indeed that he turns Locke's epistemological optimism upside-down: if all knowledge derives from sense experience, which by its nature is always particular, no necessary fixed laws can be found. Laws and causal connections are nothing but habits acquired through the repetition of similar experiences, and can have no higher status (we shall return to Hume's view of causality in Part III, see p. 323).

Adam Smith (1723–1790) was a professor of moral philosophy and contributed to that subject, but his fame and importance are mainly founded upon his *Wealth of Nations* from 1776. The fundament of modern liberalist national economy, it is also a necessary starting point for the Marxist concept of the economic structure as a relation between social classes (even this will be dealt with in somewhat greater depth in Part III, see p. 284).

In its essence the Enlightenment was a broad and far from homogeneous pan-European movement which cannot be adequately reduced to a single formula. It changed from the first generation (Montesquieu, Holberg) to the next (Diderot, Turgot, Marquis de Condorcet, Johann Friedrich Struensee), and it involved writers as different as Giambattista Vico (1668–1744), Jonathan Swift (1667–1745), Jean le Rond d'Alembert (1717–1783), Antoine Laurent Lavoisier (1743–1794), Gotthold Ephraim Lessing (1729–1781), and Immanuel Kant (1724–1804). (Some of these names will be discussed further below.)

GENERAL THEMES AND ACCOMPLISHMENT

In spite of the diversity of the staff of the Enlightenment movement, some main tendencies can be singled out—first of all that whereas the French Enlightenment made direct use of the English example in its attacks on French feudalism (a word coined by Montesquieu), Enlightenment thinkers in the rest of Europe would be inspired by the English example mainly through its French interpretation. It is thus characteristic that the "enlightened" German physician and statesman Struensee (1737–1772),[11] when accompanying the young King Christian VII of Denmark to

---

10. "Mercantilist policies were the continuation of warfare by other means" (Gay 1967: II, 346).

11. A recommendable biography of this interesting figure, written in explicit Enlightenment perspective, is Winkle 1983.

England and France in 1768, took care that Christian should become acquainted with *English industry* and *French Enlightenment philosophers.*

Independently of their nation, almost all Enlightenment philosophers believed in and argued for the possibility of science (natural as well as "moral," that is, roughly human+social[12]) and reason to improve the social world. As Diderot formulated in old age, the only preoccupations in which a high spirit should take interest are "the laws, the customs, the abuses of authority, religion, government, vices, errors, prejudices" (*Lettre apologétique de l'Abbé Raynal,* ed. Vernières 1964: 648). The justification for the conviction that science and reason *could* improve the world was offered by the triumphs of seventeenth- (and, as time passed, eighteenth-) century natural science as interpreted by empiricist philosophy; and, no less, by the absurdity of existing habits and of the prevailing social order, which application of a bit of reason could so easily expose. France and England, for example, had fought protracted wars over a disagreement which was no more important than the question whether eggs should be cut in the narrow or the broad end (namely Catholicism versus Protestantism—thus was the opinion of Swift, clergyman in the Church of England); and public office was only given to those who were hypocrite or infirm enough to see a square as an oblong (thus was that of Holberg, a Danish public official).[13]

---

12. *Moral* derives from Latin *mores* ("custom," "manners," "ways to behave"), which roughly corresponds to the meaning of the word when the eighteenth-century spoke about "moral science"; the idea is thus not too far away from present-day "behavioral science."

13. As regards the ambition to improve their world, the Enlightenment intellectuals were certainly no exception in their century; "projectors" with fanciful ideas both for technical inventions and for improvement of the body politic were plentiful enough to turn up as a recurrent laughing stock in Enlightenment writings; what Enlightenment philosophers would see as the difference between themselves and the projectors was that the latter's proposals had an all too obvious character of isolated *schemes* which augmented the disorders of society instead of correcting them by means of comprehensive reason.

While the social criticism of the Enlightenment had taken its inspiration from the achievements of seventeenth-century science, seeing in them the triumph of critical reason rather than the accumulation of specific results, the projectors were rather inspired by the wave of *inventions* which had characterized seventeenth- and eighteenth-century technology.

In a different framework, the difference between the projectors and their Enlightenment critics may be conceptualized as the difference between tactics and strategy—throughout history, military action was generally the most explicit case of instrumental rationality, and accordingly a slight generalization of the concepts as discussed in Clausewitz's *Vom Kriege* (von Clausewitz 1973: 292–294) may be useful. Tactics is the kind of instrumentally rational action which uses the means at hand and tries to achieve a given aim as efficiently as possible *on these given conditions.* Strategy, instead, is also instrumental, but it is concerned with creating the adequate means for future tactical action; it is global in its approach, whereas tactics is only locally rational.

The difference between the two highlights, and is highlighted by, Hobbes's view of the difference between the state of nature and the foundation of a society based on contract. In

As stated, the Enlightenment held the ultimate purpose of knowledge to be human utility and welfare—no less in the case of natural sciences and technological knowledge than regarding the disciplines of social and moral knowledge modelled upon natural science. The Enlightenment thus turned upside-down the traditional ranking order of knowledge as explained, for instance, by Aristotle in his *Metaphysics* (see p. 26): supreme rank was ascribed to those arts which procure the necessities of life, and the theoretical sciences derive their legitimacy from their ultimate serviceability in the same domain. Aristotle's ladder had already become shaky during the Renaissance, when the purpose of natural philosophy was seen by Thomas More, Bacon, and others as being *both* to honour God through study of his accomplishment *and* to improve the condition of the human race; few (if any), however, had ever been as radical as the Enlightenment *philosophes,* and only in the eighteenth century did the turnover of the classical stance spread widely.

It was at least in part a consequence of this understanding of the purpose of knowledge that Enlightenment philosophers rarely attempted to construct all-encompassing *systems*—even the materialists mostly argued less dogmatically about man the automaton than their seventeenth-century predecessors had done,[14] although their better knowledge of the nervous system might have incited them to be even more self-assured. As Diderot explains in the *Encyclopédie* (article "Eclecticisme"), the aim was to combine "the best from all philosophies," which of course makes no sense if you believe that these philosophies (or any of them) are really coherent systems. Another symptomatic observation is that made about "truth, wisdom, prudence" in the article "Sens commun": no attempt is made to define them precisely, nor are they however reduced to mere subjective opinion; instead, they are told "not to be purely arbitrary." Montesquieu counters an objection to his climatic theory with the remark that its author "ignores the contradiction of the human spirit," which he has in fact discussed amply in the chapter which is criticized (ed. Derathé 1973: II, 437).

---

the state of nature, men act tactically, everybody trying to kill before he is killed; the social contract comes not from any transcendental morality but from the strategic insight that a change of the situation will improve the situation for oneself as well as for everybody else.

Rousseau's disagreement with the main stock of Enlightenment writers concerning the actual uses of human intelligence (see p. 133) can be said to be a disagreement whether human "intelligence" will naturally tend to be strategical or tactical.

14. The seventeenth-century pious successors of Descartes would perform the most cruel experiments on animals with the argument that these were merely machines which could not feel, however much they screamed. The Enlightenment materialists, as we have seen, accepted that *their* machines might be happy or unhappy, notwithstanding the conflict between this experience and the consequences of the normal (mechanical) understanding of machines, and used this system-alien premise as the very foundation for their moral philosophy.

Even though the Enlightenment in general was no more inclined than Diderot and the *Encyclopédie* (cf. p. 132) to make an absolute distinction between natural and moral/social/human science,[15] the Enlightenment contributed to the creation of genuine social science and human science. Both Montesquieu and Hume formulated explicitly that they wanted to do for moral philosophy what Newton had done for natural philosophy. It is quite clear from the accompanying expositions that none of them understood much more about Newton than that he had given a supposedly exhaustive explanation of his field; but this was also enough for them to propose the ambitious aim.

An aim, of course, is one thing, and the production of actual scientific explanations another. Even on the latter account, however, the Enlightenment marks a divide, of which only the most important aspects shall be mentioned:

— In two more or less consecutive steps, the physiocrats and Adam Smith created the first genuine theory of "the societal household" ("political economy," later abridged into "economics" and again expanded as "macroeconomics").

— In their psychological philosophy, Diderot as well as the mature Hume went beyond the simple and mechanistic statements of the seventeenth and earlier eighteenth century, according to which ideas were supposed to collide, attach to, or bounce off each other as material particles. Hume and Diderot, it is true, still built on everyday experience and observations, not on systematic observation or experiment, as the psychology of the nineteenth and twentieth centuries did. But systematic thought about everyday psychological experience and observation (accepting the inherent ambiguities of the psyche), instead of building on mechanistic metaphors or on principles derived from a general postulate about human nature, was itself a decisive turn.

— Montesquieu, Vico, and Adam Smith integrated sociological patterns in their understanding of historical processes, thus originating the perception of history as a developmental process governed not by immutable laws of general validity[16] but by historically determined quasi-regularities.

> There may seem to be a conflict between the Enlightenment belief in *Reason* as a seemingly abstract and suprahistorical principle and the

15. The Italian Vico is an exception to this rule. Like Hume, he is a skeptic regarding our knowledge of the natural world. But since we are ourselves part of the social world and know human motives from within, a much more certain science of history and human action can be established. Because of this atypical point of view, certain historians of philosophy do not count Vico as an Enlightenment figure at all but rather as a forerunner of the Counter-Enlightenment. His presentation of his views and results as a *Scienza nuova* (1725), however, shows him to share the aims of the early Enlightenment.

16. That had been the position of Machiavelli, who saw no difference between the situation of Moses and an Italian Renaissance prince—cf. p. 96.

> recognition that different societies and social epochs induce different psychologies and attitudes. Yet Montesquieu does assert that a Christian baby when put into a Turkish crib will develop a Turkish conscience; similarly, Holberg claims in *Niels Klim* that males who, like European housewives, are forced to stay at home outside general social interaction will develop the habits and psychological characteristics of women (and, in *Jeppe paa Bjerget,* that the drunkard peasant has been forced into drinking by the treatment which a poor peasant receives). The contradiction evaporates when one observes that the Enlightenment belief in absolute Reason is an invention of later interpreters whose own (positive or negative) obsession with philosophical system building has made them read the open-minded eclecticism of the Enlightenment as yet another system.

Even though the Enlightenment was subversive with relation to existing regimes and social structures (with England as a partial exception), the perspective was still restricted by the horizon of the time. A good example of this is Holberg's play *Don Ranudo* (written c. 1723). The fools of the play—an elderly married couple—embody the high nobility, proud beyond measure but also impoverished beyond hope. Their reasonable counterpart is a double personification of the progressive forces: the prosaic peasant, with his common sense, and the enterprising territorial magnate. This was also to be the (restricted) perspective of the physiocratic school: the productive classes are, without distinction and without perceived conflict, *those who own the land* and *those who work on it.*

More generally it can be said that certain antagonisms were obvious, while others would only become visible as a result of later social developments. The latent conflict between labour and capital, in particular, was veiled, not only for those who—like Holberg—might perhaps be characterized as associates of the bourgeoisie and the entrepreneurial nobility, but even to those who formulated the points of view and the interests of the working classes. Only in England had the conflict materialized to such an extent that Adam Smith was able to formulate the modern class-based analysis of the social structure, distinguishing "those who live by wages," "those who live by profit," and "those who live by rent"—that is, the working class, the capitalist class, and the land-owning aristocracy which leased its land to farmers investing their capital (*Wealth of Nations* I.xi.iii, ed. Smith 1974: 356f).

It might be objected that Rousseau's denunciation of the normal use of human "intelligence" shows awareness of some of the latent conflict of the Enlightenment utopia more than a decade before Smith pointed this one out (without seeing what huge impact it was going to have). This is quite true—but Rousseau's awareness had been that of the prophet rather than that of the analyst. Rousseau's criticism thus belongs to the same class as Swift's bitterly satirical castigation of the ultimate consequence of economic rationality, which he set forth in *A Modest Proposal* in 1729: since Ireland was hit by deep economic crisis, the most rational way to help poor parents would be to prepare their children as food for the well-to-do. Both exemplify that certain Enlightenment writers were aware that their utopian *reason* was not an automatic consequence of the generalization of instrumental *rationality*.[17] But the Enlightenment was unable to go beyond the mere recognition that the problem was there. Only as some of the political aims of the Enlightenment were achieved in the Revolution would the conflict born from their womb come into the open.

It was this veiled character of the conflicts implied by the Enlightenment project that until 1789 permitted a number of absolute monarchs to make alliance with Enlightenment philosophy as a means to achieve political rationalization and modernization of their backward realms—thus in particular Friedrich II of Prussia, Catherine the Great of Russia, and Joseph II of Austria.

---

17. From this point of view, the distinction between "strategical" and "tactical" instrumentality forsakes; Swift's parody is in fact aimed at strategical, not punctual instrumentality, arguing that the "modest proposal" will benefit the rich no less than the pauper parents—not to speak of avoiding "those voluntary Abortions, and that horrid practice of Women murdering their Bastard Children" and of the advantage accruing from "the Addition of some thousand Carcasses in our Exportation of Barrel'd Beef" (ed. Eddy 1973: 22, 28). It illustrates that the difference between the two is only relative (as already in the military domain).

For further confrontation, Max Weber's distinction between *Zweckrationalität*, "instrumental rationality," and *Wertrationalität*, "value rationality," may be useful (at least as it came to look in Weber's later writings, his usage is not consistent—cf. Parsons in Weber 1947: 115 n. 38). Instrumental rationality is only concerned with how to achieve a given goal (with already given means, or by means to be constructed); the goal itself is outside discussion. Value rationality encompasses the reasoned discussion of our ultimate aims, of their attainability, mutual relations, and consequences.

The presuppositions of instrumental rationality are kindred to those of the "geometric method," namely what the Enlightenment would see as naive overestimation of *a priori* judgement: whether Euclid's postulates are really to be trusted can only be decided when all their consequences within the General Theory of Relativity have been analyzed. Similarly, whether we *really* want to achieve a given aim can only be decided when we have analyzed the consequences of achieving the aim; in the military analogy, tactical as well as strategical rationality are concerned with how to attain victory, this political aim being in itself taken for granted; the question of value rationality is whether we really want *victory at any price*, perhaps whether we want it at all given (for instance) the prestige and influence that will fall to the armed forces that achieved it?

Weber's ideas borrow from Kant, who in 1785 formulated the principle that "you should act in such a way that Humanity, as represented by yourself as well as by any other person, is always used as an aim, and never as a mere means" (*Grundlegung zur Metaphysik der Sitten*, BA 66–67). Kant, on his part, is inspired by Rousseau. A direct road thus leads from Rousseau's prophetic feelings to the terms in which contemporary philosophy discusses the dilemma.

PHILOSOPHY REDEFINED

The Enlightenment *philosophes* were presented above as the leading intellectuals of the bourgeois public sphere. They were thus philosophers in the sense of the pre-Socratics rather than in the sense which has been current since Aristotle. We might claim them to be amateurs with regard to philosophy, but it would be more pertinent to notice that much of what they did contributed to the detachment of specific fields of knowledge from the grasp of the broad field of philosophy as understood till then, and to a transformation of them into separate sciences. The Enlightenment gave up the notion of "experimental philosophy" while developing the approach covered by the term into a variety of experimental sciences;[18] and the Enlightenment began speaking of "moral sciences," at the same time as it started to sever economics, political science, and sociology from each other and from their common origin in philosophy.

One important Enlightenment figure, however, made his most important work within philosophy in the narrow sense—a sense which had become narrower because so many fields of knowledge took their own way): Immanuel Kant, who actually held the philosophy chair in Königsberg. (But Adam Smith was also employed as a professor of moral philosophy; the position alone was thus not sufficient to create a "real" philosopher).

The works which primarily define as Kant as a philosopher in the strict sense are his three *Critiques: Critik der reinen Vernunft* (1781), *Critik der practischen Vernunft* (1788), and *Critik der Urtheilskraft* (1790), together with a number of affiliated shorter works from the same years. These are also the works which more than any other define what "philosophy in the strict sense" came to mean. In a way, Kant did to philosophy precisely what Adam Smith did to economics, detaching it from the common mother discipline and establishing it as a particular type of knowledge (though in this case usurping the name).[19] After Kant, the main current of philosophy came to inquire into the *conditions* for knowing and judging, leaving the acquisition of actual knowledge to the single scientific disciplines. Moral and aesthetic judgement, on their part, have tended to be disconnected from the world of science and scholarship (in which aesthetic judgement had never possessed full citizenship) and to find their main professionalized soil in politics and art criticism.

The conditions for knowing are the theme of the *Critique of Theoretical Reason.* Roughly speaking, the aim of the work can be explained as an appropriation of the

---

18. Thus Robert Boyle, regarded as "experimental philosopher" *par excellence* in his own times, was characterized by d'Alembert in the "Discours préliminaire" to the *Encyclopédie* (vol. I, p. xxviii) as the "father of experimental physics."

19. One may object that Hume had already launched the analysis of the conditions for knowing, thus foreshadowing Kant's undertaking (as actually admitted by Kant in several passages); but the difference is as great as between the physiocrats and Smith.

Humean rejection of rationalism and too easy empiricism (cf. above, p. 135), but reshaped in a way which permits Kant to avoid the skepticist conclusions which Hume had been forced to draw.[20] That knowledge cannot be derived from Cartesian "evident truths," thus far Kant agrees with the empiricists. Nor can experience, as rightly seen by Hume, tell us about necessary causal connections in the world we observe, or lead us by necessity to the truths of mathematics. But neither causality nor time and space can be reduced to mere mental habits or subjective expectations; they are, indeed, the indispensable prerequisites (*a priori* conditions) for knowing about the physical world. We have no possibility to ascribe these attributes to "the thing in itself," it is true; but *we* are unable to grasp things without using this framework. In a similar way, the truths of mathematics (which Hume had held to be mere logical identities) are *a priori* prerequisites for any scientific-theoretical reason.

The *Critique of Practical Reason* (in the first part of which Kant returns to the "geometric method") pursues similar goals in relation to moral knowledge. Moral philosophy ("pure practical reason") does not tell whether one action or another is morally justified, but asks for the criteria which by necessity must characterize any directive if that directive is to be considered a moral command, that is, a "categorical imperative"[21] (cf. the discussion below, Chapter 20). What pure practical reason does tell is merely that "you should act so that the rule governing your will may always be valid as a general law" (A 54). Since such action presupposes freedom to act, the rule follows never to treat one's fellow beings as mere means, thus depriving them of their freedom.[22]

---

20. Kant himself explains that his critique of pure reason "was prompted by the Humean doctrine of doubt, yet went much further" (*Critik der practischen Vernunft* A 89ff, quotation A 92).

21. Kant opposes *categorical,* that is, absolute imperatives ("thou shalt not kill") and *hypothetical* imperatives, that is, imperatives conditioned by instrumental rationality ("if you do not want to go to prison/Hell you should abstain from murder"). Obviously, only the former constitute moral rules; the latter amount to nothing but prudent advice.

22. Kant's formulation is beautiful enough to be quoted in the original:

> Der Mensch ist zwar unheilig genug, aber die *Menschheit* in seiner Person muß ihm heilig sein. In der ganzen Schöpfung kann alles, was man will, und worüber man etwas vermag, auch *bloß* als *Mittel* gebraucht werden; nur der Mensch, und mit ihm jedes vernünftige Geschöpf, ist *Zweck an sich selbst.* Er ist nämlich das Subjekt des moralischen Gesetzes, welches heilig ist, vermöge der Autonomie seiner Freiheit.
>
> (A 155f)

Once again, we see, the "freedoms" of particular groups are rejected because they encroach on that *freedom* which belongs to every human being (and every being provided with reason).

What is morally right should thus not be derived from knowledge of Good and Evil, as moral philosophy since Plato had held. Instead, the discussion has to start from the "principle of moral conduct," which is an *a priori* condition for the determination of the will (A 110).

Critique of Judgement, in its first part, attempts to define the foundation on which aesthetic value judgements can be made in a way that avoids both the regulation by rules known from French classicism and that subjectivism which had largely replaced it. In the second part, it takes up the problem of how to speak of apparent appropriateness in nature (a favorite theme in "natural theology"), where (once again) *purpose* in nature is seen as a product of "the particular constitution of my capacity to know" (A 329 / B 333). What keeps the two parts together is a new fundamental insight: the essential characteristics of the category of judgement, it is true, are displayed most clearly in the case of the aesthetic judgement—namely, nonreducibility to strict proof from first principles, but concomitantly a necessary assertion of validity which allows argument; but judgement is also an essential presupposition for the application of both theoretical and practical reason as discussed in the first and second *Critique.* Both the construction of general concepts from particulars (Hume's problem) and the decision under which general rule a particular action falls are instances of judgement with the same characteristics. (We shall return to this problem in the end of chapter 22.)

Before Kant, and also for his Enlightenment precursors, philosophy had told or analyzed the True, the Good, the Beautiful, and the Cosmical Order. Kant redefined it as the *investigation of the possibilities for human reason* to make such descriptions and analyses. Globally, the *Critiques* constitute a critical examination of the Enlightenment project. Regarded in one way, Kant stood aside by being a professional philosopher and thus engaged in making this investigation systematically; otherwise seen, however, he argued philosophically what the Enlightenment had suggested through its practice: that the grand philosophical system explaining everything True, Good, and Beautiful cannot be constructed.

ENLIGHTENMENT AND REVOLUTION

It is much too simplistic to see the French Revolution merely as a logical consequence of the Enlightenment movement—if only for the reason that all-encompassing social processes like revolutions are never mere "logical consequences" of one or two isolated factors, and in particular not of a set of ideas. Concretely, the immobility of political and economical structures were important; so were political scandals and even meteorological circumstances resulting in a bad harvest. None the less, the Enlightenment is an important part of the background of the Revolution, and many of the revolutionary parties, from the moderate to the most radical, took their ideas from one or the other component of the Enlightenment current. Of special importance in this connection is the belief in science, education, and utopian "reason," which eventually led to the quasi-religious worship of, and building of temples for, Reason during *La Terreur*, the radical phase of 1793–1794, and to thorough innovations in the educational system from 1794 onwards (cf. below, p. 149, on the École Polytechnique, which is one prominent example).

At the same time, the Revolution exposed the fissures and actualized the potential cracks and contradictions in the Enlightenment utopia. Temple-building notwithstanding, Robespierre's policy of terrorizing the enemies of the Constitution into obedience turned out not to be rational; the rationality of Napoleonic warfare, on the other hand, was not conducive to a world of human reason; and much of what happened was neither reason nor rational.

As a consequence of the Revolutionary adoption of Enlightenment ideals no less than because of the shortcomings of these ideals, the Revolution came to be seen by its opponents as proof that the Enlightenment project was to be condemned as false prophecy. Friedrich Schiller and Johann Wolfgang von Goethe, one-time sympathizers of the Enlightenment, were scared.[23] In England, already the early phases of the Revolution led the (utterly moderate) Whig politician Edmund Burke to formulate *conservatism* as an explicit political program and philosophy: human reason is frail and much more likely to err dangerously when it tries to change everything at a time than such well-worn institutions as church, nobility and monarchy that have been tested and have learned from their mistakes for centuries.[24] Similar though less consequential philosophies were formulated by French emigrants

---

23. Goethe, however, who had joined the Prussian army, was perspicacious enough to tell his companions after the victory of the revolutionary Paris paupers over the well-drilled and well-equipped Prussians at Valmy in 1792 that "Here and today, a new epoch of world history begins, and you may tell that you were there" (*Kampagne in Frankreich 1792,* 19.9.1792; *Werke* XV, 305).

24. In Burke's own words (ed. O'Brien 1969: 183):

> We are afraid to put men to live and trade each on his own private stock of reason; because we suspect that this stock in each man is small, and that the individuals would do better to avail themselves of the general bank and capital of nations, and of ages. Many of our men of speculation, instead of exploding general prejudices, employ their sagacity to discover the latent wisdom which prevails in them. If they find what they seek, and they seldom fail, they think it more wise to continue the prejudice, with the reason involved, than to cast away the coat of prejudice, and to leave nothing but the naked reason; because prejudice, with its reason, has a motive to give action to that reason, and an affection which will give it permanence.

This is different from that routine continuation of old institutions and habits against which the Enlightenment philosophers had fought. It comes closer to Polybios's defense of religion (see p. 40, n. 22); but whereas Polybios's upper class was sufficiently cynical and sufficiently

during and after the Revolution. The Romantic movement in Germany is the third important constituent of the *Counter-Enlightenment,* the broad intellectual movement reacting to—and mostly against—the Revolution and its supposed origin in an unfounded trust in Reason.

A further consequence of the Revolution was that not only intellectuals but also the higher bourgeoisie gave up their Enlightenment sympathies, and did so much more thoroughly than the intellectuals. After the *de facto* social (though not political) victory of the new capitalist class, intellectual criticism of outdated and irrational power structures could be dispensed with: the critical potential of utopian reason had become a threat rather than a weapon. Strictly speaking, this happened already during the Napoleonic era in France. A group of intellectuals pursuing Enlightenment-inspired critical analysis of the origin and development of ideas—the *idéologues,* as they called themselves, referring to their critical investigation of prevalent ideas and their origin—became a favorite aversion of the emperor himself, who preferred nobody to disturb him when he made the necessary compromises with the church, the wealthy bourgeoisie, and the returning *émigrés.*

Not all European intellectuals who reacted to the French Revolution in the end of the eighteenth and the beginning of the nineteenth century did so negatively. There are pro- as well as anti-revolutionary Romanticists—as representatives of the former class we may recall the names of William Blake, George Byron, and Johann Gottlieb Fichte.[25] After a couple of decades, several currents arose which in their own ways continued the Enlightenment movement and the intentions of the *idéologues* while learning from their shortcomings. Some, like the Comtean positivists (cf. below, p. 156), developed what in recent terminology could be called its "scientistic" trend, that is, its belief in the possibility of solving technical and social problems by means of systematic application of science, and thus to produce sciences (modelled on the natural sciences) which were suited to serve this purpose efficiently. We may say that they equated reason with instrumental rationality, accepting the ways of the projectors and rejecting that identification of reason with moral responsibility that Kant had presupposed (cf. n. 22). Others related more directly to the radical Enlightenment, not least to the currents that had emerged during the peak of the Revolutionary period, and revived the humanist Enlightenment ideal on the conditions of the new

---

detached from the populace to allow itself to be enlightened while controlling the masses by means of superstition, Burke was aware that in his times those who found the conservation of "prejudice" socially useful would have to believe in the message themselves.

25. "In my view, the French Revolution is important for humanity as a whole," namely as a "rich canvas about that great text: Human right and human value"—thus the initial passage of Fichte's "Contribution to correction of the appraisal of the public of the French Revolution" from 1793 (ed. Buhr 1989: 39).

political and social scene. Merging it with Romanticist insights, they became radical democrats (like Heinrich Heine) or utopian socialists (some of whom were actually quite close to Auguste Comte). Even the intellectual superstructure of the emerging labor movement is an heir to the Enlightenment tradition.

CHAPTER 7

The Nineteenth Century

The nineteenth century brought much more than a continuation adapted to changed conditions. It innovated in many respects. If the intellectual superstructure of the incipient labor movement was a continuation, organized labor itself was an unprecedented phenomenon, only faintly heralded by the victory of the Parisian paupers over the Prussian army in the battle of Valmy.[1] It certainly also innovated thoroughly in scholarly and scientific life. However much has been said above, for instance concerning the study of language, natural philosophy, or mechanics in earlier epochs, it is only in the early nineteenth century that linguistics and physics emerge as *coherent fields with traditions and institutions which, while growing immensely in insight as well as complexity and manpower, continue into our own days.*

THE INSTITUTIONALIZATION OF UNBOUNDED SCIENTIFIC QUEST

Science as we know it nowadays, as *systematic, ever-continuous research,* is indeed a child of the nineteenth century.[2] Of course, science understood as *socially organized and systematic search for and transmission of coherent knowledge* is much older. As we

1. This example is chosen not only because it continues the end of the previous chapter, but because it may have had a heavier impact than any other innovation on the worldview of everybody. As Friedrich Nietzsche observed on the phrase *Wir sind alle Arbeiter,* "We are all workers," a late nineteenth-century German ruling class cliché, it would have been regarded as an obscenity by the nobility of the *Ancien Régime.* "How close is now even the most idle among us to work and to the worker!" (*Die fröhliche Wissenschaft,* III, N° 188, 1982). Work, the ordeal of most and the aversion of the happy few a century before, had become the meaning of life.

2. I shall remind of a statement from p. 7: the term *science* is used in these pages, and in particular from this point onward, in a wide sense corresponding to *scientia/Wissenschaft.*

have seen, the seventeenth and eighteenth centuries had produced revolutions in many domains of knowledge which underlay a number of present-day natural and human sciences, and they had initiated "scientific" organization of knowledge in many other domains. Yet the Enlightenment had tended to perceive knowledge as something limited. Certain sciences were already close to having solved all their central problems—"There is but one universe, and it can happen to but one man [*viz.* Newton] in the world's history to be the interpreter of its laws,"[3] according to Lagrange, perhaps the most eminent mathematician of the late eighteenth century. Lagrange also believed that he and his contemporaries had left little but applications to future generations of mathematicians. In other fields (not least the "moral sciences"), work had just begun, and they were further from the goal. But completion was still the goal, and Montesquieu and Hume can be taken at their word when they express the aim to do in the moral sciences what Newton was supposed to have done within natural philosophy—that is, to say the last word of real importance.

This attitude of scientific modesty may astonish us when we think of the immediate scientific experience of the eighteenth century, which looks to us like acceleration and impressive expansion rather than exhaustion. The seeming paradox is at least partially solved, however, if we remember the utilitarian orientation of the Enlightenment and the equally utilitarian institutions which produced scientific knowledge; when you are mainly looking for results which can be used, you care less about the possibility that your new results may generate new theoretical problems and open up quite new scientific vistas.

In any case, and in spite of isolated thinkers who had formulated similar ideas in earlier times, generalized belief in the unbounded growth of human knowledge only materialized in the nineteenth century. As we have seen it in connection with other thorough transformations of thought, even this was a reflection of institutional innovations. Ultimately, the new mood had its roots in the rise of the modern state and of modern society as they resulted from the technological and political revolutions of the late eighteenth and early nineteenth centuries.

The link between the general social structure and overall social needs on one hand and the changing structure and conceptualization of the scientific enterprise on the other is constituted by the need for manpower able to carry responsibility for working the new technical and administrative machinery, and thus for educational institutions where this key personnel could be trained. In different countries, different types of institutions emerged, affecting the development of sciences and scholarship in different ways.

The first important institution is the French *École Polytechnique* from 1794. It was founded in order to provide future civil and military engineers with a fundamental scientific training—in fact two years of basic studies in mathematics under

3. Quoted from Kuhn 1963: 353.

the best mathematicians of France. After two years the students were transferred to other institutions where they would specialize in mining, road- and bridge-building, or other fields.

One reason for the importance of the École Polytechnique is that it represents the first appearance of the *engineer in the modern sense:* a practitioner trained in the scientific knowledge of his own days, and not just in the ways of other practitioners combined with third-hand knowledge of scientific results and methods created a hundred years or more ago. Today, as we know, engineers in this general sense not only build bridges and construct machines. Social planners and practicing economists, for instance, make "engineering" decisions based on social science; consultant psychologists and professional communicators use the insights of the humanities correspondingly. Another reason for the significance of the school is a consequence of the historical context within which it was created: the Revolutionary conviction that public utility, scientific rationality and utopian reason coincide. The teachers were obliged to publish their courses in print in order to make this supposedly useful learning available to everybody. As a result, the school became a center of mathematical *research*—not least because the teachers were recruited from among the best mathematicians at hand who used the opportunity to teach and publish their own results. The original design survived not only the transfer of the school to the Ministry of War in 1804 but also the Restoration. When a Romantic philosopher (H. C. Ørsted) made the discovery that an electric current influences a magnet, Polytechnicians (Biot, Savart, Ampère) were responsible for the transformation of this astonishing *fact* into an *element of scientific theory.* Even the very creation of physics as *one discipline,* encompassing mechanical physics, heat, light, electricity, and magnetism, is mostly a product of the École Polytechnique.

> Only around 1830 was it becoming clear that the highbrow research orientation of the École Polytechnique might not be the best way to train engineers for practical work. From that time on, the school lost its importance as a center for scientific development. It remained an engineering school, but even in this domain it lost its prominent position as other institutions modelled on the same pattern but oriented toward the civilian domain were erected. In German territory, this creation of (scientifically less ambitious) Technische Hochschulen soon became a widespread phenomenon.

In England, a reform movement at universities in the 1820s brought some renewal, firstly by introducing the results of seventeenth- and eighteenth-century research into the teaching and adopting, for example, Newton's mathematics in French interpretation into the curriculum of Cambridge; secondly by gradually causing *research* to become a natural part of university life. Oxford and Cambridge were too dominant, however, and too much oriented toward the training of clerics, to allow the English reform movement any strong impact. On the other hand the Mechanics Institutes, intended to supplement the education of skilled workers in industry, were too much tied to the improvement of workmen's practice to enter a direct alliance with scientific research.

For the humanities, the central development took place in Germany in the wake of the Prussian university reform of 1809, which we shall discuss in some detail below, and which soon spread from Germany to other countries.

In spite of their diversity, all these developments had the same ultimate background: the increasing need for qualified manpower. They also had a common effect: science and research returned to educational institutions, primarily the universities, which thus earned that characterization as "research institutions engaged in teaching at the highest level" which they try to defend today.

THE GERMAN UNIVERSITY REFORM AND THE HUMANITIES

At the surface of historical events, the German development as well as the explicit integration of teaching with research had its beginnings in the Prussian reform of 1809 (whatever the outcome, research had not been an explicit institutional aim of the École Polytechnique). An important element of the immediate inspiration for this reform was the course of the Napoleonic wars: at the battle of Jena (1806), the Prussian and Saxon armies had been beaten decisively by Napoleon; in the succeeding Treaty of Tilsit, Prussia was reduced to half its former size. In the context of a still feudally colored and absolutist Prussia, whose most progressive element was a "bourgeoisie of officials" rather than an industrial or mercantile bourgeoisie, the response to this "Sputnik-shock" was a claim for *spiritual renewal*.[4] Another reason for this orientation of the claim for renewal was evidently the existence of the Romantic movement, and in particular the Romantic response to the French Revolution. (The two explanations are not independent, since the orientation of the German Romantic movement was itself correlated with the social composition of the German educated elite).

The central idea of the university reform was the need for the German elite to be freed from that sluggishness which resulted from its education in dogmatic and fossilized universities and—before it got so far—in a secondary school whose teachers, themselves coming from the universities, were no better. Therefore the quality of the teachers of the *Gymnasium* had to be raised, morally as well as regarding their scholarly level. This should be done by improving their level in the *Geisteswissenschaften* (the "sciences of the spirit") considered identical with the *Altertumswissenschaften,*

---

4. The response of the *polytechnicien* Sadi Carnot to the French defeat a few years later constitutes an illuminating contrast: in the introduction to his work *On the Motive Power of Fire,* which marks the birth of thermodynamics, he argues that France had no lack of either skilled scientists nor brave soldiers; but the English had the advantage of *industry.* The fundamental need for France was therefore *more steam engines, and more efficient steam engines.*

the "sciences about antiquity"—that is, Hebrew, Greek, and Latin philology, history, and mathematics.

The name of the program is *Neohumanism;* it was, in fact, close to the Renaissance interpretation of antiquity, especially to German post-Reformation Humanism as formulated by Melanchton. In spite of many changes in the content of *Gymnasium* teaching, it remained the ideological backbone of German secondary education until 1933, and was eventually resurrected in both Germanies between 1945 and 1960.

The only place where future *Gymnasium* teachers could be taught the *Geisteswissenschaften* was in the Arts Faculties of universities. Since the Middle Ages these had been the preparatory school of universities whose main task was to train priests, lawyers, and physicians; but in the post-medieval period the Arts Faculties had lost and never regained that central intellectual position which had been theirs during the thirteenth and fourteenth centuries. Now, however, they were given the status of "lucrative faculties," in the terminology of the Middle Ages, and students were to be given a complete education at the "Arts" or "Philosophical Faculty." The final level for students should be one of independent research, reflected in a dissertation—and in order to make sure that the quality of university professors was sufficient to bring the students to this level, they would have to be appointed on the basis of their own scientific work, not according to family relationships or sociability as judged by future colleagues from other disciplines (since there was in principle only one professor from each discipline, future colleagues from the same institution would normally be unable to make a scientific evaluation). The aspiration was not only to provide the *Gymnasium* with a staff whose members had once made a single piece of independent research for their thesis. Also after entering their career, *Gymnasium* teachers were expected to use part of their time on research; articles in the yearbooks of many gymnasia show that quite a few teachers actually did so.

Research was not meant as an aim in itself. The overall purpose of the enterprise was moral improvement as provided by the *unified humanities,* in agreement with Neohumanist ideology and with the integrative and organic worldview of the Romantic movement. But the undertaking was so efficient in creating new knowledge that unification became more impossible than ever. The totality of humanities was soon splitting up into disciplines, and these into subdisciplines, each possessing greater and greater knowledge of its own domain but also less and less understanding of neighboring areas.[5] As a result, even the gymnasium teacher in the humanities

5. The tendency to break up disciplines into subdisciplines was furthered by a peculiarity of the university system: growing numbers of students in a particular field at a university might make it necessary that another professorial chair was created to share the teaching; but since each discipline was only entitled to have one ordinary chair, creation of an extra chair had to be argued from the existence of a new discipline.

stopped being an all-round humanist, developing instead into a specialist with some, but rarely all-encompassing, knowledge of other fields.

In spite of discipline formation and specialization of single scholars, however, general attitudes to the subject-matter and to the aim of the humanities developed which expressed the common research experience of the whole field and hence cut across the single lines of interest. This was not an exclusively German phenomenon, even though the rapid progress of humanistic research in Germany makes the phenomenon most conspicuous here. One of these attitudes is the *regard for the factuality of the material.* History is not (or not primarily) there to be used in moral and political reasoning—not to speak of strategic planning à la Machiavelli. The first task of the historian is to find out *what actually went on*—"wie es eigentlich gewesen," as it was formulated by Leopold von Ranke (1795–1886). This approach has been labelled "positivist" by critics for whom this term is negatively loaded. However, what Ranke intended by the phrase was first of all that history should get beyond conventional fables and look instead at the sources; since Ranke's main purpose, moreover, was to find the general pattern *through,* but not in ignorance of, the confusing mass of details, the term is undeserved in its pejorative sense.

Another closely related stance is the *historicist* attitude: the world is continuously changing. We should not believe that our own reason and worldview possess general validity and are suitable for suprahistorical explanation of events belonging within other historical contexts. Historical material should be explained as something specific, on its own terms. We may go from the historically specific to general regularities and patterns, but inference in the opposite direction is not legitimate. In this respect, the conditions of the historical sciences were seen to differ from those of the physical sciences: Newton might look at another apple, and Ampère repeat Ørsted's observation of the magnetic deviation produced by an electric current. But once Napoleon was defeated there could be no other battle at Waterloo with the same outcome and the same consequences; history is characterized by *Einmaligkeit,* "once-ness" (Ranke again).

We observe that the approach of nineteenth-century history differs fundamentally from Machiavelli's use of Moses, Caesar, and Cesare Borgia as illustrations of the same, ever-valid principles (cf. p. 96), and in its radical formulation even from the Enlightenment belief in sociologically determined quasi-regularities. Due among other things to inspiration from Georg Wilhelm Hegel (behind whom we find both the Romantic movement and Vico), history was seen not as a mere sequence of events (of which the historian should write a chronicle) but as an evolutionary process. This point of view is of course in virtual conflict with the radical interpretation of the *Einmaligkeit* postulate: if every event stands completely on its own and is unconnected to any other event, nothing but chronicle-writing is left to the historian. When forced to choose their side, most scholars would opt for evolution and historically determined quasi-regularities and against radical *Einmaligkeit.*

Evolution, indeed, was a very widespread idea, accepted not only in history proper (the "history of events") but also (and perhaps more unconditionally) in other fields: linguistics, cultural history, and history of ideas and philosophy, and even in anthropology, geology and biology. We might say that history was the hegemonic science, and that history was integrated in other sciences as their central perspective. History of the single sciences also came to occupy the role of philosophical justification of their accuracy and legitimacy.[6]

> The inclusion of anthropology, geology, and biology shows that the evolutionary orientation was not an exclusively German affair. But whereas British evolution (represented by Charles Lyell in geology and Charles Darwin in biology) tended to be materialistic and "uniformitarian," referring to unchanging natural forces and mechanisms,[7] German humanistic scholarship was predominantly bent toward idealism: evolution was not seen as the product of material social processes but (by Hegelians) as the gradual unfolding of the World Spirit or (by Romanticists and their offspring) as the product of the spirit of specific epochs or nations (*Zeitgeist* and *Volksgeist*). It may go without saying that the latter orientation was often coupled to nationalist and, increasingly toward the end of the century, racist persuasions. But this was not the starting point of the *Volksgeist* idea;[8] nationalism and racism, moreover, were certainly not German specialties but rather pan-European phenomena.

---

6. This discovery of the general importance of history finds an evident parallel in general nineteenth-century experience. Admittedly, rapid change had characterized Western Europe at least since the late Middle Ages; but the general view of change had either been that it provided a restoration of lost values (the Renaissance and, on the level of piety, the heretic movements and the Reformation); or the present moment was seen as the final victory of the new over the old (*Le siècle de Louis le Grand*); or one would see oneself as engaged in the battle which, if won, should lead to the final victory of reason (the Enlightenment). Only the nineteenth century discovered the present as a passing moment in the midst of continuous change, not only in the sciences (cf. above, p. 148) but as an all-pervasive situation. Only at the onset of the nineteenth century could Faust get the idea to accept damnation in that very moment which he wanted to remain without change: "Werd' ich zum Augenblicke sagen: / Verweile doch, du bist so schön! / Dann magst du mich in Fesseln schlagen, / dann will ich gern zu Grunde gehn!" (Goethe, *Faust,* 1699–1702). Goethe himself was skeptical enough as regards this expression of the modern mind to prove Faust wrong: in the poem, Faust is saved precisely when he wants to stay in the moment instead of running away from it.

7. Lyell, in his *Principles of Geology* from 1830–1833, had formulated "uniformitarianism" as the principle that the surface of the Earth had been shaped by erosion, sedimentation, volcanic eruptions, earthquakes, and similar forces still to be seen in action; Darwin, in the same vein, claimed in *The Origin of Species by Means of Natural Selection* from 1859 that species are formed naturally by that same selection process which is used deliberately by human breeders in order to produce improved races of cattle.

8. Gottfried Herder (1704–1803), the creator of the *Volksgeist* notion, declares the equal standing of all nations forcefully in his writings. A striking example is offered by his strongly

In the historical, textual, and linguistic sciences, the factuality, historicist, and evolutionary orientations were the basis for new approaches. In the historical and textual sciences, they were responsible for the creation of systematic source criticism and textual criticism. Glimpses of these techniques can be seen in earlier epochs, both in the Hellenistic era, in the twelfth-century confrontation of authorities, and in late Renaissance and Early Modern Humanist studies. But since the aim had then been to restore particular classical texts, to find the correct interpretation of an ancient authority, or to expose forgeries, the techniques had never developed into a general method, and certainly had never been seen as the defining qualities of history and textual studies. This only happened when texts were read systematically as expressions of their time and *Zeitgeist.* In linguistics, the break was even greater. Until 1800, linguistic can on the whole be described as *grammar*—mostly grammatical descriptions of single languages, at times also as a search for general grammatical structures or semantic categorizations (especially in thirteenth- to fourteenth-century scholasticism and in seventeenth-century "general grammar"—cf. p. 118). The limited and often specious aims of the proponents of the sixteenth- to seventeenth-century etymological school and "Scythian theory" (cf. above, p. 120) had prevented this approach from gaining influence and from systematic continuation. To Friedrich Schlegel, the Grimm brothers, Rasmus Rask, and Franz Bopp, grammar was only one of several resources used in comparative linguistics, the object of which was *understanding of the evolution of specific languages* and *description of their family relationships.* To them, etymology could not be "a science where the consonants count for very little and the vowels for nothing at all," as a scornful Voltaire is reported to have maintained about the method of Goropius Becanus and his successors (Ellegård 1973: 662a): as important as grammar was the integrative investigation of the details of phonology and vocabulary, which allowed to put etymological studies on a healthy basis—and even grammatical structures had to be analyzed more closely than in the traditional formulation of rules and exceptions.

"POSITIVE KNOWLEDGE"

The integration of higher education with research and the ensuing explosion of research activities and results were not restricted to the humanities. The natural sciences, too, received new impetus even more than the humanities. Whereas the Enlightenment *philosophes* could look back upon an accretion of epoch-making

---

polemical *Neger-Idyllen,* where precisely those virtues and that high mind are ascribed to the slaves which it had been customary since the Renaissance to detect in ancient Rome. Unlike Rousseau, moreover, Herder does not present these virtues as expressions of "noble savagery." A black prince who has been caught by treason and sold as a slave is no less prince, and no less civilized, than any princely peer of his.

discoveries in natural philosophy which they might still see as essentially once-only events, the humanities of the mid-nineteenth century were forced to see natural science as a neighbour in continuous and ever-accelerating development toward greater knowledge based on increasingly precise and certified empirical foundations.

On one hand, this inspired the creation of a new humanistic discipline that (initially) was quite different and segregate from those based on texts and sources: experimental psychology. Since Aristotle, the "philosophy of the soul" had in principle been a branch of natural philosophy (from which, as we have seen, most other branches had deserted in the wake of Newton's *Principia*). In practice, common-sense psychological considerations had long formed part of pedagogical philosophy (and of Aristotelian rhetoric); both Locke and (to a lesser extent) Hume had based their theories of knowledge on (dubious) psychological postulates; Diderot, Stendhal, and other authors had made acute psychological insights central themes of their works but not made a specialized field of study out of them. In so far as psychology existed as a scholarly field, it was thus part of theoretical and practical philosophy, and only empirical to a limited extent.

Shortly after the middle of the nineteenth century, however, a new approach appeared: "man" was regarded as a living being with a sensorial apparatus which could be investigated experimentally, as can other characteristics of living beings. "The soul," or at least its manifestations, could be measured and counted by methods not fundamentally different from those used by physicians to investigate human metabolism—it is certainly no coincidence that the first practitioners of the field were physiologists. Even though, as a rule, early work along these lines regarded only sensory psychology,[9] they laid the foundation for one of the main trends in late-nineteenth- and twentieth-century psychology.

Only one trend, however. Toward the end of the century, a counter-movement set in, inaugurated by Freud's psychoanalysis. Originally, no counter-movement to the prevalent physiological approach was intended; Freud's starting-point was also a medico-biological view of human nature, combined with hypnosis therapy. But through work with this technique, and especially through its failures, Freud was led

---

9. Anecdotal history—thus R. Thomson 1968: 50–52—often mentions as the beginning of experimental psychology the discovery and investigation in the 1830s and 1840s of astronomers' "personal equation" (that is, the fact that the times t_A and t_B in which the same phenomenon is observed by astronomers A and B differ by an approximately constant amount *c*, t_A-t_B = *c*). No less important than such studies of reaction times were, however, the seminal investigations of the relation between impression and stimulus strength, leading to the "Weber-Fechner law," according to which the minimal increase in, for example, sound level that can be perceived is a constant percentage of the existing sound level. (The law is indeed approximately true for all kinds of perception within the "normal" range of intensity; for sound, the minimal increase which can be perceived is always of the order of one decibel, corresponding to an increase in energy density of c. 25 percent.)

to psychoanalysis as *interpretation* and as a midwife for the patient's *own understanding.* Through this integration of meaning and interpretation into its field of interest and its methodology, psychology (or at least this approach to psychology) was brought into contact with the main trend of the humanities, and thus emancipated from medical science.[10]

On the other hand, the new triumphs of the natural sciences inspired Comte to formulate his *Positive Philosophy* (*Cours de philosophie positive,* 1830–1842, ed. Serres et al 1975). An expression of aftermath scientistic Enlightenment, positivism was erected into a philosophical system and became a theoretical partner of utopian socialism (inspired by the *idéologues*). According to Comte, every field of knowledge passes through three phases. In its beginning, it is integrated in religion and myth; next, it passes through the metaphysical phase; finally even metaphysical and philosophical notions are found to be superfluous, and they are replaced by a science built exclusively on "positive" knowledge, that is, firmly ascertained empirical facts.[11] Social evolution follows the same scheme. In the sciences, the scheme provides the framework for theoretical progress; in social evolution, for social and moral progress. The view that the Enlightenment is "as totalitarian as any system" because "for Enlightenment the process is always given from the start" (Horkheimer & Adorno 1972: 24), parodically mistaken as it is when applied to Voltaire and Diderot (or the early Greek natural philosophers), is hence less obviously fallacious when applied to Comte's thinking.[12]

> Before identifying Comteanism with latter-day positivism, one should be aware that the nineteenth century was going to produce a very different positivism: the "empiriocriticism" of Ernst Mach and Richard Avenarius. This current did not share Comte's scientistic optimism. On the contrary, its central claim was that the only certain facts are *sense impressions.* Whether something in some "outer world" corresponds to these

10. Certainly without ever forgetting the biological basis for mind and behavior, as argued in depth by Frank Sulloway (1992). In Part II we shall return of some of the ambiguities arising from this delicate alliance.

11. An illustrative example, which runs better in German (*not* borrowed from Comte, but useful all the same): At the religious stage, the misfortunes of life lead to, and are explained with reference to *Der Böse* ("the Fiend"); during the metaphysical phase, this mythological figure is replaced by a hypostatized metaphysical concept, *das Böse* ("Evil"); positively, it is recognized that "Evil as such" is only a way to speak of single evil actions and inclinations of people, and a way to avoid understanding their real background and nature.

12. But still misleading, it might be argued, because it treats as a theoretical assertion the strategic optimism of a fighter for (what he considers) progress—mistaking, so to speak, a general encouraging his men that "somehow we are going to win this battle" for an arm-chair strategist purporting to know in advance the results of the war.

impressions is a problem which for reasons of principle cannot be answered and which therefore makes no sense.

Whereas Comte's positivism can be characterized as a radical epistemological optimism, for which reason it inspired utopian-political radicalism, empiriocriticism became popular in politically reactionary circles in Wilhelmian Germany because of its rejection of the (critical) possibilities of reason. If science and reason are unable to tell us anything definite, the way is open to religion and religious authority.[13]

The "logical empiricism" of the twentieth century is yet another brand of positivism, within which currents of epistemological optimism as well as pessimism can be found (cf. p. 266).

Comte's positivism was probably not very influential in the natural sciences—their progress was determined by other forces.[14] But it provided a tool for those inside the humanities who reacted against German historicism because they saw it as an expression of Romanticism (or just because it was German).

Of great and long-lasting influence was Hippolyte Taine's "positivist" theory of literature.[15] According to Taine, every artistic product—and most forcefully the best—expresses the psychology of the culture within which it is produced.[16] This psychology of the culture is a combined product of "*la race,*" "*le milieu,*" and "*le moment,*"—this attempt at causal explanation provides the actual link to Comtean positivism. "The race" is the ethnic fundament of the culture; "the environment" refers to the geographical circumstances in which the particular segment of the race has ended up, not least the climate—the Aryans who ended up in the swamps of the North developed quite differently from those who settled in Greece or Italy (the inspiration from Montesquieu is obvious and explicit); "the moment" stands for the actual history which shaped the culture (in the case of the English, Saxons by race, the Norman invasion turns out to be a main aspect). It is, according to Taine, the task of the historian (in particular the historian of literature and art) to decipher from the particular artistic production the psychology of the culture in which it was

---

13. This counter-enlightenment use of Empiriocriticism (amply documented in quotations) is the reason for Lenin's diatribe against the movement in *Materialism and Empiriocriticism* from 1908.

14. Even the eminent physiologist Claude Bernard, whose formulated philosophical opinions were "positivist" in the sense that he rejected any pretensions to find the "essence" or "primary causes" of things, distanced himself strongly from Comte's ideas (Grmek 1970: 31); his philosophy was obviously mainly indebted to his own research experience.

15. Formulated in the introduction to his *Histoire de la littérature Anglaise* (1863–1864) (Taine 1877: I, iii–xlix).

16. The *Zeitgeist* and *Volksgeist* ideas are near at hand, and Taine does indeed refer to Germany as the place where his approach was first developed, precisely since Herder, who is explicitly mentioned (p. xii) together with Goethe and others. Taine is neither anti-German nor anti-Romantic.

created, and to trace how this particular psychology has been produced by race, environment, and moment.

No less important was the influence of the sociologist Émile Durkheim, who claimed to base his sociology on objective "social facts" placed above human interpretation (a question to which we shall return on p. 209, n. 15). Durkheim was one of the channels through which Comtean positivism influenced twentieth-century sociology and anthropology. Another channel was more direct, though probably less important when it comes to methods and subject-matter: Comte indeed invented the very term *sociology* for that "social physics" (another term of his, shared with the utopian socialist Henri Saint-Simon) which he tried to develop (see König 1968: 202b).

POPULARIZED SCIENCE AND POPULAR SCIENCE

The "positive approach" to human nature and to human culture—be it physiological or Comtean—had a background and a sounding board in more general moods and broader cultural currents. The nineteenth century gave rise to the multifarious phenomenon of *popular science,* of which only a modest detachment (which we might call *popularized science*) aimed at broad diffusion of the results and approaches of academic science. Mostly, popular science was constituted by movements of "parallel science" or even "counter-science" emerging around figures who had created a doctrine of their own, inspired by some feature of academic science but often curtailed, simplified beyond recognition, or distorted when seen in the academic perspective.[17]

In the user perspective, such movements served the purpose of self-assertion, as "our own science" in a social world which was irrefutably dominated by science and by technological change purportedly derived from science; in contrast to "popularized science," this type thus fully deserves the label *popular science.*[18] The parallel to the "enlightenment" role of early Greek natural philosophy and twelfth-century

---

17. The characterization of this kind of "popular science" as a nineteenth-century phenomenon should *not* be read as a statement that a "sound" stem of "real" science had existed since (say) Greek antiquity (or 1100 CE, or 1543), and that "popular science" then arose as a parasitic outgrowth around 1800. Much (not all) of what is considered "science" in earlier epochs is just as close to the model of "popular" as to "academic" nineteenth-century science; the interest in medical science in the High Middle Ages and the sixteenth century presents us with particularly clear examples. But with the professionalization of scientific research in the early nineteenth century, the gap between the two models became obvious, and it became just as important for the "academics" to prove themselves different from the "quacks" as for the latter to prove the opposite. As we shall see on the example of eugenics, the difference might sometimes only manifest itself on the level of social standing and scholarly mores.

18. "Popular," but neither automatically nor predominantly "lower class" science. Most of the public was middle class, and part of it belonged to the educated classes.

astrology is unmistakable, and the first manifestations were indeed produced during the "low Enlightenment" of the 1780s. Jean-Paul Marat (1743–1793), physician and future spokesman of radical revolution, was by then deep in optical investigations on his own, and his undertaking was related in spirit to much of what other leaders of popular science movements did;[19] yet he never gained any perceptible public audience on *this* account. Immense success, on the other hand, fell to the physician Franz Anton Mesmer (1734–1815), who taught the doctrine of "animal magnetism" (mostly referred to nowadays as "Mesmerism"), a phenomenon which was later reinterpreted as hypnosis and thus made a concern for academic psychology.

The failure of Marat and the triumph of Mesmer illustrate an important characteristic of the main body of nineteenth- (and twentieth!-) century popular science: it had to be immediately relevant to human existence. At the same time, its self-asserting role vis-à-vis the dominance of natural and related science demanded that natural and medical science should provide the model for its humanly relevant insights. Popular science of this kind thus tended to be overtly scientistic. (Then as now, even the popularizers of the insights of academic science would often tend to oversell the product, making new achievements more certain, more broadly consequential, and more meaningful for the public than warranted, thus tending to make this popularization just as scientistic).

Many different examples could be mentioned: patent medicines and patent cures;[20] anti-masturbation machines; healing transformed into "Christian Science"; spiritism and psychical research; and so on. Several of these, as could be expected for movements giving meaning to human existence, served as alternative religion.

---

19. C. C. Gillispie (1980: 290–330) gives a detailed and sensitive account, both investigating Marat's actual work and analyzing similarities with and contrasts to Mesmerism.

20. Mark Twain's portrayal (1986: 53f) of Tom Sawyer's Aunt Polly is a picturesque illustration:

> She was one of those people who are infatuated with patent medicines and all new-fangled methods of producing health or mending it. She was an inveterate experimenter in these things. When something fresh in this line came out she was in a fever, right away, to try it; not on herself, for she was never ailing, but on anybody else that came handy. She was a subscriber of all the "Health" periodicals and phreneological frauds; and the solemn ignorance they were inflated with was breath to her nostrils. All the "rot" they contained about ventilation, and how to go to bed, and how to get up, and what to eat, and what to drink, and how much exercise to take, and what frame of mind to keep one's self in, and what sort of clothing to wear, was all gospel to her, and she never observed that her health-journals of the current month customarily upset everything they had recommended the month before. . . . .
>
> The water treatment was new, now, and Tom's low condition was a windfall to her.

The examples which were just mentioned were too far from the academic sphere to produce much influence that way or to illustrate the expectations and norms which prevailed in the vicinity of the academic environment. Others, however, had influence on academic science or are at least illustrative of its surrounding moods.

One discipline which illustrates the existence of an effective cultural demand for a "positive" approach to man modelled on medicine and natural science is *phrenology.* Even though the phrenology movement developed institutions and periodicals fashioned after the academic norms, it is uncertain whether it was directly academically influential, except in the sense that academic science, in its attempts to keep a healthy distance, got an extra impetus to develop and stabilize its own institutions—making more explicit, for instance, what distinguished an *academic* journal.[21] In view of the *new* character of early nineteenth-century science, it is obvious that its institutions were still unstable.

The phrenological doctrine had been developed by the Viennese physician Franz Joseph Gall (1758–1828) around 1800, and its basic tenets have been summarized as follows:

> (i) the brain is the organ of the mind; (ii) the brain is not a homogeneous unity but an aggregate of mental organs; (iii) these mental organs or physical faculties are topographically localized into specific functions; (iv) other factors being equal, the relative size of any one of the mental organs can be taken as an index to that organ's power of manifestations; and (v) since the skull ossifies over the brain during infant development, external craniological means can be used to diagnose the internal state of the mental faculties.
>
> (Cooter 1984: 3)

The phrenological doctrine bears some resemblance with the ideas of the Enlightenment materialists but has totally different implications. They, most explicitly Helvétius, had taught that the environment determined the function of the human machine, that is, that *education* was all-decisive (cf. p. 134). The phrenologists, on their part, held that everything was a question of *inborn biology.* To Helvétius, men were thus fundamentally equal (and potentially good, if only educated in the right way); to the phrenologists, they were unequal beyond educational repair (and their moral quality fixed in advance).

The phrenologist creed, not least the belief in inheritance and the conviction that external measurement of the skull provides exact information on a person's intelligence and psychical makeup, became immensely popular, in particular in Britain, and stayed so until well after midcentury. By then it mingled with and was

---

21. A similar outcome of the encounter with "popular Mesmerism" is analyzed by Alison Winter (1994).

gradually crowded out by the eugenics movement and Social Darwinism, spiritually related doctrines which explicitly held the upper classes to possess the better inheritance and which obtained indisputable influence in academic science. Eugenicists and Social Darwinists were both inspired by Darwin's theory of evolution by natural selection—a process which, as pointed out by Darwin, had results analogous to the effects produced by means of the *artificial* selection that was used by breeders to produce better races of grain and cattle. The champions of both movements would certainly have protested vociferously if they had been classified with phrenological and Mesmerian "quacks," but the dynamics of the process by which their tenets became fashionable were much the same.

Herbert Spencer, originator of Social Darwinism, summed up its view of the Darwinian process as "survival of the fittest." Darwin, who clearly saw the circularity of the argument ("fittest for what? Fittest for survival!"), espoused it in the end because it would ease the spread of his teachings in a public which had already taken Social Darwinism to heart. The central idea of the movement was that *social survival* (that is, property acquisition) was understood through the image of survival, and that the *better fitness* that appurtenance to the propertied classes was evidence of was equated with better (moral) quality. The doctrine, which gained important influence in later nineteenth-century sociological thought, was an unmistakable justification for economical inequality.

The eugenics movement was carried by people (most prominent among whom was Francis Galton, Darwin's cousin) who knew too much about social statistics and biology to accept the tenets of Social Darwinism: since lower-class people on the average produced more surviving children than their social betters, application of Darwinian standards would show that the social fitness scale was an inversion of its biological counterpart. *Fitness* being understood by eugenicists according to the social scale, the program of the movement was (though certainly not told in these terms) to undertake artificial selection on a societal scale, and thus to improve society by eradicating the ignoble heritage of the socially inferior.

Ideas similar to these went into a particular applied science of man, criminology, in particular as developed by Cesare Lombroso (1835–1909) in his work on the "born criminal."[22] Still other versions entered an unclear symbiosis with the more simplified versions of socialist theory that were spreading in working-class environments and were in part deliberately used by the Social Democratic parties in their agitation; in these cases, it was the turn of the upper class to be degenerate.[23]

---

22. *L'uomo delinquente* (first published in 1876), inspired *inter alia* by Comtean positivism and Darwinism, and famous for identifying the born criminal from anatomical and physiological features, in particular skull measurements and facial asymmetries (see Wolfgang 1968).

23. August Strindberg, the uncomfortable ally of the Swedish Social Democrats, was among the proponents of this idea; it is visible, among other places, in his play *Fröken Julie.*

These simplified theories are themselves instances of the phenomenon of popular science asserting itself as "*our* science." Scientistic popular science serving as an underpinning for social identity and legitimacy and in the formation of a worldview was a widespread characteristic of a century where God had come to be at work only on holidays (if at all) and on holiday when everybody else was at work.

ACADEMIC AND NONACADEMIC HUMANITIES

The Prussian research-oriented university model spread quickly to other countries, and it was soon regarded as self-evident. The Battle of Jena and the ensuing quest of national moral resurrection, though certainly "immediate inspiration" (as formulated above) can therefore represent nothing more than the surface of historical events. As already suggested, the underlying cause of what happened was the general socioeconomic transformation of Europe (and soon the United States), which gave rise to an increasing demand for efficient and well-trained officials, administrators, technicians, and teachers in a society in constant change. If this need had not been urgent, the German reform would probably have been abortive—if only for the reason that the Prussian government would probably not have been willing to pay for the many new positions needed for its realization; no new chair was indeed created without extensive bargaining with the ministry. Even outside Prussia it was the demand for manpower (largely the demand of the state, either directly or via deliberate technology policies) that made public authorities willing to implement and finance educational reforms in agreement with a model which had proved successful.

General public needs, however, even if a necessary background, do not provide the complete explanation. The process soon became self-accelerating in all fields where the research orientation became effective. Systematic work created new results and new understanding, which either (in the natural and technical sciences) increased the utility of (and hence the demand for) scientifically trained manpower, or (in the case of the humanities) opened the way to a specialized and technical approach to the intellectual realm, which then came to be seen as a necessary qualification (and at times an element of national prestige). To this comes the tendency of any similar environment (discussed both in connection with the Fara and Old Babylonian scribes and with the medieval masters of arts) to connect status awareness (and pride!) with the probing of professional tools.

The prosperity of the new alliance between high-level teaching and scholarly research does not mean, however, that all professional intellectuals were absorbed by the academic environment, nor that work of lasting importance within the humanities was undertaken only in universities. Many writers remained outside, and others wrote from within to the outer world, participating in or even creating general intellectual debate. Not everybody lost themselves in scholarly specialization; and increasing and fertile specialization did not prevent many of the participants in general debate from contributing to the development of both the general intellectual climate and the scholarly specialties.

Among Danish writers of this kind, one may mention Søren Kierkegaard and (toward the end of the century) Georg Brandes, the prophet of Taine's historicist positivism and the leading figure in the introduction of modern currents in a somnolent provincial environment. Of genuine pan-European importance were the utopian socialists, who influenced and received influence from the early labor movement. Later in the century, intellectuals within the mature labor movement—not least Karl Marx, Friedrich Engels, and August Bebel—were no less important in the formulation of the worldview of this new social force than the Enlightenment philosophers had been in the formulation of the perspective of the early bourgeoisie. (In neither case, of course, is the relation between class and intellectuals described exhaustively in this simple formula; cf. pp. 127f about the "bourgeois public sphere.")

CHAPTER 8

Toward the Present: Scientific Humanities

It would lead much too far to investigate the development of the institutional contexts within which the humanities have developed in the twentieth century. We shall restrict ourselves to the observation that the trends which began in the nineteenth century have persisted. The number of universities and similar institutions integrating high-level teaching and research has continued to grow on an increasing scale; specialization has proceeded and resulted in greater accuracy and depth, though still in interplay with the formation of interdisciplinary tendencies; and the reactions of the general public as expressed in periodically recurrent surges of "popular science" have not changed much in character. The academic environment has not swallowed all activity in the humanities, but the numerical balance between academic and nonacademic practitioners has shifted decisively toward the academic side. The development of an ever-swelling stratum of "humanistic engineers"—consultant psychologists, public relations and other communication experts, and so forth—is probably the most important innovation of the twentieth century within the humanities. Only if one makes the dubious decision to regard university teachers as "free" and disregard the "engineers" is it possible to claim that twentieth-century humanities are carried on by an environment of "free intellectuals."

As a result, work within the humanities has become "scientific" in the sense presented above (p. 7) to a degree which has probably never been equalled before. This situation, and in particular its novelty, has raised the question of *how to secure—or how to justify—the scientific character of the humanities,* in a world where some 300 years of indubitable triumphs have endowed the natural sciences with the status of sciences par excellence (it is no accident, indeed, that the English term *science* acquired its specific meaning of "natural" or even "physical" science in the later nineteenth century), and where it has become a commonplace to speak of "the two cultures"—namely the literate/humanistic culture and the culture of natural science and technology.[1]

Instead of attempting to fill out the generalities contained in the first paragraph of this chapter by particular accounts (but see chapter 21) we shall look at some of the more important attempts to answer the question raised in the second, immediately preceding paragraph.

One obvious possibility has been to assert that since natural science is real science, the way to make the humanities scientific is to emulate (what one believes to be) the distinctive methods of the natural sciences. This idea was already vaguely present one way or the other in the use of the "geometric method" in seventeenth-century philosophy and in Montesquieu's and Hume's aim "to do for moral philosophy what Newton had done for natural philosophy" and somewhat less vaguely in Saint-Simon's and Comte's "social physics." In the late nineteenth and the early twentieth century, such mere postulates and programs were replaced by genuine methods and results in a number of disciplines. It is not astonishing that the first place where it happened was in experimental and related branches of psychology: being already built on exact and supposedly meaningful measurability and hence expected to be also exactly applicable (in psycho-technical tests, IQ tests, and the like), they made their experimental and "positivist" orientation more explicit. Not very much later, advertisement and propaganda studies (in due course to develop into communication studies) and other kinds of sociology took up the torch. It has generally characterized fields dominated by a technical cognitive interest, which cannot astonish: if one wants to use knowledge for obtaining a specific result or to know whether that result has been reached, it is important to be able to *measure* with some kind of precision the parameters that characterize the situation to be changed, as well as the resulting changes themselves, and the nature and intensity of the means used to obtain them.

Around the turn of the century, other disciplines, in particular the traditional sciences of culture (philology, historical sciences) formulated their scientific aim and character as *different from that of the natural sciences* as they were understood at the time. It will come as no surprise that this opposition against the conflation of *human science* with *human engineering* had its focus in the heartland of Neohumanism,

---

1. The phrase was the title of a lecture series by C. P. Snow (1964), which had as its subordinate theme the mutual fear and distrust between literary and scientific culture. The main theme was the attitude of rich toward poor countries and the importance of planned technological development, which even the literate elite should not allow itself to spurn. This main theme was never really noticed, whereas the subordinate theme immediately became a commonplace—obviously touching what both the scientific/technological and the literary elite of the rich world saw as a more urgent problem than hunger.

that is, Germany.[2] Various philosophical justifications of this segregation were proposed.

One formulation, attributed to Wilhelm Windelband (1848–1915), emphasizes the uniqueness of the single historical event and the single text (cf. Ranke's *Einmaligkeit*). According to Windelband, the aim of the humanities is to *describe the particular*—they are *idiographic.* The natural sciences, on the other hand, are *nomothetic, law-establishing,* seeking the general. Though often quoted, this view was bound to be ineffectual—any description of something particular in terms of language connects it to some more general framework simply by using nouns and verbs (cf. below, p. 340, n. 11 and surrounding text). What Windelband asks for (if taken to the letter) is that the humanities should make use only of general concepts that are so hackneyed that their general character is not noticed.

Another scheme, certainly more influential, was proposed by Wilhelm Dilthey (1833–1911). Dilthey points out that the humanities (or rather, the *Geisteswissenschaften*) are not satisfied with (causal) *explanation.* The humanities investigate human *actions,* which differ from mere movements by having a purpose, and which can only be described adequately if the purpose intended by the actor is considered. Therefore the humanities must aim at *understanding,* must use historical sources or the texts under investigation as ways to understand the *intention* of human historical action or the *meaning* which is expressed by the text.

Dilthey's thesis is akin to Vico's point of view: the humanities deal with human phenomena which we understand from the inside. But there is an important difference: Dilthey writes after the fall brought about by historicism: in contrast to what Vico claims, we have *no* direct access to the way of thinking of the actors of other historical periods. Understanding will only result from hard scholarly work and will never be definitive and exhaustive (cf. p. 184, n. 5 on Dilthey's concept of the "hermeneutic circle").

The existence of different philosophical legitimations of the same situation suggests that an underlying historical explanation can be found at a different level.

---

2. In contrast, applied sociology and psychology, though traceable even they to mid-nineteenth-century German beginnings, grew to maturity and prominence in the Anglo-Saxon world, in particular in the United States. Structuralism, to which we shall return, had foci in Geneva, Moscow, Prague, Copenhagen, and Paris (British "structural functionalism," cf. p. 209, n. 15, does not qualify as a genuine structuralism). Because *philosophies* of science have tended to reflect local *scientific* activities, it has become customary to speak of "Anglo-Saxon" and "Continental" schools in the philosophy of science, where the former encompass those which tend to see the natural sciences as paradigmatic for all science, while the latter emphasize the distinctive character of human and social science.

As one may guess, this rough division is not too precise. Most of what has happened in the Anglo-Saxon schools was thus sparked off by emigrants escaping from Austria (Popper, Wittgenstein, the members of the "Vienna Circle").

A semi-sociological explanation of the emphasis on the distinctive character of the humanities might be to see it as a defensive reaction against the explosive development of the natural sciences and against the tendency to declare that *only* the methods of the natural sciences were scientific.[3] However, the claims that the humanities should look only for the particular and not for general regularities, and that their sole object is the understanding of the actors themselves, could also be seen as a contrast and a deliberate opposition to the outlook of Marxism and the Social Democratic labor movement. According to this outlook, the working of the human spirit not only takes place on historically specific conditions in a way which may be partly or fully hidden to the actors, but it can also to some extent and in a dialectical relationship be explained causally and in general terms (according to the simplified versions that gained wide currency, mechanically and without dialectical relationship).

> In retrospect, this kind of "political" interpretation may look far-fetched. Today, the basic Marxist notion has been accepted by most historians, whether understood as Marxist, Weberian, or just common sense, and nobody will find it dangerous. No follower of Marx, on the other hand, will be unfamiliar with the notion that human actions are characterized by having a purpose (as emphasized by Marx), and few present-day Marxists will claim that an explanation can be adequate which does not involve this purpose as one determinant. Yet before the "political" interpretation is dismissed, one should remember that at least one other field changed its conceptual fundament in order to distance itself from Marxist theories—namely political economy, which abandoned the foundations created by Adam Smith and David Ricardo not least in order to avoid the consequences derived by Marx from this framework (cf. below, p. 286, n. 22). One should also take a second look at the

---

3. This explanation generalizes what Dilthey (1924: 139–240) writes about psychology. One type, aiming at causal explanation referring to "a limited number of clearly defined elements," does so in an attempt to imitate the natural sciences but can only grasp the complete mental process by entering a haze of gratuitous hypotheses that lack the foundation of experiment and measurement on which the natural sciences build. Moreover, it has to presuppose a complete determinism. Therefore, a credible psychology which also comprehends the free and creative aspects of mental life must be descriptive and analyze the mental totality of the fully developed human being instead of synthesizing a postulate from the isolated elements that can be studied by experimental psychology, and has to start from the way we *understand* ourselves and each other.

There is nothing wrong in natural sciences as such—Dilthey refers lavishly to their role in human cultural development—nor with experimental psychology in itself. Their potential for explaining human action and life, however, is limited, in part for reasons of principle and in part because the results obtained so far by experimental psychology, laudable though they are, are insufficient for such a task.

> speech Dilthey made on the occasion of his seventieth birthday[4] in 1903, and where he appears to tell precisely this.
>
> In the first place, Dilthey demonstrates by his choice of words and themes that his own natural bent is to look for explanations just as much as for mere understanding of the actors from within. "Language, law, myth, religion, poetry, and philosophy, each possesses an inner lawfulness which conditions its structure, and this again determines its development" (etc.).[5] The dismissal of "explanation" must therefore have a very precise aim. Further, in the closing paragraph, he tells that the ultimate purpose of his life-long work as a historian, and in particular of his "critique of historical reason," has been to defeat "that anarchy of convictions which threatens to break in," which in the context can hardly be anything but the convictions of the Social Democratic movement.

Around the middle of the twentieth century, a novel explication of why the humanities were different from the natural sciences materialized through the generalization of *structuralism.* Structuralism had taken its beginning within linguistics (Ferdinand Saussure, 1916), which it came to dominate from the 1930s onwards (Roman Jakobson, Louis Hjelmslev). Its first impact in the textual sciences goes back to the 1920s (Vladimir Propp's analysis of folktales, not meant by the author to be structuralist—cf. p. 301, n. 44), but only in the 1950s would it become important in this domain, due in particular to the influence of the anthropologist Claude Levi-Strauss.

The reason given by structuralists for the separateness of the humanities is wholly different from those which were advanced by Windelband and Dilthey. Structuralism sees human action as no less determined than physical phenomena. In contrast to these, however, human actions are determined by their place within a totality, *a structure,* not by a mechanical one-cause-one-result sequence (cf. below, p. 207f). Even though the model of causation is different, structuralist theories are thus more deterministic than Marxist explanations—and "vulgar structuralism" at least as deterministic as most of those variants of "vulgar Marxism" which have been produced by the various branches of the labor movement for agitation purposes.

A distinctive characteristic of many structuralist currents is the absence of historical development from view. This differentiates average structuralism from

---

4. (Dilthey 1924: 7–9). I use the opportunity to thank Kurt Flasch for directing my attention to this interesting item.

5. One may add that some of Dilthey's actual explanations come astonishingly close to Marxist views. When discussing in his *Einleitung in die Geisteswissenschaften* why ancient Greek natural philosophy was unable to make efficient use of its experiments, he refers to "the opposition of a ruling citizenry, which also cultivated science, to the slave class on which the burden of manual labor fell, and, connected with that, disdain for physical labor," and to "the lack of an industry managed by scientifically trained people" (trans. Betanzos, in Dilthey 1988: 203). In general, it is true, the work is oriented after Comte's stages.

both Marxism and from nineteenth-century historicism, and indeed from all evolutionary theories. Structures are presumed to be static, and their single elements seem to be permanently fixed by the totality in which they participate. Since historical change cannot be declared to be nonexistent, it is relegated to the region of the twofold unaccountable: that which neither can be nor should be explained scientifically.[6]

In some cases, the dismissal of change results from theoretical system-building: the theoretical construction seems to leave no place for change. In others, it expresses the rejection of alternative theories or approaches such as Marxism or the particularistic philosophies of Dilthey and Windelband. (Of course, theoretical system-building and a priori rejection of alternatives are not mutually exclusive). All in all it must be concluded, however, that although the formulation of certain variants of structuralism may have to do with the aspiration to provide a semi-political demarcation, it is hardly possible to give a single explanation of this kind for the whole movement, which is much too diverse for that. It is even possible to point at several important variants of structuralism which are strongly committed to a dialectical explanation of change (thus Roman Jakobson's linguistic theories; Piaget's structuralist psychology; and some of the Marxist varieties of structuralism): structures, through their functioning in discursive, cognitive, or social practice, engender tensions which erode their stability; in the end, the old structure is exploded, and a new structure crystallizes.

Better than external sociological explanations of the structuralist current within the humanities is the explanation via the subject matter: as the humanities moved away from and beyond mechanistic psychology and from the Neohumanist selective interest in ancient texts and history, the understanding of complexity became important. The evolutionary and historical approaches to both linguistics and texts were unable not only to solve but even to formulate many pertinent questions—or, when trying, they were forced to recognize that change was conditioned by the totality inside which it took place.[7] The structuralists took up that aspect of their subject matter which was traditionally left aside in the historical and evolutionary

---

6. This position repeats a traditional Aristotelian notion much more loyally than supposed by the less learned among the structuralists (cf. above, p. 42). Even Aristotle had supposed that science (*epistēmē*) had to deal with *necessary truths,* which could only be derived from the immutable forms. Single events and change—indeed everything dealt with by the humanities according to Windelband and Dilthey—were things which *needed not* have happened: they were *contingent,* accidental, and thus outside the scope of science.

7. This is what happened in later nineteenth-century linguistics and is particularly evident in Saussure's explanation of the sound shift laws (1972: 198–210)—notwithstanding his explicit declaration that the structural and the historical descriptions are incompatible (1972: 193–195).

A similar argument could be made about the "structural functionalism" of British anthropology (cf. p. 209, n. 15).

approach—and in a first approximation they would concentrate on the aspect of synchronous complexity, leaving the historical aspect to the tradition and the integration to a later generation (which has indeed taken up the job in recent decades—cf. p. 319).

During the last three decades or so, the dogmatic first approximation has been left behind in many fields. Whether as part of academic career strategy or for other reasons, the change has been given high-flown names (*deconstructivism,* to mention one example) by some of those who want to lead or to participate as pioneers in a scientific revolution or to be the Popes of a new dogma. Others, more eager to till their field than to put up fences around it, tend to make judicious use of all the insights inherited from earlier generations—trying, like Diderot, to use "the best from all philosophies" but being perhaps more aware of the dangers which inhere in the combination of theories whose basic ideas contradict each other. It may be difficult to distinguish nowadays a Marxist who has read Weber and Levi-Strauss thoughtfully from a Weberian or from a scholar who started out from structuralism with similarly open-minded readings.

For the sake of clarity, chapter 11 will start by putting up some fences between different approaches to the subject matter of the humanities. The gist of the message will be, however, that grazing is unhealthy if it is restricted to a single and dogmatic approach.

CHAPTER 9

Bibliographic Essay

The material on which Part I is based is too extensive to allow anything approaching adequate documentation in a supplement—just as any attempt to give full documentation within the running text would have made it illegible. What can be done in a bibliographic essay is to list a number of readable or important titles dealing with central subjects covered in the text. Works on single historical figures will not be included, nor are editions of sources except for a few anthologies of illustrative excerpts (both genres can be looked up in specific bibliographies).

As usually, references are made according to the author-date system, where the date refers to the edition listed in the bibliography.

Much valuable material, including rich bibliographic information, can be found in the five-volume *Dictionary of the History of Ideas* (DHI in the following; 1968–1974). The 16 volumes of the *Dictionary of Scientific Biography* (DSB; 1970–1980) contain biographies with mostly extensive bibliography of all major figures who can somehow be connected to the history of (natural) science (the gauge is quite liberal; not only Aristotle and Hume, but also Plato, Thomas Aquinas, Voltaire, Marx and Wittgenstein are included). Topics which can be related to the social sciences (once again according to a liberal gauge) are covered in the 17 volumes of the *International Encyclopedia of the Social Sciences* (IESS; 1968; equally with bibliographies for further study). Each of the three encyclopedias contains a most helpful index volume.

Randall 1962 can be recommended as a broad interpretation of the history of philosophy in socio-cultural context—in fact, of philosophy regarded as a series of responses to problems raised by this context—from 1100 to 1800. Bowen 1972 is a broadly oriented history of education, of educational ideas, and of educational institutions—and since so much of what relates to the humanities has also been related in some way or the other to education, it can be consulted with profit on many issues dealt with above.

Chapter 1: The two "modern classics" on general state formation theory are Fried 1967 and Service 1975, to which most subsequent discussions of the subject refer. More recent collections of studies applying the theories to single cultures are Claessen & Skalník (eds) 1978; Gledhill, Bender & Larsen (eds) 1988; and Herrmann & Köhn (eds) 1988.

Classics on the importance of the shift from oral to literate culture (somewhat overstating their points according to the opinion of the majority of recent workers) are Havelock 1976; Ong 1967; and Goody (ed.) 1968. More recent and more balanced are a number of important publications by Jack Goody, in particular 1977, 1986 and 1987.

Intensive work on the first beginning of Mesopotamian literacy has been done during the last two decades; a convenient summary is Nissen, Damerow & Englund 1994. A documented synthesis along the lines presented here will be found in my 1994: 45–87, 296–306, which also traces the relation between state formation, transformations of the social structure, and scribal culture through the mid-second millennium BCE.

A number of path-breaking Soviet works on the developments of the third millennium will be found in Diakonoff (ed.) 1969. A readable account of the social and ideological characteristics of Old Babylonian society is given by Horst Klengel 1980. Lucas 1979 and Sjöberg 1976 describe the scribal school.

Chapter 2: The characterization of ancient society as a *slave holders' society* and in particular the claim that this social structure conditioned the development of ancient thought has been the subject of much discussion. Most objections, however, have been aimed at a rather mechanistic use of the concept read (wrongly) into Farrington's classical statements of the connection between class structure and philosophy (1965; 1969). An overview of the controversy (until 1960) can be gained from Finley (ed.) 1968. Austin & Vidal-Naquet 1977 combines a survey of central issues in ancient Greek economic and social history with select excerpts from ancient sources.

A concise and stimulating discussion of the connection between the appearance of philosophy and the emergence of the Greek city state is contained in Vernant 1982. A more thorough discussion of the roots of Greek philosophy and rationality along similar lines is Lloyd 1979. G. Thomson 1961 is a controversial classic of broad scope, written from the perspective of a creative and historically informed historical materialism.

The standard work on Pre-Socratic Greek philosophers is Kirk, Raven & Schofield 1983 (revised from Kirk & Raven 1957). Guthrie 1962 is an extensive history of Greek philosophy until Aristotle (6 volumes), whereas Guthrie 1967 contains a concise survey of the same subject matter. Mean proportionals (as to extension) are Seidel 1989 (a university course) and Jürß 1977 (on Pre-Socratic philosophy only). The history of Greek philosophy from Plato onwards, including

Christian philosophy until 1100 and Islamic philosophy until c. 950, is dealt with in Armstrong (ed.) 1970. Seidel 1988 is a briefer presentation of ancient philosophy from Aristotle onwards. Greek "rational medicine" in its relation with natural philosophy has been treated by Longrigg (see especially 1989; 1995).

A classic on Greek educational ideas, with particular emphasis on the sophists and Plato, is Jaeger 1973. Actual education is dealt with in Marrou 1956, Clarke 1971 (on "higher education") and Bonner 1977 (on the Roman world).

Much information on ancient culture and philosophy can of course be taken directly from the original authors, many of whom (from Plato through Augustine and Cassiodorus) are accessible in modern translations with introductory explanations.

Chapter 3: A general, very readable (and beautifully illustrated) introduction to the broad features of medieval economic history is Bautier 1971. Works concentrating on the formation of feudal structures (and seeing them in different ways) are Anderson 1974, Duby 1973, and Gurevič 1979—all footing on the classic Bloch 1965. An insightful account of the interaction between technological innovation and social change in the Middle Ages, from the adoption of the stirrup onwards, is White 1962. Ovitt 1986 deals with the transformation of the attitude to labor and technology brought about by medieval Christianity.

A description of the passage from Antiquity to the incipient Middle Ages on the level of culture is Brown 1971 (no less beautifully illustrated than Bautier's book). The interpretation of the (literate) Middle Ages as an "age of renaissances" is set forth in my 1988. Nonliterate medieval culture is the subject of several works by Gurevič, for example 1986, and also dealt with in Richter 1994. A recommendable history of the medieval Church arranged thematically rather than chronologically is Southern 1970.

Education and literate culture in the early and Central Middle Ages are dealt with in Riché 1962 (English translation 1976) and 1979 (both include the Carolingian age and educational effort); more oriented toward the products of literate culture is Laistner 1957, while the perspective of McKitterick (ed.) 1990 is more attentive to the tension between oral and literate culture. McKitterick (ed.) 1994 contains similarly oriented studies of the Carolingian period. The cultural situation of the centuries following upon the Carolingian failure is part of the subject of Southern's stimulating 1953.

The thesis of a "twelfth-century renaissance" was formulated in Haskins' classic 1976. More recent but inspired by the same approach are Brooke 1973 and, to some extent, Southern 1970a. An eminent exposition of rationalist and other philosophical and theological currents from c. 1050 to 1200 in social context (connecting the topic to the rise of towns and to the specific character of the urban environment) is Werner 1976. Related in spirit are Chenu 1966 and his brief 1977 (an avowedly historical-materialist analysis written by the Dominican Father who was quoted on p. 82, n. 3). Further references to discussions of the cultural meaning of astrology and naturalism in the High Middle Ages can be found in my 1994: 123–146, *passim.* Haskins 1924 remains a central work on the twelfth-century wave of translations.

Much has been written on the early history of universities, and on that of single universities. A classic which remains astonishingly vigorous is Rashdall 1936, which was reprinted at least as late as 1964 (my copy). Among recent books on universities and university culture, Pedersen 1979 and Piltz 1981 can be recommended (both have much to tell about the scholarly traditions behind the universities), along with Cobban 1975 and J. W. Baldwin 1971. Thorndike 1944 is an anthology of excerpts from original sources in English translation.

A recommendable and extensive survey of thirteenth-century philosophy is van Steenberghen 1966; a more concise introduction to the topic is his 1955. The prohibitions of Aristotelian natural philosophy are the subject of Grabmann 1941; these and related conflicts are also dealt with extensively in Zimmermann (ed.) 1976. Much material can of course be found in the vast literature dealing with Thomas Aquinas, Albert the Great and other scholarly churchmen. Significant but mostly rather specialized articles on single aspects of scholastic philosophy will be found in Kretzmann, Kenny & Pinborg (eds) 1982.

A famous classic on the predominant moods of late medieval culture is Huizinga 1955. A more recent treatment of the many disparate facets of the fourteenth-century intellectual and spiritual scene is Leff 1976. Social conflicts and revolts are the theme of Mollat & Wolff 1970.

Anthologies of illustrative medieval texts in English translation are Herlihy (ed.) 1968, 1971, M. W. Baldwin (ed.) 1971, and Ross & McLaughlin (eds) 1977a.

Chapter 4: Since modern practitioners of the humanities indulge in thinking of themselves as the legitimate heirs of Renaissance Humanism, no soil has probably been cultivated more intensely by historians of ideas than the Renaissance. As a consequence, to include the Renaissance in a brief bibliographic essay is a hazardous affair. First of all I shall therefore point to Burke 1972, the initial chapter of which is a fine survey of research traditions and approaches to the cultural and social history of the Renaissance since the publication of Jakob Burckhardt's seminal *Kultur der Renaissance in Italien* in 1860. Further on, Burke's very informative book revolves around writers and visual artists, their products and their cultural significance, and the particular and general social conditions of artists and Humanist writers.

The two main approaches to Renaissance Humanist culture are probably those presented by Kristeller (for example, 1961+1965, and the collection of articles in his 1956), concentrating on the relations of ideas to other ideas and to professions, and Garin (for example 1965), more oriented toward the Renaissance as a general civic movement. The latter approach is also prominent in Martines 1983, which sees Humanism in the context of Italian city states as these had developed since the twelfth century, and Bec 1967, which explores its relations to the commercial development of Florence. An original attempt to characterize Renaissance society and culture as a coherent structure is Batkin 1979. In his provocative 1950, Haydn characterizes much of what is normally considered part of Renaissance culture, including Elizabethan literature and the empirical scientific tradition which established itself in the sixteenth century, as a *Counter-Renaissance.*

Eisenstein 1979 explores the importance of the "Gutenberg revolution" for many facets of late Renaissance society and culture. Important and conflicting views on Renaissance art are found in Gombrich 1971, Panofsky 1960 and the pertinent parts of Hauser 1962. The way from Renaissance *studia humanitatis* to humanistic scholarship is dealt with by Grafton & Jardine 1986, while the emergence of historical scholarship in a fairly modern sense is dealt with by Kelley 1970 and Huppert 1970. Grafton 1992 analyses the impact of the discovery of America and other "new worlds" on a world view hitherto based on ancient texts.

Mathematics as part of the concept of 'humanity' via its civic utility and mediated by the image of Archimedes is dealt with in my 1992. The "scientific Renaissance" of the sixteenth and early seventeenth century is in itself an immense subject; more or less at random I shall point to Debus 1978, a concise survey more open to the occult aspects of Renaissance science than those traditional "Whig histories" which seek present science in the past; Zilsel 1976, a collection of articles originally published 1940–1945, and emphasizing the relation between "higher artisans" and the new science; Heidelberger & Thiessen 1981, a more recent book applying the same perspective; and Schmitz & Krafft (eds) 1980, which deals with the interconnectedness of the scientific and the Humanist movement. In her now classic 1964, Yates insisted that there is a strong tie between magic and Renaissance science, founding however the argument on philosophers and phantasts who did not participate in the scientific movement. Equally classic is Thomas 1971, a multidimensional investigation of the decline of magic over the seventeenth century emphasizing the role of the new science. A useful survey of Ramus's work (though blemished by adulation and a strong "Whiggish" orientation and lacking in deeper analysis of Ramus's argument) is Hooykaas 1958, which may be supplemented by Mahoney 1975. Randall 1962: 230–255 locates Ramus within the general search for "method," partly in parallel and partly in contrast to Bacon, while Ong 1958 concentrates on Ramus's conceptions of dialectic and method.

Anthologies of illustrative Renaissance texts in English translation are Ross & McLaughlin (eds) 1977 and Cassirer et al (eds) 1956.

Chapter 5: A concise exposition of the correlation between courts and artistic forms (Mannerism, Classicism, Baroque) is Koenigsberger 1979. The early history of the Academy-concept is described in the first part of Pevsner 1940. A concise survey of the quarrel between ancients and moderns with further bibliography is Aldridge 1968. French classicist culture is the subject of Hauser 1962: II, 172–191. Much information on the philosophy of the epoch can be found in DHI, in Randall 1962, and in the biographical articles in DSB. The history of the man-machine model is sketched in Vartanian 1973. Metcalf 1974 describes the "etymological" school in linguistics. My 1997 contemplates the contradictory relation between the Baroque mind and the new science.

Chapter 6: The Enlightenment is another cherished subject of historians of ideas, and the literature on the theme is of exorbitant extent. Only a very superficial introduction can thus be given.

The idea of a specific bourgeois public sphere was formulated by Habermas 1968. Though original, the book of course draws upon and synthesizes a number of older observations and ideas, some of which are set forth in Hauser 1962.

A comprehensive interpretative treatment of Enlightenment philosophy is Gay 1967, the first volume of which ("The Rise of Modern Paganism") deals with the tension between Enlightenment and Christianity, whereas the second is adequately characterized by the title "The Science of Freedom." The two volumes include bibliographic essays of 133 and 135 pages, respectively. Another large-scale synthesis, emphasizing the new world view and the "revolutionary spirit" of the *philosophes,* is Wade 1977. Reformist and utopian Enlightenment ideas are dealt with by Baczko 1978 and Venturi 1971. A brief presentation with select bibliography is Pappe 1973. A fairly recent book on the specific Scottish Enlightenment is Daiches, Jones & Jones (eds) 1986.

An annotated selection of texts highlighting the Enlightenment philosophy of knowledge (starting with Locke, and heavily biased towards English and Scottish authors) is Berlin 1970. The attitudes of the Enlightenment philosophers to science and to single sciences are dealt with by Kiernan 1973, whereas the situation of the sciences seen in the general Enlightenment context is treated by Hankins 1985. A selection of texts illustrating the impact of Enlightenment philosophy on the educational reform ideas of the French Revolution is contained in Baczko 1982. A brief introduction to the Counter-Enlightenment is Berlin 1973.

Chapters 7–8: The continuity from Enlightenment thought over utopian socialism to Comte is the topic of Manuel 1965. The Prussian university reform, its background and consequences for the emergence of systematic academic research in the natural sciences as well as the humanities have been dealt with by Turner in a number of papers (1971, 1974, 1981). The historicist attitude, from Vico and Herder over Ranke until the early twentieth century, is traced in Iggers 1973. Specific developments within the single branches of the humanities are best looked up in studies of the history of single disciplines (a valuable first approach to many of these is provided by articles from IESS and DHI); a broad survey of nineteenth-century humanities (including certain social sciences and some early twentieth-century developments) is found in Störig 1957: 589–752. Darnton 1968 and Cooter 1984 offer broad perspectives on Mesmerism and phrenology as instances of "popular science."

PART II

Human Science and Human "Nature"

CHAPTER 10

Cognitive Interests

In chapter 1, Habermas's concept of *Erkenntnisinteressen* ("cognitive interests," that is, incentives for searching knowledge) was introduced succinctly; in subsequent chapters, these interests have been referred to repeatedly. Habermas's categorization is inspired by other philosophers, but the total scheme is original, as is his way to connect three fundamental conditions of human existence with three *Erkenntnisinteressen* and with three prototypes of scientific knowledge. Original (and problematic) is also the denial that curiosity in itself may be a fundamental condition of our existence, and that we therefore might find knowledge in itself to be interesting.

Any comprehensive theoretical investigation of *knowing* as a human activity should ask the question why we want to know. Since Habermas's scheme is in itself rich in facets, and since discussion of its inherent problems will exhibit further aspects of the question, it may provide an adequate starting point for that systematic approach to the knowledge of the humanities which is the necessary complement to the preceding historical account.

First in Habermas's scheme comes *work.* As human beings we work, that is, we produce according to a conscious plan in agreement with knowledge of materials and tools. Work provides us with the material fundament for our existence. To work as a condition of human existence corresponds the *technical* cognitive interest, *knowledge as power,* knowledge of how means can be used to attain a goal we have (already) set ourselves.

Ultimately, this understanding of work is inspired by Marx (and, behind him, by Aristotle and Plato). But the actual formulation of the idea owes much to Max Weber's idea of *Zweckrationalität,* "instrumental rationality," which is precisely the rational choice of the best means to achieve an already given goal—a goal which itself is not subject to discussion (cf. p. 140, n. 17).

Next we have *interaction.* As human beings we participate in a human community built not least on communication. Communication, however, presupposes the possibility of mutual *understanding.* The corresponding cognitive interest is *practical:*

the quest for *knowledge as a guide for practice,* that is, for life together with others within an existing social order and community. One will notice, firstly, that a distinction is presupposed between *that which is good for something (else),* corresponding to Greek *technē*, and *that which is good (in itself),* corresponding to Greek *praxis* (whence the terms *technical* and *practical*).[1] One observes, secondly, that "life together with others within an existing social order and community" implies that the framework of this community is taken for granted as the self-evident fundament for discussion. For this reason, the practical cognitive interest automatically becomes legitimizing, a search for mental assimilation of the existing order as it *ought to be because it is* (namely because it is necessary for our participation in the community); the knowledge we get out of this search tells us that *we are OK.*

Since its beginning, this legitimization of ourselves has always been an important purpose of the writing of history—be it in the message of the Old Testament about the Select People, or in Herodotos's lessons about the contrast between the Greek *polis* and the despotism of Oriental societies). In recent years it has again been taken up by conservative thinkers as the main purpose of history as a school subject: it is important that Danish children learn about the Danish colors (the *Dannebrog*) falling from heaven during a battle against the heathen Estonians in 1219—important because it tells us about the specific (and, given the story, superior) qualities of being Danish.[2]

The third condition is *emancipation.* The given circumstances which determine our existence as human beings are historical products and not perennial. We can thus free ourselves from their compulsion (though not from the need for *some* social framework)—not arbitrarily, but if we know to distinguish necessities to which we have to bow from fetters which are not (or no longer) necessary. The corresponding

---

1. We remember that a *techne* or "art" was, in the opinion of Plato, Aristotle, and other ancient philosophers, knowledge about the vulgar and morally inferior production of useful goods (with "the art of medicine" and "the art of rhetoric" as exceptions). *Praxis,* on the other hand, was the subject of the most elevated branch of philosophy, the one dealing with the GOOD.

2. This purpose has important implications for the teaching of history. If history teaching is primarily meant to inculcate awareness of what it means to be a Dane, it is less important whether the story is objectively true or (as I was told by a subversive teacher in 1956) an international anecdote imported from Portugal together with the Portuguese-born queen Berengaria. There is hence no reason to oppress the children with the tools of scientific history such as historical method and critical sense. At best they are superfluous, at worst they undermine the *real* purpose of teaching the subject: the affirmation of national identity.

The story of *Dannebrog* is not to be learned by children as an coincidental objective fact but *as a meaningful message*—they are to *understand it as Danes* (I shall leave the translation into other national idioms to the reader—parallels should not be difficult to find). History is no science, but belongs at the same level as poetry (more precisely, at the level of poems like *Britannia Rule the Waves* and *Deutschland, Deutschland über Alles*).

cognitive interest is *emancipatory.* Whereas the technical cognitive interest can be correlated with Weber's *Zweckrationalität,* his *Wertrationalität* or "value rationality"—the "reasoned discussion of our ultimate aims, of their attainability, mutual relations, and consequences"—is thus associated with the emancipatory cognitive interest.[3]

> Habermas's scheme is often regarded as a reaction to Marx's social theories (which are then claimed to focus exclusively on the aspect of work). It has also occasionally been presented in this way by Habermas himself. In fact, Marx's concept of social practice encompasses all three domains: work, the process by which men reproduce their living conditions, always takes place within a social community, within and as part of the historical process. It is thus more accurate (as done by Habermas on other occasions) to see the threefold scheme as a detailed specification and schematization of ideas that are found in Marx's writings but are forgotten by many of his interpreters—Marxists and anti-Marxists alike.

According to Habermas, three prototypes of scientific work correspond to three fundamental cognitive interests:

— The *natural sciences* are regarded as mere means for technology: engineering physics, agricultural chemistry, agricultural and medical biology, mining geology—such "applied sciences" provide the model. The method of this type of science is claimed to be experimental and empiricist (on occasion hypothetical-deductive)—summed up as "positivist."[4]

— The *humanities* are the disciplines that try to understand texts from other historical periods or other cultures, or the expressions of the mind of others, *on their proper conditions which we have to accept,* in agreement with the radical historicist ideal. The method is hermeneutic, that is, "dialogue" with the subject-matter.

> The reason that practical cognition must make use of this method is quite simple: at our first approach to a foreign text (in the wide sense, that is, to any spoken, written, sculptured, painted, or similar expression of meaning), we interpret it in agreement with our own presuppositions and prejudices, which are in fact our only access to the universe of meanings. But if the foreign text does not fit our interpretive expectations on all points (which it rarely does), and if we investigate the points of noncompatibility seriously, we will be led to revise our prejudices—so

---

3. The aspirations of emancipatory knowledge is illustrated by an old prayer (attributed at times to St. Francis of Assisi): "My Lord, give me strength to do what ought to be done; patience to suffer what cannot be otherwise; and wisdom to distinguish one from the other."

4. This use of the term is not directly derived from either Comte nor Mach nor from the "logical empiricism" of the Vienna Circle; it relates to Comte's original usage through a series of misunderstandings by social scientists claiming to emulate the objective methods of physics—and to polemics against these formulated by other social philosophers (especially those belonging to and inspired by the "Frankfurt school").

> to speak, by "listening." The revision will enable us to understand the presuppositions of the foreign mind (or understand them better) and hence even to understand ourselves from the foreign point of view. Understanding the other leads us to better insight into our own expectations to universes of meaning, and hence allows us to approach the foreign text (and other foreign texts) with better prejudices. This "hermeneutic circle," as it has been called by Martin Heidegger and Hans-Georg Gadamer,[5] is thus a generalization of the ideal dialogue between equals.

— *"Critical" social theory* is not satisfied by knowing "how it is" but also asks the causal question "why it must be so," namely in order to answer the ultimately decisive questions "which change is possible?" and "how can it be brought about?"[6]

---

5. The term is older, and was originally used by the Romanticist theologian Friedrich Schleiermacher (but still by Dilthey) about the relation between the single parts of a text (in the first instance the Bible) and its totality, which follows a similar pattern: only understanding of the totality allows us to choose a particular interpretation of the single sentence—but on the other hand, the meaning of the totality has no other source than the interpretation of its single constitutive parts.

If we see the meaning of the totality not as an objective horizon inside which the single sentences are formulated but as *our* construct, as a framework which *we* have to build up and presuppose in order to be able to understand (certainly a sensible view of a text which, like the Bible, was produced by many different hands over quite a few centuries), the step from Dilthey's notion to the current use is seen to be short.

6. It can legitimately be asked whether there is necessarily anything *critical* in these questions. Aren't they the questions of any *applied* science, including applied social science? Does not the economic planner in a firm ask *why* it encounters losses, and what should be done in order to change negative profits into positive ones? On the surface of things the questions are indeed those of any applied science, and it might seem that no distinction can be established between social technology and emancipation, as indeed argued by some social theoreticians with a nominalist bent—see, for instance, Gibson 1980.

The matter is intricate and may be approached in several steps. At first we may try to distinguish *different levels of causality*, turning to Hume's classical example: the billiard ball which *starts* rolling "because" it is hit by another ball (Humean causality, which, if we leave Hume's subjectivism apart, is grossly identical with Aristotle's "efficient" cause, cf. pp. 323ff)—but which only starts *rolling* (instead of falling or sliding) because it is spherical, placed on a smooth table covered with cloth (here we might speak of "structural" causality, referring to the total make-up of the situation in which the event of hitting takes place). The critical "how," "why" and "which" all concern the structural level, whereas normal applied social science only asks for efficient causes: which events should be brought about in order to engender the situation asked for.

So far the difference is more or less the difference between tactics and strategy (cf. p. 136, n. 13) (though Habermas' notion of emancipatory praxis may go even further toward value rationality as defined on p. 140, n. 17). If we see no difference between, on one hand, an inhabitant of Hobbes's state of nature who chooses the best means to cut the throats of his fellows before they cut his, and on the other, the collective of inhabitants deciding to make a

The principal method, according to Habermas, is critically reflexive hermeneutics—social structures and relations can neither be understood nor *a fortiori* be changed in a progressive way if we disregard the way participants understand their world. As Hobbes's social contract it can only be a result of cooperation and common action of those involved.

This principal method has, however, to be supported by the methods of technical cognition. The inclusion of these "technical" methods in the project of emancipation can be explained by the need to step outside the given world and make it an object of your reflection, in the same way as the technician places himself above the objects he is going to change or construct. But whereas the engineer is no part of the bridge he builds, no revolutionary can avoid being himself a member of the social world, however much he believes to be in opposition; being in opposition simply means that he is involved in its inner conflicts. Emancipatory changes can therefore only be produced in dialogue with that social community; forgetting that fact is often a symptom that your deepest aim is not really to *change the world* but only to *change your own position within this world.* Treating fellow human beings solely as *objects* is not possible; what you really do if you try is to treat them as *enemies,* trying to make sure that they shall not be able to do to you what you are doing to them.[7]

---

social contract—then there is indeed no reason to distinguish between socio-technology and emancipation. If we do see a difference in quality, the distinction makes sense.

But the parallel to Hobbes has further implications. The aim of an applied science is to identify the handle which the actor can activate in order to achieve his aims independently of the intentions of others, whose throats he may indeed want to cut in the most efficient way; the social contract, on the other hand, can only be made by common decision—it presupposes agreement and mutual understanding.

The attempt of the Enlightenment philosophers to demarcate themselves from the projectors (cf. again p. 136, n. 13) is obviously related to the present distinction between emancipatory *praxis* and socio-technical management. As pointed out by Gadamer, Habermas's very way of introducing the emancipatory interest presupposes the ideals of the Enlightenment; from Gadamer's conservative point of view, practical insight might also recognize that authority is legitimate: "what is under dispute is simply whether insight always dissolves the substantial [that is, social, supposedly oppressive] relations, or it might lead to mental acceptance" (Apel et al 1971: 72ff, quotation p. 74). The "always" of the quotation, of course, is added by Gadamer and misses or distorts his opponent's point of view: "insight" is required precisely in order to know which "substantial relations" need not—or no longer—be accepted. Gadamer's underlying insinuation seems to coincide with Burke's: that authority is *always* legitimate if only it is derived from tradition.

7. In the first instance, this is a purely philosophical observation which need not involve any feelings of enmity or hatred: an object cannot react with the intention to escape your influence (even though the bridge you build may defy your expectations and fall down); an enemy is someone who possesses a will, and who might intentionally try to frustrate your plans—which you will then have to prevent. *An object,* in other words, possesses no will; *an*

The meaning (and at the same time, the open-endedness) of this account of the conditions for emancipatory change may be made more clear by a comparison with certain theories about the strategies that must be followed by those who want to change the world. Most pertinent is Lenin's theory of the role of the revolutionary Party (as interpreted by Gramsci, not by Stalin): the Party is the "collective intellectual," the group which formulates the worldview of the working class and creates the concepts required to formulate it, but it cannot do so by erecting itself above the working class (as done by Blanquism, Stalinism, and in certain of the "terrorist" offspring movements of the student rebellion). This is the point in Lenin's *What Is to Be Done?* from 1903: that *the newspaper* is the fundament for the effective realization of the Party, not vice versa. Only if a channel exists for communication with the working class will it be possible for the Party to act as its "collective intellectual," that is, to *exist* as a revolutionary party.

Che Guevara's interpretation of the success of the Cuban revolutionary war elucidates Habermas's idea in a somewhat different way.[8] The twelve survivors from the landing on the Cuban shore were not able to conquer Batista's armed forces. But they were able to demonstrate the vulnerability of these forces, and hence to awaken a terrorized population and make it believe in the possibility of successful change. The guerilla attacks on the Batista militia were thus, in their essence, *texts,* acts of communication, expressed in *the medium* of military action—and they were read as intended by the "target audience."[9]

There are good reasons to associate Habermas's idea of emancipatory knowledge with the theoreticians of the socialist movement. Already the Enlightenment

---

enemy is somebody whom you try to deprive of the ability to exert his will (cf. the related observation made by Kant and quoted on p. 142, n. 22).

In critical situations, however, the abstract philosophical categories materialize in psychology and behavior: the torturer, in contrast to the blacksmith, often tends to loose self-control during his "work" and has to be observed by a "neutral" physician in order not to kill the victim too soon—even the professional torturer is rarely able to regard a fellow being as a mere object.

8. *La guerra de guerrillas,* in Guevara 1967:23–109.

9. The term *target audience* is borrowed intentionally from the advertisement industry, because it shows the weakness of the Guevarist approach, especially as understood by many student revolutionaries of my generation. A "target audience" is an object; you check its reactions in order to find out whether you should change the way you present your message in order to make it buy your toothpaste or your revolution; but you do not listen to it in order to find out whether it really *wants* or *needs* the product you offer (whether toothpaste or revolution). If it does not accept your message you may end up like the Italian Red Brigades and the Peruvian *Sendero luminoso* and shoot at trade union activists or elected peasant representatives. Even in these cases, the presumed object turned out to be an enemy not only from a philosophical but also from the everyday point of view.

believed in a natural alliance between knowledge and progress, it is true. As we have seen, however, the Enlightenment never arrived at formulating clearly the relation between *technical improvement* (that is, strategically rational change of society, where some members of society *know better* and improve on behalf of everybody but according to their private knowing-better) and *emancipation,* even though the contempt for projectors and the terms of Diderot's denunciation of despotism foreshadows the distinction. Furthermore, as the dilemma had posed itself (all too) clearly during the French Revolution, the reaction of the bourgeoisie and its thinkers was to aspire only to technical improvement and to opt for purely instrumental rationality. Only the radical intelligence and the labor movement would stubbornly preserve the conviction that *reason* could be used to criticize and improve the fundaments of society, and discovered that reason was also required if one wanted to distinguish possible change from utopian dreams. The first clear formulation of this historical determination of possible progress is in fact due to Marx and Engels.

Politics, however, is not the only frame of reference for emancipatory knowledge (nor is it Habermas's only frame). Another example is provided by psychoanalysis. Even psychoanalysis presupposes that the therapist, firstly, listens extensively to and understands the patient; secondly, understands more about the patient than the patient is immediately willing to admit to himself; and, thirdly, makes the patient understand even this.

Before we leave the problem of emancipation, it should be emphasized that Habermas distinguishes natural science, humanities, and *critical* social science, not just natural, human, and social science. A main point of his scheme is indeed that much social science is not critical but (socio-)technical, and that this constitutes a philosophically and morally illegitimate mistake of category: a science which *should* deal with emancipation (because its agents are themselves part of the "object" of their science) is misconstrued as a technique for the exertion of power and for the management of existing social relations.

Useful though it is, Habermas's scheme still calls for various comments and critical observations. One has to do with its relation to Marxism. As it should already be clear, the scheme is inspired by Marxist thought; it differs from Marxism,

---

In order not to conflate Che Guevara's ideas with those of the student revolutionaries, one may observe that these latter never took the care to assimilate the warning pronounced in the first page of Guevara's book (Guevara 1967: 27f): that if a government is the outcome of some kind of democratic consultation, even if fraudulent, and upkeeps at least some pretense of constitutional legitimacy, then the possibilities of civic action have not been exhausted, and armed rebellion is doomed to fail. In other words, as long as the possibility of basing politics on dialogue is not fully strangled, emancipatory politics cannot be achieved by the methods of the advertisement industry.

however, by being more generalizing and abstract. One might regard it as "Marxism minus the class concept," that is, as a retreat from the insights of the nineteenth century. However, another interpretation seems more adequate and more fruitful: namely that the abstract scheme is there in order to be filled out, and that social classes as well as all other historical realities will turn up (more concretely than as abstract class distinctions) as soon as we ask the questions of emancipatory, critical social science: "How is the situation?" and "Is this situation really *still* inescapable?"

Another problem is presented by the purported one-to-one correspondence between cognitive interests, knowledge types, and scientific domains. As it has been amply argued in Part I, Habermas's coupling of natural science with the technical interest and of the humanities with legitimization is not transhistorically valid even as an approximation. Admittedly, from the onset of Neohumanism and historicism until recently, the humanities have largely aimed at noncritical understanding, being thus practical in Habermas's sense. But for long periods a number of humanistic disciplines, in particular rhetoric, were technically oriented—aiming, it is true, at communication, but at technicalized communication where the receiver (the "target audience") is treated as an object (that is, according to the above, *as an enemy*) and not as a dialogue partner. The leading questions may be formulated thus: how should I use discourse in order to get it my way, how will I be able to make others see things the way *I* want them to be seen?[10] Post-1968 humanities, inspired among others by Habermas, on the other hand, tend to understand themselves as, and sometimes even try to be, emancipatory.[11] Since the rise of sophisticated advertisement and of the "scientific" use of the media for propaganda purposes, finally, sociotechnical humanities (similar in aim but broader in scope than classical rhetoric) have become increasingly important once again. After having retreated since the late Renaissance to the role of legitimizing existing circumstances, the humanities find themselves today precisely in Gorgias's dilemma as referred to on p. 33, n. 9.[12]

---

10. This basic orientation is not changed by the obvious development of Hellenistic and Renaissance rhetoric into a ritualized art of display; even a tournament (to mention a parallel), however much submitted to strict rules and aimed at display, remains combat.

11. Also Dilthey, in his birthday speech (cf. p. 169), had spoken of the humanities, or at least "the historical worldview," as the "liberator of the human mind from that last chain which philosophy and natural science have not yet broken." His distrust of the (apparently Social Democratic) "anarchy of opinions" was precisely due to his fear that it would prevent this liberation.

12. The humanities are not alone in having sometimes deviated from Habermas' prescriptions. Until the late Renaissance, we remember, the aim of natural knowledge was not technical at all but "theoretical," an expression of organized and systematic curiosity. In phases of "enlightenment" character (early Greek philosophy, and the "twelfth-century Renaissance"—precisely the phases where the interest in natural knowledge was most ardent) this search for

A third qualifying remark is that an absolute distinction between interaction and technique is philosophically problematic. A normal dialogue will rarely have as its only purpose to understand the dialogue partner; often, the conviction that *you yourself* are right is as important, together with the aim to *convince the other.* Understanding and participation on an equal footing are thus prerequisites for reaching that goal, *a means* for that technical interest which consists in making the other accept. Since dialogue may anyhow lead to the discovery that you are less right that originally supposed, however, interaction and technique cannot be separated through a simple distinction between "real" and counterfeit dialogue.

This objection does not mean that the three fundamental conditions for human existence—work, interaction, emancipation—are worthless as concepts. The crux is that they should be considered as *poles* with respect to which our acquisition of knowledge is oriented, not as *classifying boxes* inside one (and only one) of which each act of knowing belongs; so should that curiosity which Habermas disregards.

The main source of many of the problems associated with Habermas's scheme is that he builds his understanding of the single scientific domains on the interpretations of other philosophers (often normative interpretations, stating what ought to be the methods and aim of the sciences), and not on original investigations.

His description of the natural sciences as nothing but technology *in spe* is thus borrowed avowedly from American pragmatism, in particular from C. S. Peirce. His portrayal of the humanities is taken from Dilthey and, in particular, from Gadamer and other recent followers of the Dilthey tradition, and thus owes much to the German historicist and Neohumanist tradition. The concept of a critical social theory, finally, originates with the "Frankfurt school" of Theodor Adorno, Max Horkheimer, Herbert Marcuse, Erich Fromm and others, who also provide the nexus to Freud's psychoanalysis. Since these philosophers and philosophical schools fasten upon different aspects of the scientific activity, Habermas's eclectic approach causes him to miss the shared cognitive interest of the natural sciences and the humanities: the aspiration for world order and for comprehensive understanding of our *total* condition within the cosmos. This cognitive interest (which has something

knowledge as interesting in itself made it a contribution to the emancipatory movement. As Kant said, enlightenment is "Man's withdrawal from that state of minority in which he has placed himself"—"Ausgang des Menschen aus seiner Selbst verschuldeten Unmündigkeit" (*Was ist Aufklärung,* A 481). Even later on, natural science has been pursued not only as potential technology but also because of its role for the formation of a secular world picture. But the emancipation in question was always more indirect than Habermas envisioned it, thus inspiring that *courage to use one's own reason* which Kant asks for in the same address, and was no direct investigation of the possibility and conditions of social change.

in common both with the practical and the emancipatory interests and with the "theoretical" interest of Greek philosophy) could be labelled "ordering curiosity" or "critical world order construction."

CHAPTER 11

ANTHROPOLOGIES

The various theoretical and empirical objections which can be raised against Habermas's scheme do not prevent it from being a significant contribution to the understanding of the roles of the sciences, not least of the different possible roles of the humanities. Apart from this direct importance it is, moreover, illustrative of an aspect of the fundaments of the humanities (and, to some extent, of all sciences *qua* human activity): it builds on *an anthropology,* that is, a notion of what human beings are or should be, and a view of the fundamental conditions of human existence.

> It is to be observed that the term *anthropology* is used as when we speak about "philosophical anthropology"—a field that differs from and has often preciously little to do with "cultural" or "social anthropology" (≈ethnology), and which shares nothing but the word with "physical anthropology" (a discipline which investigates the variations of the human body, for example the geographical and temporal distribution of skeleton types, blood types, or genes). It is true that the empirical discipline of cultural or social anthropology presents serious challenges to many dogmatic philosophical anthropologies—but it is no less true that philosophical anthropologies too often do not accept the challenges offered by inconvenient empirical evidence.

A philosophy of the humanities as sciences dealing with the specifically human aspects of human life (as distinct from the medical aspects or the fact that our bodies are subject to the law of gravitation) must evidently contain or presuppose at least an implicit anthropology, not only because its object (like the object of any philosophy of science) is a human practice, but also because the *object of its object*—that is, the object investigated by the humanities—is human practice.

Anthropologies may be explicit instead of being tacitly presupposed. As the name *philosophical anthropology* suggests, the formulation of explicit anthropologies is an old preoccupation of philosophers, and discussions of human nature and of the nature of human society was always a fundamental ingredient in moral and political philosophy. There is, however, no reason to present established philosophical anthropologies in the context of a general discussion of the humanities. More appropriate is a discussion of the main types of (often tacit) anthropology that underlie actual scientific approaches, since it may sharpen the sensitivity to the anthropological presuppositions one makes in one's own work. Inasmuch as they remain tacit, such anthropologies obviously do not qualify as "philosophical," since philosophy, if anything, is concerned with explicitation and ensuing analysis of presupposed notions. *Examination* of implicit anthropologies, on the other hand, falls under philosophy in this sense.

A handy first division to be used in the following may be expressed in a pun, as the dichotomy between *created man* and *creative man. Created man* is the approach of the anthropologies that undertake to describe man and human action as products of some external factor. All anthropologies of this kind contradict a view of the humanities as concerned with understanding and not with explaining (the trivial reading of Dilthey). However, much human science, and even much of that which is done within the humanities in the restricted sense of Dilthey (see p. 168, n. 3) does attempt to explain; so does Dilthey himself (cf. p. 169, n. 5 with surrounding text). Moreover, many attacks on the humanities because of their "unscientific" concern with understanding and with nonmeasurable human intentions base themselves on one or the other of these deterministic anthropologies. For both reasons they have to be discussed.

The "external factors" to which the created-man anthropologies recur may refer to very different instances:

— Man may be seen as a product of his body, considered either as a physiological mechanism or as a set of biologically determined possibilities, drives, or instincts ("biologism").
— Instead, people may be viewed as products of their environment, that is, of their childhood experience, conditioned reflexes, or their social and geographical environment ("sociologism").
— Finally, single individuals may be regarded as pawns in a larger game which surpasses their understanding, and where no choice is left outside the role assigned to them ("structural determinism").

> Biologism has a tendency to dismiss states of the mind as causes of human action: "the brain secretes thought as a working body secretes sweat," in a nineteenth-century aphorism; sweat, as we all know, is a by-product and does not in itself effectuate any work. The same tendency prevails within behaviorism (to be presented below; aphoristically, "we do not run away because we are scared; we get scared because we run away"). It is not true to the same extent for sociologism and structural determinism. They accept that my actions result (or result in part) from my conscious decision. But my consciousness is determined by my past experience, my passions, and so on, and these, on their part, are determined from

> the outside by my social environment, the structure of my language, my societal conditions, my class situation, or other factors.
>
> Some brands of determinism suppose that every single move we undertake is determined, as the movements of a mechanical device (say, a clockwork or a car) are determined by its inner structure and by the external influences to which it is subjected (whether the clock is wound, or whether the car is provided with gas and somebody activates the accelerator).[1] Others are more modest in scope and only look at the average of our actions, or at the general patterns which govern people's actions in a given society. In rough approximation, total determinism is favored by those approaches which reject mental states as possible causal agents.

Creative man is the stance of those anthropologies which concentrate on man as the possessor of free will, and which see the specifically human in human action as the ability to transcend external determination. More on this below.

For clarity's sake, it will be convenient to discuss many of the anthropologies in a crude or "vulgar" version, where their distinctive features stand out more neatly. The conspicuous weaknesses of the vulgar versions certainly need not characterize the corresponding sophisticated theories—but the weaknesses of the vulgar versions are only too close at hand when anthropologies are presupposed in work dealing with other matters; sophistication (like other external adornments) is easily worn off in use.

---

1. Twentieth-century physics and information theory have rejected total predictability even in the case of physical mechanisms. Nobody will probably believe human beings to be more strictly determined than clocks, but on the other hand, most of us believe that the working and wearing of the clock is determined for all practical purposes, irrespective of the niceties introduced by quantum physics. The objections to determinism in general as formulated by modern physics are probably not the ones which first come to mind when human action is discussed.

CHAPTER 12

Theories of Created Man

DETERMINATION BY THE BODY

This type of anthropology is always an image of the technical and biological knowledge of its time. It came up for the first time in the seventeenth century, when the astounding progress of mechanical technology inspired Descartes' conception of men (and animals) as clockworks, as automata. To Descartes and his contemporaries, it had been obvious that the human automaton would have to contain a nonmaterial soul, whose relations to the mechanism was an unsolved riddle. La Mettrie and other eighteenth-century materialists, as we remember, solved the puzzle by dispensing with the hypothetical "ghost in the machine," but ran into other problems—what does it mean that an automaton is happy, and why should it be?—which they would solve by presupposing that matter itself carried the potentiality of consciousness.

The seventeenth-century mechanical understanding of life was a postulate with little empirical support beyond the newly acquired understanding of the bone-muscle mechanics, the pipeline system of arteries and veins, and the character of the digestive apparatus as a chemical plant. It was a postulate that *the functions of life had to be explained that way*—at best it was a "research program." The nineteenth-century advances in physiology demonstrated the clockwork model to be much too naïve. At the same time, however, they seemed to promise explanations of the same kind but on a higher level—and the advent of the electronic computer made still another model available, which for a while could be believed to contain the principle of the human mind. For a while, man was claimed to be a fully automatized plant governed by a central computer.

Even the computer model was a postulate without genuine empirical foundation, no less naïve as an anthropology than the clockwork model. It should not be forgotten, however, that both have been fruitful: firstly as research programs whose carrying capacity was tested; secondly and in particular as challenges, namely through the questions "through what is man *more* than a physiological clockwork?" and "what is the *difference* between the human mind and a computer?"[1]

More important in certain quarters of contemporary human science than this physiological mechanicism is the conception of man as identical with his *biology.* In one variant this notion is the fundament of the psychiatrical current that considers psychical disease an organic illness affecting the biochemistry of the brain and thus the mental state of the patient (for which reason the best cure is also chemical). Directly, of course, this regards only psychical disease—but indirectly it implies that all mental states result from chemical causation. This current thus identifies biology with physiology, nowadays mostly by searching for the connection between particular chromosome defects and specific mental diseases.

In contrast to the classical clockwork and computer models, the medico-psychiatric view is thus built on the most advanced biological research of the day and not on more or less gratuitous postulates. Seen in the context of earlier physiological determinisms, it can be claimed to fulfil some of the promises made by La Mettrie concerning genuine treatment of psychiatric disease.[2] Postulates creep in, however, when it is concluded that behavior and intelligence must be just as directly correlated with genetics as for example Down's syndrom, and when it is overlooked that even a schizophrenia is likely to be a reaction of the mind to a particular neurophysiological state of the brain that affects its way of structuring impressions from the outer world.

---

1. Many of the advances which took place in linguistics in the late 1950s can be seen in part as reactions to the failure of the naïve computer model, more specifically to the inadequacy of early attempts at computer translation from Russian into English, which demonstrated language to possess an unsuspected structural dynamic. As reported by Noam Chomsky (1979: 125–127), the originator of transformational grammar: "All these theories [inspired by the mathematical theory of communication, information theory, and the theory of automata] left me very skeptical. . . . I wanted to prove that these models were not appropriate for the grammar of natural language." In part the new developments also interacted directly with the computer model: "generative grammar" is strikingly similar to the way computer languages are defined.

A computer model for language is not in itself the same as a computer model for human behavior. Yet if language, a fundamental tool for, and a constitutive element of, behaviour, cannot be expressed in computer terms, nor can behavior in general.

2. What is said here concerning the medico-psychiatric view regards the advances made in some branches of psychiatric medication during the last ten to twenty years. Until then, philosophical inferences of the kind which are described here were no less postulates than Descartes' automata. Pills were used simply because they worked, and without any insight in their biochemistry (that is, they worked better than no pills, in the perspective of the therapist). Justifications of the method were either craftsmanlike pragmatic or vulgarly positivistic, and not analytical. For many chemical cures, they still have to be.

Quite different in character are two biologistic levels of psychoanalysis: on one hand Freud's theoretical superstructure, and on the other popularized "vulgar psychoanalysis." They do not build on actual results of biological science but on their own, prescientific concept of human biological nature as a lump of "drives": firstly the "sexual drive" (in the case of which, however, this characterization as "prescientific" only fits the "vulgar" version, cf. imminently), and secondly (in the late version of Freud's metapsychology) the "death drive."

The term itself is telling. A "drive" is something to which you feel pushed, in a way which makes it difficult or impossible to desist from doing it. It thus designates a subjective feeling; when used as an objectivistic description it belongs on the level of Comte's "metaphysical" explanations, together with the "sleep-provoking force" of opium made famous by Molière. This is at least the character of the "death drive." Even Freud's "sexual drive," it is true, is borrowed from traditional prescientific parlance. But this traditional term is transformed and given a precise meaning as a description and recapitulation of psychoanalytical empirical evidence (in a way similar to the transformation of common sense and "metaphysical" concepts like "force" and "energy" in classical physics). Experience with neurotics, in particular, who were not able to desist by simple decision (for instance, from coughing without any physiological reason), called for a scientific conceptualization of the causation mechanism that under specific circumstances would give rise to this neurotic pattern of behaviour. The "death drive," on the other hand, was appealed to as a universal key to the incomprehensibly dreadful: the mass slaughter of the World War, perpetrated and accepted by otherwise peaceful and seemingly reasonable people. The explanation of these horrors as resulting from an irreducible "drive" is nothing but the pessimistic observation that "apparently our fellow beings are unable to desist from doing it."

It should be remembered that Freud himself only uses this (pseudo)-naturalism or -biologism as an explanation of human behavior up to a certain point (in contrast to much "vulgar" or "newspaper Freudianism"). The very aim of psychoanalysis is, indeed, to give the patient (and, *a fortiori,* the nonpatient majority) so much understanding of himself that he is able to master his drives and their expression by means of reason.

Somewhat similar in character, but not supported by empirical evidence comparable to Freud's analyses, are the so-called "aggression theories" which in one or the other variant have circulated since the 1960s. They play a lesser role than psychoanalytical anthropologies in the classical humanities—for example in the interpretation of literature, where variants of Freudianism have sometimes taken over the once primordial role of *race, milieu,* and *moment* as the inevitable starting point; but they are of some importance because of their distinction within miscellaneous grandiose speculations about human society and culture past, present, and future. Making use of superficial observations of the behavior of select animal species and of superficial Darwinism, they claim that man is, because of his evolutionary past, nothing but an aggressive animal.

A favorite "empirical" argument for this principle is that sexuality need not be an irreducible "drive" but may express aggression and submission. This is a nice argument against vulgar-Freudian pan-sexualism, but is not relevant as an argument in favor of anything. It should be well known to anybody willing to open his eyes that sexuality can serve as a symbol with many functions: for example, religious—just read Pietistic religious poetry; or social—the classical myth of revolution, as exemplified in the tale of the abolition of kingship in ancient Rome, speaks of the sexual offenses of the rulers and of the just revenge of the offended.

The pan-sexualist, of course, may claim that this is just a proof that religion and social protest are nothing but (misdirected) expressions of sexuality. Similarly, the aggressionist may claim that amorous behavior and smiles are nothing but veiled aggression. In both cases the arguments are wrong. As a rule, evolution (as exemplified for instance by biological evolution) makes use of the Biblical principle of "pouring new wine into old bottles"—bottles which then in the long run are transformed so as to fit the new content better. Our auditory canal *is* no gill slit; our lung *is* no swimming bladder; and so on. The auditory canal and the lung bear witness of an evolutionary history, no doubt about that; smiling does so, too, as obvious in its relation to certain ways of grinning. But this is quite different from identity.

Since the scientific foundation for aggression theories is meagre and highly ambiguous, it seems reasonable to explain them sociologically. Like that social Darwinism which they perpetuate (cf. above, p. 161), they naturalize a specific historical situation: that of the mature capitalist epoch, where competition is no longer believed to be conducive to the equilibrium between all legitimate interests (as could still be assumed when Adam Smith wrote his *Wealth of Nations*) but as a system where those who survive and breed do so because they are morally entitled to do so *qua* the best (*viz.* best to survive and breed). In an era of militarization, the aggression theories explain that we should accept prevailing conditions and policies on the premise that "war and preparation for war are not evil, just natural, and so are social and sexual inequality."

The sociological explanation is most relevant if one considers a popular writer like Desmond Morris, probably the best-known early exponent of the aggression theorists because of his bestseller *The Naked Ape* from 1967. The Austrian ethologist Konrad Lorenz argues to a much larger extent from his own research in animal behavior. He is thus not only a popularizer on a higher scientific (and stylistic!) level than Desmond Morris, but is also to be taken more seriously as a scientist. Even *his* theories, however, must be understood as political arguments depending on their historical context—as should his reasoning in an oft-cited but rarely read large paper from 1943 that the eugenicist program should to a larger extent take *physical beauty* and not just *race* into consideration. Evolution, according to the gist of the article, has selected an aesthetic feeling in us which gives the most healthy biological inheritance a higher transmission probability by making us find their carriers more sexually attractive than the carriers of less healthy genes. In other words, granted the year and the country (occupied Poland) *inside which,* and the idiom *in which* the article is written: the *ugly,* not necessarily Jews and Poles, should go to Auschwitz.[3] Socrates, Lorenz points out quite explicitly through his choice of pictorial representations of "degeneracy," would have had to go. Physical beauty is a token of valuable inheritance.

Related to aggression theories but seemingly more firmly anchored in biological science is the more recent field of *sociobiology* (Edward O. Wilson and others[4]). The basis of the argument is supplied by the social insects: termites, bees, and ants. To a larger extent than the individuals of nonsocial insect species, a working bee or ant shares its genes with its companions and, in particular, with the queen which takes care of reproducing the group. When the genes of the ant make it behave altruistically, for example by extinguishing a fire in the ant hill by throwing its own body into it, the chances *of the genes* to survive are augmented, since they are also present in the other ants of the hill and in particular in the queen. The apparently altruistic action on the level of individuals is thus reduced to an egoistic action on the supposed real level of evolution, the level of the genes.[5] Ants, as a consequence, are nothing but the means by which ant genes reproduce (in the well-known manner in which human beings are nothing but the means by which cars proliferate).

Obviously, similar arguments of shared genes do not fit the social behavior of human beings, who share as many genes with their siblings as do other mammals

---

3. See Lorenz 1943, *passim* (pp. 397–401 of the summary provide convenient clues), cf. also Kalikow 1980: 204–206. It is a noteworthy challenge to all facile identifications of moral and political acceptability with scientific importance that precisely Lorenz's 1943 paper has become the starting point for an influential line of philosophical thought ("evolutionary epistemology") quite devoid of racist implications. On the other hand, one may observe that the basic points had already been made, without Lorenz's dubious implications, by both Friedrich Engels and C. S. Peirce—and then wonder why it is always Lorenz that is referred to as the founding father (the theory of inheritance to which he refers is that of the early nineteenth century, not the gene theory that was already well established in 1943).

4. First presented in Wilson 1975; cf. Montagu 1980.

5. In justice, it should be emphasized that the theory implies (and Wilson points out explicitly) that egoistic behavior at the level of individuals is just as much in need of genetic explanation as is altruistic action. Individuals are no more intrinsically egoistic than intrinsically altruistic. Whether one strategy or the other is most advantageous for the survival of a gene depends on concrete conditions.

But as so often in this domain since reasons of publicity made Darwin accept the meaningless "survival-of-the-fittest" slogan of social Darwinism, such fine points are lost in the television versions of the theory.

whose patterns of social behavior differ widely from theirs. Explaining human social behavior, altruistic as well as egoistic, requires analogies with other species (in particular chimpanzees). It is claimed that the leading male monopolizes access to the females of the band, at least during the period when they are fertile; in this way *his genes* ensure their optimal chance of survival. The genes of the female, on the other hand, are best protected if she ensures the protection of a strong male for her offspring. *Therefore,* the argument goes, human males ought to be polygamous, and human females monogamous.[6]

Precisely the reference to chimpanzee behavior, however, demonstrates the weakness of the program; the image of chimpanzee life which serves the argument is outdated. It was invented at a time when primatologists thrust a duplicate of their own ideal human society on the chimpanzees, looking only for sexually monopolizing dominant males and compliant females. Since then (not least since woman primatologists invaded the field!) females have appeared to possess most of the initiative as regards the choice of partners, and the sexual monopoly of dominant males has turned out to be a case of projected wishful thinking.[7] Like many similar

---

6. Evidently, this will not fit arithmetically, the numbers of males and females being roughly equal; but this paradox can then serve as moral justification for economic inequality: the strong (males) should be rich and thus able to buy themselves a harem or mistresses; the poor (males), on the other hand, unable to afford either, will not transmit their second-rate genes to future generations.

It was the constant nightmare of nineteenth-century eugenicists that their own class of people, educated, knowing about and having access to contraception and using it actively, tended to get fewer children than poor and supposedly less worthy people. Today, when easy access to contraception has come to depend quite as much on geography as on social class, the nightmare of their heirs has changed its appearance but not its contents: there are too many Non-White Anglo Saxon Protestants on this earth, and their proportion is growing!

7. Cf. for instance Thuillier 1985 and de Waal 1982. The link between the perspective and the gender (or other kinds of extra-scientific life experience) of researchers is, of course, not restricted to primatology. An eminent example is offered by an attack made a decade ago by the anthropologist Derek Freeman on Margaret Mead's work on the sexual behavior of young girls in Samoa. The young Margaret Mead had interviewed adolescent girls from outlying islands about what they actually did (and may occasionally have been told what they day-dreamed of doing). Freeman interviewed *his* peers—middle-aged males of high social standing (moreover, from the colonial center and committed to the "Christian" ideals of the colonizers)—about what young girls should do, and got wholly different answers (Rappaport 1986: 330 and passim).

Evidently, both the actual behavior and desires of young girls and the moral expectations of the aged elite are constituents of a culture, and neither of the two approaches is wrong—none of us can ask all questions at one time. But your particular situation as an individual research worker—who you are, what problems come to your mind immediately, with whom you communicate most easily—opens your eyes to certain questions and closes them to others.

undertakings, sociobiology has bought its all-encompassing synthesis at the cost of irrelevance with regard to much of that world which it wants to explain.

ENVIRONMENTAL DETERMINATION

That environment which is supposed by environmental determinisms to govern our development and our character is sometimes understood as the more or less accidental sum of events making up individual experience ("type I" in what follows); sometimes it refers to that culture or global social structure which each of us shares with many fellow beings ("type II"; "sociologism" in the strict sense); it may also be understood as that sum of planned experiences and events which make up an education ("type Ia," a specification of type I). Since the moment when the Enlightenment openly rejected the acceptance of everything *existing here and now* as *natural* and hence inescapable and timelessly valid (a tacit tendency no less inherent in Scholasticism and Aristotelianism than in sloppy common-sense thought), type Ia has been the supposition underlying much *socialization theory:* the child is taken to be an infinitely malleable lump of clay in the hands of the educator. Type II, on the other hand, is implied in Montesquieu's assertion that a Christian child placed in a Turkish cradle will grow up with a Turkish conscience (cf. p. 139) as well as in his climatic semideterminism. Somewhere between types I and II we find Holberg's claim that in a country where women force their husbands to stay at home like housewives, the males will be prone to gossiping and in possession of all the vices stereotypically ascribed to females in *his* world.

A common variant of type I from our own century is offered by vulgar psychoanalysis (often in obscure mixing with the "drive variant" of biologism): everything we do results from childhood experience and childhood traumas, usually in ways we do not recognize; the reasons we may give for our actions are nothing but *rationalizations,* cover-up stories by which we hide from ourselves and others their real, inadmissible cause.[8]

---

8. Evidently, vulgar psychoanalysis, like other "vulgar" versions of grand theories, deserves its name by taking over prominent features from the original theory. 'Rationalization,' for instance, is Freud's concept and represents an important insight; much of Freud's analytical work aimed precisely at finding the childhood experience which had patterned the psyche of his patients and made them act in ways they did not understand properly and therefore explained by such "rationalizations." But vulgar psychoanalysis is vulgar not only because it presents these features in cruder versions—this happen even in the best popularization of a theory; more important is that a few constituents of a complex and potentially dialectical network are picked out as absolutely valid, with the indirect implication that everything a neurotic (and in the next step everybody) does is neurotic. This is why vulgar psychoanalysis can be classified as a determinism (or a bastard breed of two mutually exclusive determinisms), while the original theory cannot.

In the idiom which was proposed above (p. 158), vulgar theories represent "popular," not "popularized" science.

Of greater theoretical and philosophical interest is *behaviorism.* Originally, this was simply a psychological research program of positivist coloring: which *"positive" empirical facts* are accessible to the psychologist? Not states of mind, since they are not objectively measurable, and since the statements of people concerning their feelings and thoughts are not reliable. The only thing we can observe is their behavior (John B. Watson, 1913). A supplement to this principle of epistemological sobriety came from Ivan P. Pavlov's experiments on conditioned reflexes (starting around 1900 and continued for three decades): a dog begins salivating when presented with food. If food is repeatedly offered while a bell rings, the dog will start salivating at the sound of the bell—even if no food is actually offered. And if a lamp is then turned on each time the bell rings, the lamp alone may be sufficient to provoke salivation, although it was never accompanied by food.

The idea that the Pavlovian principle could apply to learning was in the air around 1930 and constitutes one of the key themes in Aldous Huxley's *Brave New World* from 1932. Genuine scientific work in this direction took its beginnings with B. F. Skinner's experiments around 1950 (preceded by his World War II attempts to train pigeons as pilots of missiles and torpedoes—apparently a success as far as the pigeons were concerned but a failure for research-organizational reasons, see Capshew 1993). As Skinner showed, a rat in a cage that receives food when it happens to step accidentally on a pedal will end up learning to step on the pedal when it is hungry.

So far, only sound methodological reflections and experiments were involved. However, Pavlov's and Skinner's results soon led to the familiar "positivist fallacy," which can be summed up in two points. Firstly, that which cannot be investigated by means of "objective" methods *does not exist.* (Of course, this statement may simply be understood as a definition of "existence," in which case it can neither be confirmed nor disproved but only deemed appropriate or irrelevant.) Secondly, valid scientific explanations *can* be constructed which refer exclusively to positively "existing" entities. (This may, of course, be true, but it is still a fallacy in the sense that it does not follow from the premises).[9]

A further consequence has been Skinner's "social engineering" as described in his book *Beyond Freedom and Dignity* from 1971 (Skinner 1973). The first step in the argument deals with the nature of learning: all learning is assumed to follow the pattern suggested by the rat that learns to use the pedal in the "Skinner box"—that is, to be a sum of conditioned reflexes. We learn to write "4" when we see "2+2"

---

9. The treatment given to color by early seventeenth-century science may exemplify the fallacious character of the arguments. Galileo regarded color as a secondary—that is, purely subjective—quality, which science could not be concerned with; Descartes tried to explain it by mean of what he considered "existing" entities, namely from the rotation of "light particles" (cf. p. 114, n. 9). The latter approach brought no new insights, but later physical science has demonstrated that Galileo's epistemological pessimism was exaggerated: color *can be* treated scientifically, but not in terms of what "existed" for seventeenth-century science.

simply because the teacher gave us "positive feedback" (praise) when we happened to give this answer. All talk of "understanding" or "truth" in this (and any other) context is nonsense. From this follows that we cannot learn to behave peacefully and decently toward each other through "understanding." Clever experts (the psychologists) will have to construct a program for the training of other people (and themselves, for that matter) in globally adequate behavior. This implies that we should abandon the illusion of human freedom, which will be felt by many to hurt human dignity, but since the survival of the human race is at stake (and highly threatened), we cannot afford freedom and dignity—whence the title of the book.

The argumentation may seem attractive to intellectual desperados. None the less it is highly deceptive for philosophical as well as empirical reasons. Firstly, it follows from the same argument that Skinner cannot have written the book because it "is true," but only because he has been conditioned to do so (as he also states himself). In school he got praise, a piece of chocolate, or a scrap of Jesus when answering as required by the teacher; at university he got degrees, appointments, and higher salary when writing publishable books. *Beyond Freedom and Dignity* is thus simply a continuation of a behavioral pattern which was once *adequate for its author as a person* (but does not need to be so any longer—the dog's salivation continues even if the bell stops being accompanied by food). But *personal* adequacy of book-writing has not the least thing to do with the question whether the advice dispensed by a particular book is *globally* adequate.

Secondly, it is questionable whether much of human behavior can be described appropriately in analogy with the behavior of rats in a cage, and whether our complex activities can really be described as composite conditioned reflexes. "Human beings are not rats," as the objection has often been formulated.[10]

Thirdly, according to more careful empirical investigations of animal behavior, even rats are not (Skinner) rats. Skinner rats only learn to use the pedal after maybe fifty accidental releases of the pedal. But a colony of rats finding a new type of food only needs one of them to become ill a single time before they all keep away from the nourishment in question (Vloebergh 1984). At least for rats, Skinner learning is a very poor and ineffective simulation of natural learning. In other animal species, furthermore, Skinner learning has been shown to work only in connection with the instinctive patterns that are already present: doves can learn to use their beak on a colored dot, but not to step on a pedal (Gould & Marler 1987). This makes all attempts to use Skinner learning as a model for the highly flexible learning of humans

10. Neurophysiological results from the last three to four decades may be more convincing than such slogans. It turns out, indeed, that conditioned learning and learning involving conscious awareness make use of different mechanisms and different brain centres (Kandel & Hawkins 1992: 53f). Huxley the novelist, when claiming that only subconscious attitudes and not factual knowledge can be inculcated by means of Pavlovian methods, had better foresight than Skinner the psychologist.

even more dubious. Yet if behaviorist planning does not work, nor can behaviorist explanations tell as much about our socialization and the shaping of our actions and conduct as claimed by their proponents—if type Ia behaviorism is radically insufficient, then type I must also be.

SOCIOLOGISMS

Let us turn to type II, the genuine sociologisms, those theories of environmental determination which see the total constitution of the external world and not the sum of random individual experiences as determinants of human consciousness and behavioral patterns. Normally, such theories are not as much interested in predicting the behavior of single individuals as in the broad average of individual thought and actions that determines the flow of historical and social processes.

One current of this type is *vulgar Marxism.* In its most simplified form, it sees history as a sequence of stages: first comes primitive communism (the "egalitarian society" of anthropologists), to be followed by slave holders' society (at times, an "Asiatic mode of production" is inserted between the two). Slave holders' society is replaced by feudalism, itself to yield to capitalism. In the end, capitalism will have to capitulate, and socialism—to unfold as mature communism—will succeed. Within each stage, culture and consciousness are determined in full by the individual's position within the "mode of production," that is, the framework within which the exploitation of the producing class by the upper class is organized (whether the producers are owned by slave holders, personally bound to them as serfs, or free but forced by economic necessity to sell their work power to capitalists).

Two concepts which have been much used by historians of literature and mentality during the last two decades demonstrate that this conceptual straitjacket has been widely accepted as common sense: 'noncontemporaneity' and 'everyday consciousness' (*Alltagsbewußtsein*). If authors writing in the same moment of history differ in their understanding of or attitude to social life in a way which cannot be reduced to differences in class position, then they are considered to be "not really contemporaneous"; and if the working class looks at commercial TV instead of brandishing its class consciousness as (once) anticipated by academic "revolutionaries," then class consciousness must have been supplanted by the "consciousness of everyday." The two concepts allow scholars to eschew the suffocating effects of the straitjacket and to come to grips with the complexities of real mentality and history; but their wording shows clearly that the vulgar Marxist theory of consciousness is a background expectation which the users of the terms have to elude.

An oft-quoted maxim in the quarters of vulgar Marxism is an abbreviated version of Marx's sixth Thesis on Feuerbach, claiming that "man is the sum of his social relations." The full formulation is rather different in tenor and states that ". . . the essence of man is not an abstraction inherent in each particular individual. The real nature of man is the totality of social relations . . . " (MEW 3, 6 & 534). The thesis is thus a polemic against all attempts to define MAN metaphysically by ascribing to the species an essence carried by each individual as an Aristotelian form (and moreover, as Engels reveals, merely a notice for further elaboration which was rapidly penned and certainly not meant for publication—MEW 21, 264). From Marx's hand it has nothing to do with the assertion of complete sociological determinism. As stated in the third Feuerbach Thesis, "the materialist doctrine that men are products of circumstances and education, and that changed men are thus the products of other circumstances and another education, forgets that circumstances are changed by men and that the educator must himself be educated" (MEW 3, 533f, trans. Bottomore & Rubel 1963: 83). Other writings from Marx's and Engels's hands (not least their letters) make it clear that they did not accept the vulgar sociologism attributed to them by eager friends as well as ill-willed foes.[11] Vulgar Marxism relates to Marx more or less as vulgar psychoanalysis relates to Freud.

Related to vulgar Marxism, and occasionally derived from it, are various other "vulgar materialisms" (cf. the above quotation from the third Feuerbach Thesis, which attacks precisely such theories). One, classical type (going back to Montesquieu, cf. p. 131) is *geographical determinism.* Certain climatic conditions (in most varieties the temperate European climate) are supposed to force society to develop technology and civilization, whereas others make it superfluous (the bountiful tropics) or impossible (the arctic zone).

A particular Cold War variety of geographical determinism—much more deterministic than Montesquieu—is the "Wittfogel thesis," formulated in Karl Wittfogel (1957). The great "Oriental" civilizations had developed around the great rivers (the Nile, the Euphrates, the Indus, the Ganges, the Mekong, the Hwang-Ho, and the Yangtze). "Evidently," the purpose of civilization and of state formation was the management of irrigation by a centralized despotic power—and "evidently" the Oriental state (firstly the Soviet Union, secondly the Peoples Republic of China) was still despotic, and the Free West was both entitled and obliged to fight it.[12]

WEBERIAN SOCIOLOGY: AN EXAMPLE

The "vulgar" deterministic theories can be contrasted with another famous thesis: the *Weber thesis* on the relation between capitalist development and Calvinist

11. In older age, Marx refused emphatically to be labelled a "Marxist," since this term was already used by vulgar-deterministic followers.

12. I shall waste no more space on Wittfogel's version of ancient history and vulgar Marxism—but see my 1994: 47f for the reasons why Wittfogel is wrong on all factual accounts in the case of Southern Mesopotamia (that region of the world where his ideas could *a priori* be expected to fit best).

Protestantism.[13] This example is important because of its demonstration that sociological explanations need not be one-dimensional nor assertions of automatic determinism. In very rough outline it runs as follows.

In a number of European centers, not least in Italy, England, and the French Atlantic cities, the early sixteenth century had produced the beginnings of capitalist development. Religion, however, was an obstacle on the full unfolding of a capitalist economy, because the head of a business house would normally bequeath an appreciable part of his capital to the church or the poor; a large-scale merchant might have good reasons to fear his fate in the next world, according to a religion that had proscribed usury and tended to equate profits on trade with this nasty crime. A new creed which disapproved of such extravagant habits—as Calvinism did—was thus psychologically attractive, which made many accept it precisely in the proto-capitalist environment.

An essential theological theme in Calvinism is the doctrine of predestination: even before creating the world, God has decided who is to be saved and who is to be condemned, and we can do nothing to make him revise his decision. This, of course, is psychologically almost unbearable: I dedicate my life to Him, I renounce this and that in which the godless indulge; yet eternal felicity may be promised *to them,* while *I* may end up in the cosmic basement as firewood in the stove! Within a single generation a "folk level" was therefore added to the doctrine: the belief in *signs.* We cannot *know* who are chosen, but signs have been given to us which permit us to make a reasoned opinion about our future prospects: salvation will fall to those to whom it has been granted to live virtuously down here and who have success in their secular trade. A life in dissipation and sin and failure in business, on the other hand, are portents of future *definitive* failure.

To the businessman, this folk theology was a powerful psychological incentive to be thrifty in his private life, to avoid squandering his means, and to invest what could be saved so as to make sure that his business would prosper and ensure him of his predetermined salvation. To the germs of capitalist *economic* structure, the conscious and unconscious choices of human agents (the Calvinist creed and the spontaneous invention of the lore of signs, respectively) had thus added a "capitalist spirit" in harmony with the potentialities of the economic structure.

But even in many geographical regions where (for one or another reason) Calvinism did not succeed, the capitalist spirit developed, often through adaption and reinterpretation of prevailing religious doctrines. For instance, Weber is able to point to specific varieties of Lutheran Pietism which served the purpose—even though neither Lutheranism nor Pietism seems directly fit for that.

Finally, after some 200 years, the Calvinist scaffold had become superfluous, and a generation which had grown up together with an aggressive and enterprising

---

13. Formulated in 1904–1906 (Weber 1984).

capitalism was able to formulate an ideological basis for the capitalist spirit which was independent of religion (Weber quotes Benjamin Franklin).

In contradistinction to vulgar determinisms (including, by the way, a "vulgar Weberism" ascribing to Weber the simplistic point of view that *Protestantism produced capitalism*), Weber thus investigates the interplay between several factors and levels; he describes this interplay as the result of human choices, conscious as well as unconscious; and he demonstrates how these choices influence the development of "given" circumstances, frequently in a way which was neither intended nor foreseeable. The same complex stance, as we have seen, was also Marx's position—as illustrated for example by that third Thesis on Feuerbach which was quoted above.

STRUCTURALISMS

Particular formulations of the thesis of determination through the global pattern of the external world, theoretically more rigorous than vulgar materialism, are offered by various structuralisms. The principles of structuralist thinking can be illustrated by means of two simple examples taken from totally different domains.

All well-formed Danish main clauses are arranged according to the same underlying scheme: "(■) | V S A ; v s a." (■) is the "fundamental position" to which one of the other sentence members is moved unless the sentence is interrogative or imperative. *S* and *s* are nominal sentence members, *V* and *v* are verbal members, and *A* and *a* are adverbial members. One or more places may be empty or doubled. For instance:

> Jeg (*I*)$_{■=S}$ går (*go*)$_V$ nu (*now*)$_A$.
> Har (*has*)$_V$ han (*he*)$_S$ allerede (*already*)$_A$ kysset (*kissed*)$_v$ sin mor (*his mother*)$_n$ på kinden (*on the cheek*)$_{a1}$ i morges (*this morning*)$_{a2}$?
> "Vil (*will*)$_V$ løven (*the lion*)$_S$ ikke (*not*)$_A$ spise (*eat*)$_v$ sin dyrepasser (*its keeper*)$_n$, nu hvor han har været så uforsigtig at gå ind til den (*now he has been imprudent enough to enter its cage*)$_a$.

Irrespective of the meaning of the clause, its members will have to fit a sequence of fixed places.

In his discussion of the English Factory Acts in *Das Kapital* (vol. I, IV.13.9; MEW 23, 515), Marx tells about two capitalists, Mr. Cooksley and Mr. Simpson, who appealed to the Children's Employment Commission for the introduction of compulsory protection of working children. On their own initiative they had introduced certain protective measures; thereby, however, their competitive position had been undermined—Mr. Simpson even complained that "he always felt restless at night" because of the contracts he lost when he had closed his own shop while those of the competitors were still at work. Capitalists, like adverbial sentence members, have to obey the rules belonging to their position, or they will soon be out of business, their position being taken over by others.

Observations like these are elevated to the rank of ultimate principle in structuralism. Structuralists do not deny that *I, its keeper,* and *his mother* are different words which refer in all probability to different physical persons. But they claim that these accidental individualities cannot be made the objects of scientific investigation, which must deal with laws and regularities of general validity—the ultimate consequence of the nomothetic ideal and in fact another variant of the criterion by which Aristotle distinguishes "science" (*epistēmē*) from other types of knowledge (cf. p. 170, n. 6).

Structures can be of many kinds, as we see from our two examples. Structuralist explanations are equally diverse. The early Michel Foucault (1966), for instance, claims that each historical epoch has its own inner coherence (cf. p. 304, n. 50; the organization of medieval culture around relics as a core concept exemplifies the idea[14]). How the passage from one epoch to the next takes place is not clear at all, nor is this the problem with which Foucault's "archaeology of knowledge" is concerned; it rather looks for the demarcation which sets off one coherent conceptual structure from the other. In the same way as Taine's principle of *race, milieu,* and *moment* is an adaptation of the *Zeitgeist* and *Volksgeist* notions to the Comtean search for causes, thus Foucault's "archaeology" assimilates them to a structuralist thinking in which the sentence scheme can be claimed to be the model.

Mr. Cooksley and Mr. Simpson, on the other hand, provide the model for Louis Althusser's structuralist version of Marxism as presented, for instance, in the books *Pour Marx* (1965) and *Réponse à John Lewis* (1973): history is the history of class struggle, and individuals are bound to play the parts already written into the manuscript of the play. Althusser classifies his approach as "theoretical anti-humanism," individuals being in this perspective (and notwithstanding our private illusions) nothing but interchangeable actors, that is, passive intermediate links between the role structure of class society and the action of history's grand stage play.

Althusser does not deny the existence of development, nor does he claim that development falls outside the scope of scientific explanation. Quite the contrary, indeed, since class struggle, the inner conflict of the structure, is the motor of history. Another problem remains, however: the structure does not allow the possibility that the two English capitalists might introduce protective measures of their own, nor that they might appeal to the Parliamentary commission in order to circumvent the structural constraint on their freedom of action—clearly individual actions not written into their part of the dialogue between class representatives. The strict structural determination also makes it hard to conceive that Althusser, as a French professor of philosophy, should engage himself in the class struggle, albeit on the level of theory.

Admittedly, it is possible to read Althusser merely to the effect that *the essence of history* can be painted with the coarse brush of class struggle cleaned of personal

---

14. The example is my own and not taken from Foucault.

intervention. But if this is the message it implies that everything which goes on at the individual level remains individual, thus disappearing from view when the societal average is made. In consequence, class relations would have been no different today if Marx had become a rabbi and Engels remained an accountant, if Lenin's famous locomotive had run off the rail in Poland, if Hitler had died from his gas poisoning and Stalin from tuberculosis at the seminary.

Not all structuralist thought is dogmatic-deterministic or antihistorical like the examples just mentioned (which, however, are not the only ones of these types). Structures can also be seen as conditions which (more or less forcefully and with a more or less margin for coincidence and individual decision) drive development (including structural change) in a specific direction.

One example of this is Marx's thinking, as it was illustrated by the Cooksley-Simpson story. Another example is constituted by the linguistic structuralism of Roman Jakobson and of the "Prague circle," as summarized below, p. 319. A third important instance is Piaget's theory for cognitive development (presented in chapter 16) and the more general structuralist theory which he has formulated on that foundation.

FUNCTIONALISM

A final formulation of the social determination of the individual and its behavior (and, not least, of single societal institutions) is *functionalism,* which is primarily used in sociological explanation.[15] I shall not go into details but exemplify the approach by pointing to functionalist interpretations of two of the above examples.

One is the Weber thesis. A crude functionalist will assert that Calvinism arose *because* emerging capitalism needed a religion which encouraged capital accumulation instead of Christian charity. A less crude representative of the school will explain that the institutions of Calvinism and capitalism, by being mutually functional (that is, fulfilling each other's needs), stabilized each other and thereby that societal complex in which both participated.

---

15. Functionalism arose as a formulated approach around the turn of the century with the French sociologist Durkheim, who took the functions of social institutions to constitute those "social facts" which were to provide the fundament of his sociology. It was soon taken over by the anthropologist Bronislaw Malinowski, who emphasized the function of social institutions for the biological survival of the group. A different approach was formulated by A. R. Radcliffe-Brown, another anthropologist; according to his "structural functionalism," social institutions are to be explained by their function with respect to the *existing social order.* In later sociology the latter, "conservative" approach is probably more important than Malinowski's vaguely "progressive" idea.

The other example is represented by the British Factory Acts. Functionalists will assert that protective laws were introduced (and that our two friends insisted on their introduction) because English capitalism of the 1860s needed a working population whose health was not worn out too quickly.

None of these claims are totally mistaken. But there are at least three important problems to functionalist explanations of this simplistic character.

Firstly, they presuppose that the society whose institutions they explain by their functions is cast in one piece. Yet no society is: the needs of emerging capitalism in the French Atlantic cities were not identical with the needs of the paupers who were supported by Christian charity in the hospitals of the Church, nor with those of the French absolutist state which fought the Calvinist Huguenots the best it could (and won, thus showing in a way to *be* French society). Similarly, *some* industrial branches needed a stable and acceptably healthy working class, whereas others lived excellently with (and from!) extreme exploitation. In spite of the inveterate inclination of many functionalists to regard social institutions as virtually permanent, the real point of the functionalist explanation is, if we stick to Weber's example, that the mutual pressure and the possibility of a "positive feedback" between incipient mercantile capitalism and protestant theology happened to grind them into a complex of strong institutions, attitudes, and beliefs which supported each other; in other words, they made a social pattern appear within which certain components grew strong because they were highly functional with regard to each other; their mutual functionality thus stabilized the global pattern.

Secondly, an institution or social action is supposed to have only one consequence, which is identified as its function. In any complex structure this is obviously wrong. Any physician (and most patients) know that *effects* without *side effects* are rare.

In order to get around the latter point, the sociologist Robert Merton[16] has introduced a distinction between "manifest functions" (the "*conscious motivations* for social behavior") and "latent functions" ("its *objective consequences*"). The main advance stemming from this distinction is, however, that it highlights a recurrent fallacy in less sophisticated functionalist thinking, namely the mistaking of an intended effect (the "manifest" function) for an actual ("latent") function, and vice versa (that this *is* indeed a mistake is illustrated by the Calvinist example: Jean Calvin had certainly no intent to further the development of capitalism when stressing the predestination dogma). Only "latent functions" can be relevant for functionalist explanations.

A global conclusion to be drawn from the discussion of environmental determinism is then that all vulgar versions (vulgar Marxism, vulgar geographical determinism,

---

16. See his 1968: 73–138, in particular p. 114.

vulgar structuralism, and vulgar functionalism) are misleading and impossible already for reasons of philosophical principle.[17] But few of them are completely off the point, and the less vulgar mother theories are less so. There is nothing wrong with an explanation just because it is (for example) structuralist; the error comes in when (*in casu*) the structuralist explanation is claimed to exhaust the question.[18] One should only remember when trying to combine approaches of which none "are completely off the point" that each of them rests upon a set of underlying tacit assumptions, and be aware that these may be in mutual conflict.

---

17. Cf. what was said above (p. 203) on the self-defeating claims of Skinner's *Beyond Freedom and Dignity.* Similar objections can be raised to all absolute determinisms, environmental or otherwise.

18. Foucault, in a late interview, states that what he has said or written should not "be seen as laying any claims to totality," and that "what I don't say isn't meant to be thereby disqualified as being of no importance" (Burchell et al., eds., 1991: 73). This is apparently meant to cover his "structuralist" as well as his "post-structuralist" phases, with which the preceding pages have not dealt.

CHAPTER 13

Humanity as Freedom

Let us return once more to our two English friends, Mr. Cooksley and Mr. Simpson. They served the purpose of introducing the notion of structural determinism, that is, a fixed pattern of social roles which individuals have to fill in a manner already defined by the role in its relation to other roles. But the way in which they demonstrated the constraint of their role was by stepping a bit outside it. They *did not* identify blindly with the role but tried (however hypocritically) to act morally in spite of it. Just as much as the constraint of the structure, they can thus exemplify the principle that human beings are *free,* not completely fettered by the *already given*—by biology, nature, social roles, language. They are free in the sense that they will regularly encounter alternatives among which they have to choose. If we claim that people's actions are causally determined, we must as a minimum acknowledge that this can be true only if we allow their own consciousness (their own values, their own shaky rationality, and their wishes) to be a (co-)determinant of their actions. Whereas the various determinisms discussed above under the heading *created man* aim at "explanation," the acceptance of *human creativity* as (part of) the object of the humanities entails that "understanding" must be (part of) their approach. In order to confront the question *how* this is to be done, we may look at a philosopher—namely Jean-Paul Sartre—who in his earlier works has come to embody more than anybody else the anthropology of freedom, and who in later years struggled with the elusive connection between freedom and explanation.

THE EARLY SARTRE: FREEDOM AS AN ABSOLUTE PRINCIPLE

Sartre's early philosophy accepts at face value the freedom which becomes visible in the action of the two capitalists, and draws the full consequence of this acceptance: man is *identified* with his freedom—man is, he claims, not human through *what he is* but through *that which he chooses to be,* that is, through *that which he is not* (yet). Man is pure negation of the already given (*la facticité*), and only exists through his incessant separation from the given. *Man is* by being conscious of that which *merely is.*[1] (cf. p. 3, n. 4, on the multivalued gender of "man," also valid for French "homme".)

Man, furthermore, not only has *the possibility* to choose. He is *forced* to do so, even when he attempts—in *mauvaise foi,* dishonesty—to hide behind apparent objective necessities and nonchoice.

Choice, however, is even more. According to *L'Existentialisme est un humanisme* (originally a lecture from 1945), choice is a *moral obligation,* namely the obligation to make a choice which you would accept as a *general principle.* Any choice which you will accept to be such is, on the other hand, legitimate. The freedom to choose between diametrically opposed possibilities, moreover, is *always* present, "dans n'importe quelle circonstance," "under any circumstance whatsoever"—even in a concentration camp and when in the hands of the torturer.[2]

This moralistic and abstract interpretation of human freedom exposes some of the inherent paradoxes of Sartre's existentialist philosophy: how can that which is inescapable be an obligation? And why should moral double standards be *philosophically* illegitimate while consistent fascism (as claimed by Sartre the antifascist in 1945) is not?

To the first question, an indirect answer is given in Sartre's stage plays from the same years. They demonstrate that his central concern is *responsibility; freedom* is only secondary, but essential because it is impossible to speak of responsibility unless you have at least a theoretical possibility to evade it, that is, of choice (this primacy of responsibility echoes Kant's argument for freedom, we observe; cf. above, p. 142).

Concerning the illegitimacy of double moral standards, it will be observed that the request that choices should have general validity is nothing but a repetition of Kant's determination of the categorical imperative: a rule can only be a moral command if it is of general validity, that is, if it can be asserted irrespective of actual conditions and of the identity of the persons involved. Yet Sartre's argument is paradoxically Platonic: MAN (as "Cuphood") does not exist along with or as a model for single individuals; but then each of us *is* MAN, *is* the human counterpart of Cuphood, and we should honor that obligation by acting only in ways which fit man in general.

In other respects the philosophy is radically anti-Platonic, for which reason the argument advanced in favor of generalizability of course becomes untenable. Sartre soon discovered that, and he was highly displeased by the popularity which fell to

---

1. The main exposition of this early philosophy is the monumental *L'Être et le néant* from 1943 (Sartre 1976). The booklet *L'Existentialisme est un humanisme* (Sartre 1946) is a valuable introduction, Sartre's later disapproval notwithstanding—not least because the briefness of the argument makes the problematic points stand out clearly.

2. In 1969 Sartre commented upon this statement with the words "it is unbelievable; I really thought so."

his booklet (no doubt the most widely read of all his writings). During the later forties he wrote voluminous drafts for a treatise on the foundations of morals. However, he was unable to finish them, and they were only published in 1984, years after his death.

What Sartre formulated between 1935 and 1945 is thus an abstract philosophy of freedom. Freedom only *is* as "annihilation of that which is." Everything is formulated in abstractions. These Sartre filled out with everyday illustrative *examples,* and his plays and novels suggest *possible concrete interpretations* of the abstractions; but Sartre's illustrations and concrete interpretations do not *lead* to the abstractions.

At times they lead away from them. One example of this is the story of the young man from *L'Existentialisme est un humanisme.* He addressed Sartre in order to be told whether he should join the Resistance or stay at home to take care of his mother. Sartre refused to make his decision, thus forcing him to choose by himself. In the very last sentence of the discussion following upon the lecture, Sartre says, "By the way, I knew what he would do, which was what he did"—a clear confession that abstract freedom is an empty concept which badly describes the real process of human choice.

Beyond being abstract, the philosophy is characterized by an extreme orientation toward the isolated individual. Human consciousness is not described in its emergence through interaction with others, through the upbringing, through common action ("work") or communication. "The other" is somebody by whose "gaze" (*regard*) you discover to be fixated—a gaze which makes you see yourself as the object of a foreign subject, and which makes you objectify yourself and *feel ashamed.*

The innermost core of this partly contradictory philosophy can be approached in different ways, all of which elucidate it (though the third less than the first two), allow us to interpret its intentions, and thus permit us to use it as the starting point for further reflections.

Firstly, Sartre's ideas can be seen against their *philosophical* context. When Sartre formulated his early philosophy in 1935–1945, France was Catholic, and its bourgeois class was prudish ecclesiastical in orientation. The prevailing philosophical attitude was a kind of Christian Platonism, according to which the order of the World was guaranteed by the Platonic ideas that were present in the divine mind: *essence,* the idea of things (Cuphood), is thus prior to *existence,* the concretely real, the singular (the particular cup), that which men produce through their actual choices. This whole theory is then turned upside-down by Sartre; *existentialism* becomes *a humanism* by being a*n anti-theism.*[3]

---

3. Originally, Sartre did not use the term *existentialism;* he adopted it as it was becoming modish towards the end of the War, in order to enforce "the right" interpretation of a philosophy to which "everybody" professed allegiance.

Sartre's early philosophy is thus a (nondialectical) reversal of a heteronomous anthropology (in this case, divine determination).

Secondly, the *political context* of the formulation of the ideas may be taken into account. Sartre, indeed, was politically active, and combined philosophy with political action. An interpretation along these lines is thus in agreement with the Wittgensteinian *dictum,* "Don't ask for the meaning; ask for the use"—that is, for the actual use in that particular historical situation where the philosophy was formulated.

It should be observed that this "political" interpretation is only relevant for the formulation given by Sartre to his philosophy after the outbreak of war and after his experience as a prisoner of war. In contrast to the philosophical and "private" interpretations, this one does not touch the philosophical treatise *L'Être et le néant* (published 1943) nor the novel *La Nausée* (1938) directly—although the novel discloses a disgust for the *honnêtes gens* which suggests that Sartre knew quite well how they were going to behave during the occupation.

The political interpretation should be seen on the background of the war and the defeat to Germany, after which the vast majority of the French upper class threw itself into the arms of the Nazis. "Not choosing" meant the choice of collaboration, dishonestly disguised as an inevitable necessity. References to the pressure of society, or to the given (*la facticité*) meant to shun your responsibility by claiming to "act according to order."

To Sartre (and to many other French intellectuals), we should remember, the Resistance was no obvious continuation of an antifascist class struggle or of the Spanish civil war; nor was it carried by the "antiteutonic" chauvinism that characterized some quarters of the political Right. It was a choice of something which *according to his personal conviction was morally right* irrespective of prevailing social norms (a *completely personal* conviction rejected by many members of his social class, experience would tell him).

Thirdly, the philosophy can be interpreted with relation to *Sartre's private history,* as reflected, for example, in the childhood memoirs *Les Mots* (1964). On one hand, they demonstrate that the concentration on the isolated individual, which recurs in all phases of Sartre's philosophy, is in fact a formulation and a continuation of his chronic childhood experience. On the other hand, the equally recurrent anti-Platonism turns out to be a rejection of the young Jean-Paul's belief in *security provided by a higher meaning* of his life. The rejection of any higher meaning is indeed so violent a theme precisely in the early *La Nausée* that it suggests an origin deep in private experience.

THE ELUSIVE CONNECTION: FREEDOM VERSUS EXPLANATION

The "empty" concept of abstract liberty thus turns out to be provided with substance as soon as we see it in contrast to a background. As long as this background remained actual, the seemingly abstract concept remained the centre of Sartre's philosophy. Yet when that background was *no longer actual*—namely because the philosophical confrontation had been brought to a victorious end—Sartre's philosophical development demonstrates the need for *a new substantiation.* It was formulated in various ways, all of which include the *facticité* as an *inescapable aspect* of any choice, always in interplay with freedom.

One formulation of the new approach is found in a comprehensive, psychoanalytically grounded biography of the nineteenth-century novelist Gustave Flaubert (whose *Madame Bovary* is often considered a starting point for literary modernism), written in 1960–1971. The reliance upon psychoanalysis is remarkable—nothing could stand in more violent contrast to the previous denunciation of references to "the given," that is, to those circumstances which condition our consciousness and our actual choices, as plain dishonesty and bad excuses.

A different expression is the philosophy formulated in another large-scale work, the treatise *Critique de la raison dialectique* (Sartre 1960) and summed up in an introduction (*Questions de méthode*) first published separately (1958). The work is meant as *a critique* in the Kantian sense: an attempt to put "dialectical reason" (Marxism) on a firmer basis. The starting point is the historical materialist conception of history as a sequence of modes of production, each of which has its own characteristic philosophy. Thus Enlightenment philosophy is the "living philosophy" of early, still progressive capitalism, whereas the "living philosophy" of the epoch of mature and overripe capitalism is Marxism—the philosophy which formulates the worldview of that working class which *by capitalism* is brought into position as the carrier of a new society.

So far the point of view is a very orthodox Marxism—more programmatic perhaps than Marx and Engels would have formulated themselves. But the critique also becomes *politically* critical (in the everyday sense) by pointing out where the Stalinist interpretation becomes mortally dangerous (in a quite literal sense): namely when the group in power is so sure about the long-term perspective of history that it disregards all those particularities which do not fit its vision. If, instead of seeking the general in the particular, you take to the "intellectual terrorist practice" of "liquidating the particular" *in theory,* you end up all too easily in "la liquidation physique *des* particuliers" (physical liquidation of particular *persons*) (Sartre 1960: 28 n.1).

What remains of "existentialism" in this phase is thus primarily *anti-Platonism,* the insistence that CUPHOOD *is derivative from cups* and the perspective of history from people, not vice versa. Another reminiscence of the early philosophy is the basis in the isolated individual, not in the *individual belonging irreducibly to a social community.* Analyzing the mobilization of the Paris population July 13–14, 1789, Sartre claims that only fear of a common oppressor was able to produce common action and common struggle (namely the taking of the Bastille, with all that followed). Otherwise a plurality of people is, to this no less than to the Sartre of 1940, nothing but a *sequence of elements.*

Even this phase of Sartre's philosophy can be seen against the background of Sartre's political commitment and activity. After World War II Sartre had devoted much of his intellectual force to resisting the Cold War, cooperating among others with the French Communist Party. In the second half of the 1950s, furthermore, he was one of the most strongly outspoken intellectual opponents of the French colonial war in Algeria. His opposition to the repression of the Hungarian rise in 1956 never made him forget that what went on in Budapest was modest compared to the one million corpses produced by the French in their southern department—with the decisive difference, however, that those who were responsible for the oppression in Budapest belonged on the side where they *ought* to behave differently, whereas nobody should expect anything but ruthless brutality on the part of the colonial power.

The continued concentration on the isolated individual is probably best understood against the background of Sartre's private history. One thing, of course, is the persistence of childhood imprinting and the tenacity of personal inclinations. But Sartre's actual experience from the period was certainly not likely to change his experience of isolation: too many of his old companions from the resistance joined the ranks of the colonial army morally, or at least refused to dissociate themselves clearly (Albert Camus is one well-known example).

This interpretation through private history notwithstanding, we should remember that *nothing private is completely private.* Sartre's extremely individualistic "phenomenology," grounding the understanding of being an individual among others on the immediate experience of being observed in a moment of scratching oneself indecently (or whatever hidden reason Sartre's individual has for feeling ashamed when being unexpectedly observed), *is* part of common human experience. This experience of being the object of a foreign subject is not the totality of human experience of interaction with others, yet it is no less real than the "complementary" phenomenology: the *pure subject-subject experience* of ideal dialogue. Together, the two phenomena are aspects of a total interaction situation, which always (except perhaps in the limiting cases of psychosis) includes *both* objectivization of the other, each participant being primarily himself and outside the other; *and* a subject-subject relation, because dialogue and any other communication (even the tactical communication of advertisement) is meaningless without this presupposition (from p. 185, n. 7 we remember the difference between the torturer and the blacksmith : the torturer, instead of doing a "technical" job, "answers" the screamed or suppressed protest of the victim by supplementary acts of violence).

CHAPTER 14

Toward Synthesis: Human Nature as Dialectic and History

DIALECTIC

Taken at bare face value, Sartre's early philosophy was a nondialectical response to the various determinisms. But looking at the total course of his philosophy and seeing its different phases in context and as responses, we discover that the postulate of freedom is not abstract and empty but substantiated through contrast, and we find in the later phase a (preliminary) suggestion of independent substantiation.

The challenge with which the total course of Sartre's philosophy presents the humanities as investigation of the particularly human can be summed up as follows: central importance must be given to the authenticity of human choice—human beings produce their own history, and so does the human race as a whole; but we must transcend the empty abstraction, which ultimately reduces human freedom to an uncommitted cliché, and the one-sided concentration on the isolation of human individuals which leaves no space for grasping the possibility of dialogue. With regard to the first, existentialist phase, one must understand choice and freedom as being always relative to a given historical, social, and personal factuality, instead of treating the two aspects of human existence as irrelevant to each other. And with regard to the second phase, we must attempt to substantiate the relation between freedom and necessity not only in principle but through understanding of the interplay under concrete historical circumstances.

We may supplement with a short fable from everyday Denmark (an oblique reference to Holberg's play *Jeppe paa Bjerget,* see p. 139):

Thursday Morning, we meet Jeppe in the public assistance office telling his sad life to the social worker Nille. He has been a habitual drunkard for years and would like to quit alcohol. At the same time, however, his mind and his mouth are full of excellent excuses, marvelous reasons that Jeppe should drink. Of course, Nille may start by being kind and demonstrating her sympathy through such interspersed remarks as "Yes, I understand" and "Certainly." But she has chosen the easy and irresponsible way out if she contents herself with showing compassion and understanding and does not go on to explain that in spite of all excuses, "Only *one* person can stop drinking, and that is *you. You* must pull yourself together, because nobody but Jeppe can pull Jeppe together."

Friday afternoon, Nille participates in the weekly meeting of the Social Welfare Committee, trying to get special permission to pay Jeppe's rent while he is in long-term treatment in a home for inebriates, in order that he may not find himself in the street at his return. Here, the Baroness from party X insists that the only solution to Jeppe's problem is that he pull himself together—everything else is nothing but squandering of tax money. In this context it is Nille's professional duty *not to repeat* what she said to Jeppe the day before; here she must make sure that Jeppe is given the material possibility to pull himself together, not only while he is in treatment but permanently, and insist that the reasons for Jeppe's drinking are real and massive. Any solution proposed by the Committee which does not provide Jeppe with substantial reasons *not to drink* is hypocritical moralizing and liable to fail.

A short poem by Bertolt Brecht may serve as supplementary illustration of the problem:

> General, dein Tank ist ein starker Wagen.
> Er bricht einen Wald nieder und zermalmt hundert Menschen.
> Aber er hat einen Fehler:
> Er braucht einen Fahrer.
>
> General, dein Bombenflugzeug ist stark.
> Es fliegt schneller als ein Sturm und trägt mehr als ein Elefant.
> Aber es hat einen Fehler:
> Es braucht einen Monteur.
>
> General, der Mensch ist sehr brauchbar.
> Er kann fliegen und er kann töten.
> Aber er hat einen Fehler:
> Er kann denken.[1]

The immediate reaction to the fable of Jeppe and Nille might be the word *complementarity:* Jeppe's free choice and the circumstances of his previous life (or his

---

1. (*Gesammelte Werke,* vol. 9, 638). For copyright reasons, I cannot bring a translation. To paraphrase, the poem addresses a general, pointing out that his tank is a powerful chariot, which knocks down a wood and crushes a hundred people. Similarly, his bomber is strong, flying swifter than a hurricane and carrying more than an elephant. Man, finally, is very useful, being able to fly and being able to kill. But each has its characteristic flaw: the tank needs a driver, the bomber needs a mechanic—and man can think.

hereditary biochemistry?) that made him an alcoholic are *two aspects of the same matter*, which can neither be separated nor reduced to one. Brecht's poem might be seen as calling for the same interpretation. Still, simple explanation by complementarity is insufficient and amounts to little more than an admission of defeat. Saying to Jeppe and the Baroness alike that "well, of course you are right, but on the other hand" will only make each of them persevere; so will, even more clearly, agreeing with Jeppe on the issue of Necessity and with the Baroness on the issue of Freedom. If the "complementary contradiction" is to be *made productive*, it is important to see, firstly, that each of the two aspects can only be meaningfully applied in a particular *practical* perspective—Jeppe's freedom in the perspective of *Jeppe's practice*, and his conditioning in the perspective of that Committee whose practice is going to *determine his conditions*. Secondly, it is essential to understand the relation of the two aspects to each other: how can adequate material conditions be created which will permit Jeppe's pulling himself together to develop into an inveterate habit (which, as discussed in connection with the problem of emancipation on p. 185, will hardly be fruitful if not decided in dialogue with Jeppe)? Turning to Brecht, we may feel satisfied by using the poem simply as consolation through paradoxes. But if we want to cash the check of consolation, we will have to focus our interest on the questions, *when* people begin thinking beyond and at cross-purposes with their "usefulness," and *how* they are brought to this decisive point.

The established term for this productive complementarity, in which something qualitatively new is generated from contradiction, is dialectic. A main point in dialectic can be formulated in an aphorism (which should be taken precisely as an aphorism, not as full-fledged philosophical theory): "Insight is never final and definitive. Insight emerges in process in a world in perpetual development—which, when it comes to the development of the human world, is often development determined in part by the preliminary insight attained by the participants in each moment of the process; *new* levels of insight (with new types of answers) are only made possible through new development which generate new kinds of questions and new conceptual tools."

This somewhat opaque formulation is illustrated by the hermeneutic circle (cf. above, p. 184). Both in the Schleiermacher/Dilthey and in the Heidegger/Gadamer variant, *we as observers* undergo the change, not the dialogue partner, which is a fixed text—in contrast to what will hopefully happen to Jeppe and the mechanic. In relation to the hermeneutic circle, dialectic is thus a tool for theoretical insight, not for practical change.

A third illustration will show dialectic at work in a broader historical process, where not the observer's but the participants' insight is involved. For convenience I shall refer to Sartre's formula of a "living philosophy" for each historical epoch (but the point should not be understood as depending on that particular example, which is just one particular expression of the idea, unduly simplified in order to facilitate the exposition). The living philosophy of the early bourgeois world was the

Enlightenment; it conceptualized a world which was only emerging, and it assisted in its unfolding. No philosophy could do more before the taking of the Bastille and the battle of Valmy; it would have had to describe and reflect upon a world order even the germs of which did not exist. Only the breakthrough of industrial capitalism and the appearance of organized labor made possible a new level of philosophy—which took the form of Marxism, the worldview of an organized working class which had undertaken to change the world and to abolish the capitalist relations of class power (and thus, according to an essential Marxist tenet, *all* class power).[2] Marxism, on the other hand, is (and *must be*) just as unable to predict the contents of the philosophy or world view with which a future society will respond to its problems[3] as was the Enlightenment philosophy to foretell the philosophy of a revolutionary working class whose appearance on the scene was itself totally unexpected.

The principles of classical logic are often summed up in the formulas "*A* is *A*" and "*A* is not *non-A*." Dialectic, in contrast, as it was said by one J. Macmurray, deals with "an *A* in the midst of being transformed into a *B*."[4] Seen in this light, dialectic is the framework for understanding *change in spite of continuity* and *continuity across change,* which has much to do with the role of the insight acquired at each moment in the process; but it also opens the horizon on the role of those historical forces which transcend the insight of participants, either because they are too self-evident to be submitted to reflection, or because every human action has consequences beyond our foresight and even our comprehension. We may continue the example just given: utopian socialism, and thus eventually Marxism, did not start from scratch, but as a continuation of central aspirations of Enlightenment philosophy.[5] Yet in spite of this continuity, the total picture that emerged was quite new, both because a new social situation had permitted the formulation of new dilemmas, and

---

2. Marxism is evidently not the only mid-nineteenth-century philosophy which could not have been formulated in the context of the eighteenth century—which is one of the vulnerable points in the idea of *one* living philosophy of the epoch. Though less momentous outside Denmark, Grundtvigianism is a parallel example: "organized farming" was no less of a novelty than organized labor, and it produced its own intellectuals and its own philosophy (still with repercussions as far removed as Appalachian people's schools in the 1930s and Paolo Freire's strategy for popular education in Brazil of recent decades).

3. At present, more than a century after Marx's death, we are able to see that it will not least have to respond to problems of global survival which were not, and could not be, anticipated concretely in the middle of the nineteenth century.

4. Retranslated from the Swedish quotation in Ljungdal 1967: 22.

5. In synthesis, by the way, with strains of popular culture and political demands going back to the High Middle Ages. Through the very choice of its name, the *Commune de Paris* of 1871 took up not only the institution of the Sansculottes of 1793 but also the glorious banner under which the popular movements of the twelfth century had claimed (and often gained) autonomy from feudal power.

because old themes had come to mean something new within a new social situation, whether those who formulated them knew so or not. And conversely: a socialism which (forgetting about dialectic—concretely, about the actual continuity of many necessarily repressive structures) considers the civil rights formulated in the early bourgeois epoch as *merely bourgeois and hence irrelevant for the working class* will (according to sad historical experience) be no effectual expression of the power and authority of working people. Instead of the continuity of human rights, we end up with the continuity of actual repression, in spite of changes.[6]

SUMMING UP

Dialectic is, in this formulation, no precise method and even less a universal key. It is a suggestive pattern of thought and nothing more—but still a pattern in agreement with fundamental conditions of human existence—or at least in agreement with structures which necessarily turn up when *we* try to describe human existence. Whether their necessity follows from "reality in itself" or merely from our need to describe (certain aspects of) reality in absolute but mutually exclusive terms is parallel to Kant's problem whether time, space, and causality characterize "things in themselves" or merely constitute necessary prerequisites for our conceptualization of things. As the Enlightenment physicist d'Alembert resolved in the "Discours préliminaire" to the *Encyclopédie* (vol. I, p. v), we may decide that as long as the use of the framework seems unavoidable, it does not really matter whether the spectacles are to be counted as part of external reality or as part of our own equipment, and for convenience we may consider dialectic an aspect of human reality. Less pragmatically, we may also observe that *we* are part of the human historical process and contribute to it with our understanding and our will; therefore dialectic, by being a prerequisite for understanding, is *eo ipso* part of social reality "in itself."

In order to transform the dialectical generalities into at least a rudimentary guide for work within the humanities, we may return to the problem of anthropologies and mention some issues which should be included in a dialectical understanding of anthropology and history.

All the anthropologies considered up to here make a point. That is the reason they have not only been formulated but have also evoked a certain response. But none of them when taken absolutely is satisfactory (which would astonish few of their originators—most of them were formulated as part of a polemic and in order

---

6. These observations are no rationalization of the events of October to December 1989. Apart from the explanatory passage "concretely . . . repressive structures," they are translated literally from Danish lecture notes used long before the collapse of Soviet and Eastern European socialism.

to counterbalance other prevailing opinions). Montesquieu, for instance, would hardly have believed that a shotgun placed in a Turkish cradle would develop Turkish morals. That aspect of human nature which we might call "the potential for developing morals" is taken for granted, as something not worth discussing.

As regards those anthropologies which understand man through his relation to society, it is important to remember that "society" is no simple entity but a complex relationship made up by social groups, institutions, ideologies, and habits in interaction and conflict. Dialectical explanations involving our social existence will thus become *at least as complex* as Weber's interpretation of the interplay between Protestantism and capitalist development. In this connection one may observe that the common counterargument to social determination, "How should society be able to generate opposition to society?" is about as bright as the corresponding "How should a stick of dynamite be in possession of characteristics permitting it to destroy itself?" Both questions presuppose (in the kindest interpretation) the trivial functionalist belief that all characteristics of a system serve to conserve or protect the system; in a less kind interpretation, they build on the tacit anthropomorphic assumption that "society" and the dynamite cartridge are quasi-conscious beings which ("who"?) would never *get the idea* of destroying themselves—a bland assumption even regarding the members of the human race.

As a guiding principle we may thus state that *people create their own existence and history,* individually and collectively; but they do so on given historical conditions, involving both a material fundament and a complex web of institutions and ideologies which structure their relation to each other and to the material fundament. Rarely do they have more than rudimentary insight into these conditions. All the more often their creation of history produces unanticipated results.[7] Our theories and our understanding of the world are, indeed, created under the conditions of *that world which is already there,* and in response to *the world which is already there to the extent that we have come to know it.* In the likeness of generals, we can only plan our strategy according to the experience of the previous war. When we create something new (or merely something outside the range of the familiar) it is far from sure that traditional winning strategies will have the anticipated outcome. As formulated by the American historian of philosophy John Herman Randall (1962: 10):

> History is made by men, by groups of men living in a natural environment, partly intractable and inescapably there, partly lending itself to human efforts at its reconstruction. Those efforts are always particular

7. These principles could be legitimately attributed to Marx. Yet Marx, of course, was not the first to discover them. Both the ancient Babylonians and the ancient Egyptians spoke about the future as "that which stands behind your back [ready to attack you]," whereas the past was "that which is in front of you."

> and piecemeal. But they have consequences, and those consequences, even when not intended or even envisaged, are as inescapably there as any other part of man's environment. Men do something in their need, and then find that they have to do something else. In solving one problem they find they have created others. In learning how to grow more grain and better wool, they find they have undermined a whole culture, and have to create a new science, a new ethics, and a new theology. In a different jargon, we can say that changes in the instruments of production demand ultimately the creation of new ideologies—because they have changed the character of man's experience. The architects are men, and there is much in the structures they build that is the product of what Aristotle calls chance. But the purposes for which they are built, the needs they are to serve, the materials that are employed and the tools that are used, are not due to chance, though they are equally beyond human control. History is a human achievement; like everything human, one within natural limits, but nevertheless an achievement.

The one who has really digested these words is well equipped to avoid the traps presented by the many vulgar anthropologies discussed so far, and to let himself be inspired by their mother theories for much less vulgar purposes.

More precise rules of methodological conduct in this domain cannot be given, since, as another sage summed up many years' experience,[8] *dialectic is the logic of the unpredictable.*

8. L. Gudmund Damgaard (private letter).

PART III

The Art of Knowing

An Essay on Epistemology in Practice

CHAPTER 15

INTRODUCTORY OBSERVATIONS

Parts I and II of these pages have dealt primarily with the *distinctive characteristics* of the humanities—such features as constitute their particular identity. Part III concentrates on the complementary aspect: those features which the humanities share with other sciences, the features that characterize the humanities as well as biology, physics, economics, and sociology as *sciences.*[1] On their part, scientific knowledge and cognition share a number of characteristics with other kinds of human knowledge and cognition—not least the quality of being a less direct rendition of that reality which we know about than we tend to think in naïve everyday life.[2] Critical understanding of the general properties and categories of knowledge, moreover, will throw light on certain problems belonging more or less specifically to the humanities.

In brief survey, Part III is built up as follows. Chapter 16 introduces the broad problem of what knowledge is via a presentation and discussion of Piaget's "genetic epistemology." Chapters 17–19 develop what could be called a "philosophy of science" in the habitual sense, starting in chapter 17 with a general presentation of some basic categories and some classical points of view—Platonic and Aristotelian realism, empiricism, and Popperian falsificationism, instrumentalism, and the

1. I recall the definition of the concept of a 'science' which was set forth on p. 7: "a *socially organized and systematic search for and transmission of coherent knowledge* in any domain." Speaking in the following of the humanities as *sciences* is thus not meant as an implicit claim that they should emulate the natural sciences, but rather that *all* sciences share a number of qualities of which some have mostly been discussed in relation with the natural sciences and others in the context of the humanities or the social sciences. Illustrations will, accordingly, be taken from all three domains.

2. It should be noticed that this distinction between scientific and other kinds of knowledge presupposes that *knowledge* is more than explicitly formulated theory. Knowledge, we may say briefly, is *any kind of conscious or subconscious acquaintance of the surrounding world allowing some kind of adequate action*—communicative action included.

"demarcation problem." Chapter 17 also comprises a first confrontation with what I have chosen, in accordance with the philosophical tradition, to designate a "materialist" notion of truth (but which many contemporary philosophers would call instead a "realist" concept). Chapter 18 contains a presentation and critical discussion of two main approaches to the problem of scientific *development:* Lakatos's concept of dynamic "research programs" and Kuhn's theory of progress through a sequence of "normal science" phases separated by "scientific revolutions." Chapter 19, footing on the discussions of chapters 17 and 18, confronts and connects three classical core problems of the epistemology of science (and of epistemology in general): the question of truth, the question of objectivity, and the notion of causation.

Already chapter 18 looks at the acquisition of scientific knowledge as the product of a scientific community. Chapter 20 unfolds some consequences of this approach, in particular under the perspective of norm or value systems ("morality") as regulators of the functioning of social communities. In the process of doing so, it develops a general view on what norm systems are.

Chapter 21 takes up the production of scientific knowledge under the historically concrete conditions of the "scientific-technological revolution." Starting out from a discussion of the relation between "theoretical" and "applied" knowledge, it returns to the Kuhnian cycle, viewing it now specifically as a description of the development of scientific disciplines, and contrasts it with the inherent interdisciplinarity of applied knowledge.

Chapter 22 appears to abandon the philosophy of science altogether. Its central problem is indeed whether, and in which sense, *art* constitutes a way of knowing, drawing for this on the epistemology which was developed in previous chapters. The conclusion, however, is that art plays a central role as training of the skill in synthetical judgement without which analytical thought is useless, in science as in any other domain.

It goes without saying that the exposition owes much to many precursors, from Aristotle and Kant onwards. Some of those by whom I am inspired I have read in original, from others I borrow indirectly (and certainly often without knowing so). It must be emphasized, however, that the essay is not meant as a survey of the views of select philosophers and philosophically minded historians, psychologists, and sociologists. Instead I have attempted to formulate a personal synthesis while keeping it so open that readers will still be allowed to get a broader view of influential opinions and important problems and to agree or disagree with the single strands of the argument.[3]

---

3. No doubt my readings of the philosophers and other authors that enter the argument are so-called "productive interpretations," selecting what I find fruitful for the current argument—neglecting what I find irrelevant—and trying when several readings are possible to select the most interesting one; attempting, however, to be faithful when criticizing. In terms

The argument is indeed a complex network containing many threads and open suggestions. Even though the underlying thought is certainly rationalist, the ideal of rationality which forms its basis is that of dialogue and not the absolutist ideal of the strict proof.

PHILOSOPHY AND THE PROBLEM OF KNOWLEDGE

At least since the pre-Socratic philosophers Parmenides and Zeno, the "problem of knowledge" has haunted philosophy. Zeno, in particular, is famous for a number of paradoxes meant to show that our naïve everyday "knowledge" cannot correspond to genuine reality—we "see" the arrow reach its target and "know" that Achilles takes over the tortoise. But the intellect demonstrates clearly, according to Zeno, that this cannot possibly be true (cf. p. 32, n. 7 and preceding text). Plato, probably following Socrates on this account, argued that we cannot come to learn what we do not know already, and developed his doctrine of ideas on this foundation. Aristotle tried to put things straight by distinguishing different kinds of reality ("particulars" and "universals," in the language of his medieval followers—to be explained in more detail below) and different kinds or levels of knowledge. Thereby Aristotle set the stage for the discussion as it took place until Kant, in his "critical" approach, introduced a distinction between the *conditions which delimit and determine our possibilities of knowing* and the properties of *that reality which we strive to know about,* the famous "thing in itself" as distinct from "the thing as it appears to us": *We* cannot know (material) reality without categorizing it into objects, time, space, and causality. Whether *reality in itself* is structured that way will forever remain undecided and undecidable.

Epistemology (the theory of knowledge, of *epistēmē;* alias *gnoseology,* theory of *gnosis*) does not end with Kant; nor does Aristotle's "setting the stage" imply that philosophers followed his doctrines until the late eighteenth century. But some way or other all philosophical discourse in the field has been concerned with the relevance or irrelevance of Plato's, Aristotle's, and Kant's categories, concepts, and doctrines.[4] Mostly, it has also been "philosophical"—that is, highly sophisticated in its relation to earlier contributions to the philosophical tradition but at the same time common-

---

of the motto of part I, I exploit previous discussions as "a bank to be raided, not a church to worship in; but . . . also not a useless museum." Who wants to know more precisely the thought of a Piaget or a Popper (et al.) should therefore—needless to point it out, I hope—read their own works.

4. An introduction to epistemology based on the historical discussion is Losee (1972). One critical observation should be made in connection with this otherwise recommendable work: Hegel, and all approaches somehow derived from Hegel (including all Marxist views), are absent from Losee's universe.

sensical and often naïve[5] in its appeals to empirically established knowledge about the processes of knowing. Aristotle, it is true, based his doctrines upon profound understanding of the best standards of scientific knowledge of his time; even philosophers like Locke and Hume, however, were far from understanding the real intricacies in Galileo's and Newton's scientific methods and standards, notwithstanding their claim and belief that *they* expounded the true sense of the feats of these heroes (cf. Casini 1988).

Practical scientists, on their part, however sophisticated the methods they apply in the acquisition of knowledge about the field they investigate, are often highly naïve when it comes to understanding the philosophical implications of these "critically." As formulated by Imre Lakatos (1974a: 148 n.1), "scientists tend to understand little more *about* science than fish about hydrodynamics."

For this double reason, a direct and immediate dialogue between classical epistemology (or, more specifically, classical "philosophy of science") and actual scientific practice is not likely to be very fruitful. An introduction to the problems of knowledge which intends to further critical reflection among "practitioners of knowing" should therefore rather take its starting point in approaches developed during the twentieth century, approaches that are based on thorough empirical observation of the processes of knowing and which require that their philosophical framework be able to grasp these empirical observations.[6]

---

5. Here, as in the following, I use the term *naïve* in opposition to a generalization of Kant's notion of a "critical" approach, and not broadly as "gullible." The "naïve" attitude is the one which accepts things for what they seem; the "critical" approach is the one which investigates whether, why, and in which sense the naïve attitude is justified—"examination of its possibility and limits," in Kant's words (*Critik der Urtheilskraft,* B III).

Evidently, most of our practical life has to build on "naïve" foundations and does so without problems. As formulated by Ogden Nash (quoted from Thorkild Jacobsen 1988: 123): "O, Things are frequently what they seem/ And this is Wisdom's crown:/ Only the game fish swim upstream./ But the sensible fish swim down." Often, critique does not tell us that we were *wrong* when being naïve, but only that we were right, without knowing why, or as right as we could be; this is, for instance, the outcome of Kant's "critique of pure reason."

No critique is ever definitive. What seemed at one moment to be an absolute underpinning (be it Euclid's proofs that the methods of practical geometers were right, Kant's critique of Newtonian physics, or Marx's critique of Smith-Ricardian political economy) turns out with historical hindsight to make other "naïve" presuppositions, which in their turn can be "criticized." We choose a misleading metaphor when we speak about establishing "a firm foundation" on which we can safely build. We always build on swampy ground; what critique can do is to hammer the piles on which we build through so many layers of mud and clay that they are not likely to move significantly.

6. This preliminary description, as "naïve" as anything could be, should not be read as a claim that these approaches are *nothing but* empirical. One of the main points of the following (derived, in fact, from the empirically oriented studies, and agreeing in this respect with

(n. 6 continued) Kant) will be that no knowledge is based on empirical observation and experience alone—*a fortiori* not knowledge about knowing, which certainly remains *philosophy*, asking so to speak about the conditions for its own existence from within the horizon it tries to describe. At least when read as philosophy, the "empirically oriented" approaches "refuse to be in a place where in fact they are not [*viz.* outside the world of experience] but . . . claim, if wanted and possible, to move through the whole region of *common* experience and then to land in *other*, rarely visited places, yet, at least supposedly, not utopian and still, somehow, belonging to an experience" (as formulated by Emilio Garroni 1992: 262).

CHAPTER 16

A Piagetian Introduction to the General Problem of Knowledge

One approach to the problem of knowledge is through individual cognition. This is even, one might reasonably claim, the most obvious approach, since knowledge is always knowledge known by somebody, however socially conditioned and organized it may be. Even knowledge embodied in books or databases has been put down by somebody and, more decisively, is only transformed from black flecks or magnetic traces into *knowledge* when interpreted by a mind.

A number of twentieth-century psychological schools have set forth doctrines or theories about the nature and construction of human cognition. Some of these doctrines, like the claims of behaviorist school,[1] are mainly postulates about how cognition should be explained. Of particular interest in the present connection are the Soviet Vygotsky-Luria-Leontieff school and the Swiss Piaget (1896–1980) and his collaborators. I shall base the present chapter on Piaget's work while being to some extent inspired by the former school in my own reading of Piaget.

Piaget is most widely known as a "child psychologist," which is true in the sense that he contributed decisively to the understanding of child development, and has been amply used/misused in the planning of curriculum reforms; otherwise, however, it is a misleading statement. He started out as a biologist specializing in molluscs but with a strong interest in metatheoretical questions, not least in the logic and philosophy of science. This led him to accept in c. 1920 the task to standardize Cyril Burt's IQ test for French children, which then led him to discover his own approach.[2]

---

1. According to which, briefly spoken, all human knowledge (like behavior in general) is to be explained as a web of conditioned reflexes (cf. pp. 202ff).

2. See the introduction to Rotman 1977.

The principle of the IQ test can be described as *kinematics,* "movement-description." A number of problems are presented to the experimental subject, and correct and wrong answers are taken note of. At the same time it is known (from the "standardization") which tasks are solved by average children at a given age. In this way, the "mental age" of the subject can be determined, or the "intelligence quotient" understood as the "problem-solving capacity" expressed as a percent of the capacity of the average subject of the same age. The central concept is thus the dichotomy "correct"/"wrong" answer, and the central tool is the timetable telling at which age various types of problems normally will be solved correctly. Questions concerning the motive forces behind the process of intellectual development, and even the very idea of a *process,* do not occur.

Piaget soon noticed that the "wrong" answers were not only wrong but also systematic. We may illustrate this by an example borrowed from his later research. A girl of five,[3] asked whether there are more girls or more children in her kindergarten, may answer "more girls," "more boys," "equally many" (*viz.,* girls and boys), or "I do not know." The one answer you never get is the "obviously correct" answer "more children."

I shall return to this experiment below. For the moment we shall only observe that the consistent deviation from adult thought must correspond to a different way to conceptualize and think about quantity, not to mere absence of conceptualization and thought. This was also one of the conclusions drawn by Piaget, who set out to find the *dynamics,* the active forces and processes, of the development of cognition. The other conclusion was that things like the Kantian categories (object, number, space, time, causality) cannot be inborn, and thus that cognition has a genesis and results from a development process.

The titles of some of the books which Piaget published over the following two decades read like as many empirical tests of Kant's categories: *Judgement and Reasoning in the Child, The Child's Conception of Physical Causality, The Moral Judgement of the Child, The Child's Conception of Number.* Part of this research was based on the "typical answers" of many children to "indicative questions," for example, questions of the kind that are used in IQ testing. As his own children were born, Piaget took advantage of the possibility to observe and interact with the same children over several years, thus interpreting their mind "from within" in a more hermeneutic manner.

---

3. A "normal" girl, that is; here and everywhere in this chapter, age indications are only approximate and subject to large variations. But if your own five-year old niece gives the "grown-up" answer, you can be sure that she would have given the "five-year" answer at ages three or four. In the language of technical statistics, ages constitute an "ordinal scale," and are only numerically true as averages (and even this only for that geographical and social environment where they were established—whence the need for restandardization of British IQ tests in France).

During the following decades, Piaget re-emerged from his submersion in empirical child psychology and formulated his general "genetic epistemology," that is, a general theory of human cognition as the outcome of a process. It is this mature theory (which, it should be noted, remained a living research program and was never a fully finished doctrine[4]) on which I draw in the following, and which is the starting point for certain further reflections.[5]

SCHEMES AND DIALECTIC

Let us start from an example. A child of (say) 1½ years is familiar with a variety of balls: white, blue, red, variegated; made from cloth, from rubber, from leather; with diameters ranging from 3 to 6 cm. It knows that you may push them, make them roll, throw them from the chair where you are sitting and thus make your patient big sister pick them up for you. If a new ball of some color and within the familiar range of weight and diameter gets into its hands, the child will demonstrate in practice that it knows how to behave with balls. We may say that the child possesses a *practical concept* or (in Piaget's language) a *scheme* for balls. This scheme is not present in conscious thought but is *sensori-motor,* that is, derives adequate movements/actions from sensual perception.

If now some fully unfamiliar ball is presented to the child—say, a football or, even better, a ten-pound leaden ball, certain familiar acts are impossible. It is too heavy to be thrown, and if you roll it from the table onto your sister's feet, her patience will certainly be gone. But accident or deliberate experimentation may produce the experience that even this object can be pushed and rolled along the floor. In this way, the leaden ball is *assimilated* to the scheme. The scheme, on its part, is changed and made more flexible by encompassing even this unfamiliar ball: it *accommodates* to the wider field of experience.

This illustration introduces a number of key concepts from the Piagetian theory. Firstly, the pair 'assimilation/accommodation.' Secondly, it illustrates the concept of *equilibrium,* which according to Piaget is the central aim of this twin mechanism. Equilibrium, in this terminology, is no static condition, but to be conceived in the likeness (and according to Piaget indeed as a special case) of the dynamic equilibrium of living organisms, for which reason Piaget prefers in some places to speak of "equilibration." Irrespective of the surrounding temperature and the precise nature

---

4. Although, so rumors tell, Piaget was so dominating a personality that the group of researchers at his Institute in Geneva never developed into a genuine "school" but remained a circle of "Piaget's collaborators." Theoretical innovation seems to have been the privilege of the master.

5. It should be pointed out that some of the illustrative examples and experiments referred to in the following are borrowed from Piaget; others are my own observations, mainly made on my own daughters.

of its food, a dog will conserve approximately the same body temperature and the same organic structure; to keep up this equilibrium is the task of its metabolic processes. Extreme conditions which destroy the equilibrium will, at the same account, kill the dog. A newborn puppy, on the other hand, can stand fewer variations in its living conditions. In this sense, the equilibrium of the mature dog is more stable than that of the puppy. And in the same sense, the accommodated scheme for balls constitutes a more stable equilibrium than its predecessor, since it is able to grasp without difficulty a wider range of different balls.

Thirdly, the inescapable duality between assimilation and accommodation makes Piaget's theory of knowledge a dialectical theory. It makes sense of Plato's claim that you cannot come to know what you do not know already, and it gives substance to Aristotle's statement that you may, in different interpretations of the word, "know" and "not know" something at a time: you will only discover that the leaden ball is a ball if you already know (in the case of the infant, "know" practically) what balls are. But by discovering the leaden ball as a ball, you also come to *know more about what balls are,* and this is the key point where Piaget's dialectic really goes beyond these classics. In this way, every act of knowing is at one and the same time an assimilative interpretation in terms of an existing *cognitive structure* and an accommodation of this structure to an enlarged range of reality; although they are analytically distinct, neither of the two processes can take place in the absence of the other.

Two further observations can be made on the above example. Firstly, if nothing of what the child usually does when playing with balls fits the unfamiliar object, it will not be assimilated to the ball scheme; nor will it of course be assimilated to any other scheme if it does not match in some way. In most cases, this will result in a practical rejection of the experience in question—what makes no "sense" is not noticed or is quickly forgotten.[6] Secondly, assimilation presupposes attentiveness; it is therefore more likely to result from deliberate experimentation than from accidental events, behaviorist pedagogical theory notwithstanding (cf. p. 203).

Let us then look at another example. During the first months after birth, the baby's world is

> an objectless universe, composed of perceptual tableaux which appear and disappear by a process of resorption, and an object is not sought when it is hidden by a screen—the baby, for example, withdraws his hand if he is about to grasp the object and the latter is covered by a

---

6. This is no absolute rule. Human beings, not least children, are curious, and a puzzling object or class of objects *may* provoke intense experimentation or investigation (inquisitive "play" in the case of the child), and in this way provide the basis for the construction of a new scheme; but if not even the single sensual outcomes of experiments "make sense" with regard to the existing cognitive organization of sensory experience, this constructive process is not likely to take place.

> handkerchief. At a later stage the child begins to look for the object, lifting the handkerchief at A where it has just been covered; and when the object is displaced to a position beneath B (for example to the right, whereas A was on his left), the child, although he has seen the object being placed at B, often looks for it at A when it disappears again; that is, he looks in the place where his action has been successful on an earlier occasion. He therefore does not take account of successive displacements of the object which he has nevertheless observed and followed attentively. It is not until towards the end of the first year that he looks for the object unhesitatingly in the place where it has vanished.
>
> (Piaget 1972: 11f; cf. 1950: 108ff).

At this age the child will thus remove a blanket under which a coveted toy has been placed. Still, if a Basque beret and not the toy turns up below the handkerchief, it abandons the pursuit. Only in a following phase will it remove the beret and—triumphantly perhaps—find the toy, whose permanency as an object firmly located in space is no longer subject to doubt (Piaget 1973: 11ff).

Balls are not necessary ingredients of the universe, and the construction and stepwise accommodation of a scheme for balls cannot be cause for wonder. But permanent objects do seem to us to be necessary and unavoidable fundaments for knowledge of the world. Piaget's investigations show, however, that even this, as well as all the other Kantian categories without which no knowledge of the physical world is supposed to be possible, is the product of a development, going through a sequence of accommodative extensions and equilibria.

In one of his publications, Piaget (1973: 66) defines the *scheme* of an action as "the general structure of this action, conserving itself during . . . repetitions, consolidating itself by exercise, and applying itself to situations which vary because of modifications of milieu." On p. 114 of the same work, a scheme is defined more briefly but in the same vein as "what is generalizable in a given action." Both variants of the definition may call forth the legitimate question, whether the scheme is really part of the mind of the knowing and acting child or only a construct, a generalization or summary made by the observing psychologist for his convenience.[7] If the scheme is nothing but a psychologist's construct, the whole idea of accommodation becomes dubious, and the question of the status of the schemes must therefore be addressed before we go on.

Normally questions of this kind are undecidable and hence, one may claim, meaningless pseudo-problems. In the case of schemes, however, a decision can be

---

7. This question is indeed the normal positivist reaction (see below) to any idea of a *general structure* or *universal* encompassing and determining particulars—is it THE HORSE in general which determines the characteristics of individual horses, or is the universal concept (the species) merely a shorthand in which zoologists sum up their knowledge of similar individuals?

reached—not, it is true, in the precise case of the scheme for balls, but if we look at *grammatical schemes.*

The past tense of "hunt" is "hunted," the past tense of "reach" is "reached,"and so on. These forms, too, are formed according to a scheme, which can even be seen to apply "itself to situations which vary because of modifications of milieu" (for example, the *e* is pronounced when coming between *t* and *d* but not between *ch* and *d*). Even in this case, moreover, we may ask whether the scheme is a grammarian's shorthand or really present in the mind of speakers before they have been taught grammar.

The answer is "really present" and is provided by young children below the age where they can understand any grammatical explanation. They may have learned the forms *hunted* and *reached* by hearing them spoken by grown-ups, and might in principle just store these forms individually. But they have never heard the form *goed,* since all mature speakers say "went." None the less, their first past tense of *go* is *goed*—and when they discover that this does not agree with adult usage, they opt for a while for the compromise *wented.*[8] These forms can only result from a subconscious general scheme which the child has constructed on the basis of forms like *hunted* and *reached.* This scheme, which deviates (in extension) from the grammarians' scheme, must be present in the child's mind; it can be no mere observing psychologist's construct. If the existence of schemes is established, on the other hand, their accommodation (in the case of verbal conjugation in the steps *goed* —> *wented* —> *went*) is also real.

THE PERIODS

"Every act of knowing is at one and the same time an assimilative interpretation in terms of an existing *cognitive structure* and an accommodation of this structure to an enlarged range of reality," as stated above. But as a scheme tends toward equilibrium, the accommodative aspect of the process becomes less prominent. The different schemes available at a given moment also tend to form a coherent system and to share a number of basic features and limitations (to be exemplified below). This is precisely what justifies the idea of an overall cognitive structure.[9] Most of the time, the gradual accommodation of schemes brings about a maturation of this structure, increasing its coherence, "stability" and functionality. At certain moments, however,

---

8. "Gåede" and "gikkede," respectively, in my own observations. But English linguists confirm the English forms given here.

9. Piaget, who loved mathematical metaphors and believed them to be more than metaphors, would often bolster up this explanation by referring to a number of mathematical and (apparently self-invented) pseudo-mathematical concepts in this connection. This need not bother us here.

new mental abilities (one of them being the emergence of language and conscious thought) destabilize the structure; within a relatively brief period, new schemes arise, old schemes accommodate thoroughly, and schemes are integrated in new ways. This lays the foundation for a new cognitive structure, organized at a higher level, which is going to assimilate increasingly large ranges of earlier as well as new experience.

These changes of the overall cognitive structure demarcate the *developmental periods,* of which Piaget counts four (less thorough transformations of the structure—for example, the one allowing the child to remove the beret to find the toy—delimit stages within the period):

— The *sensori-motor* period, extending from birth until the age of 1½ to 2 years.
— The *pre-operatory* period, extending from 1½/2 until c. 7 years.
— The period of *concrete operations,* from c. 7 to c. 12 years.
— The period of *formal thought,* from c. 12 years onwards.[10]

During the sensori-motor period, the development of intelligence is characterized by increasingly effective coordination between sensory perception and motor activity (practical action). Starting off from a number of separate and unconnected "sensory spaces" (a sucking space, a visual space, a tactile space, an auditory space), a unitary picture of the surrounding world is gradually achieved, where a noise will provoke the child to look around for the source, and an interesting object coming within the visual space may make the child move towards it and grasp it.[11] The categories of space, time, permanent object, and cause are developed as practical (but certainly not conscious) categories. What this means in the case of the permanent object was already elucidated above. Having a practical category of space means possessing (among other things) the ability to plan a composite trajectory through a room, picking up a toy at point A and going on directly (and without new spatial planning) to point B, where the toy is to be used for some purpose. Possessing a

10. I remind of the observation made above that the age indications correspond to averages as established in the (urban European) environment where Piaget established the sequence. Others have made similar investigations in others surroundings, finding (with a restriction concerning formal thought in certain societies, to which I shall return) the same relative sequence but a considerable time-lag, for example, in an Iranian rural district where children were "not only without schools but also without any toys other than pebbles and bits of wood" (Piaget (1973: 154), citing Mohseni's research); it is important to note that this delay only concerned the onset of the operatory stage, whereas tests based on preoperatory thought showed no difference between these children and children from Teheran.

11. These acts of intelligence should be held strictly apart from certain reflexes which seem to mimic them. The grasping reflex, for instance, makes the new-born infant grasp whatever comes within its hand and cling to it; intelligent grasping is an intentional act which the child may choose or not choose to perform.

practical category of causation implies, for example, getting the idea to draw a table cloth upon which a coveted object is placed outside your range.

Implicit in the formation of the unified sensory space and of the practical categories is a gradual *decentralization,* or "miniature Copernican revolution," as Piaget calls it (1967: 79). The child itself is the only possible center of the sucking space, the tactile space, and so on; these spaces are "egocentric."[12] In the integrated space, on the other hand, the child only occupies one position among many possible and equally valid positions, which means that it can plan a trajectory passing through a sequence of positions.

The sensori-motor cognitive structure reaches maturity around the middle of the second year of life, when integrated space, practical categories and decentralization are attained. At this moment the child behaves as adequately as is possible without the intervention of conscious thought, given its physical and sensory equipment. At that moment it has more or less reached the intelligence of chimpanzees. It also stands at the threshold to a new period, characterized by the first emergence of language and conscious thought.[13]

One element of the new cognitive structure is created through the maturation of sensori-motor thought. Being able to plan a trajectory, or to stop at a problem, reflect and suddenly know how to go on after an "*aha*-Erlebnis,"[14] are indications that the schemes for action have been "interiorized," have become an element of thought, which can be anticipated before (or without) becoming actual action (cf. Piaget 1973: 57).

Another element, and a symptom that thought is emerging, is *symbolic play.* Sensori-motor children "play with" (for instance, with a ball, or with their toes), whereas older children "play that" (for example, that they are parents while the doll

---

12. If the concept of an egocentric space seems queer, one may think of those sensory spaces which are never really integrated into the unified space: the gustatory "space," the olfactory "space," the pain "space," the heat and cold "spaces." All of these are immediately perceived where *you are yourself,* and they are often felt as *moods* rather than as information about the world around you. Sounds, in contrast, are not perceived as heard in the ears or anywhere else in the body but at their presumed source (barring exceptional cases), and light as a rule not at all as light but as objects seen at a distance.

This does not (or not alone) depend on the nature of the stimulus: dogs, so it appears, integrate their olfactory space into the unified sensorial space, and perceive smells as linked with objects, not as mere smells.

13. Chimpanzees seem indeed to stand at the same threshold—so precisely there, in fact, that the same experiments can be interpreted by some workers as proof that they possess the ability for language and by others as proof that they lack this ability.

14. This term was first coined by Wolfgang Koehler (also known as one of the founding fathers of gestalt psychology) as a description of the same moment in chimpanzee behavior.

is the child). Such symbolic play presupposes the symbolic function, but it also provides this function with substance through interiorization of the play.

The most conspicuous element, of course, is language, which has been prepared through extensive sensori-motor play with speech sounds ("babbling") but only becomes language and a tool for thought in the moment when a string of speech sounds functioning as phonemes is used symbolically for something beyond itself and is interiorized with this function.[15]

The (embryonic) emergence of a new cognitive structure does not abolish the achievements of the preceding period. Children of three are no less sensori-motor efficient than they were a year before. Still, the immediately interesting point in a discussion of the cognition of children between two and seven is of course the *new* level, the characteristics of *thinking.*

The thought of "pre-operatory" children (a term to be explained in a moment) is characterized by Piaget as follows:

— It is egocentric.
— It is centering.
— It is irreversible.

That thought is egocentric (not the same as egoistic!) means that the child does not perceive its thoughts as *its own thoughts* which others may not share; in other words, it comprehends its point of view as the only possible point of view, which makes it difficult to entertain a genuine dialogue with the child unless you already know what it thinks. This may be illustrated by means of a conversation with a girl of two who had been to a zoo with her creche:

Adult: "What did you see in the zoo?"
Girl (with enthusiastic emphasis): "That one!"
Adult: "What is 'that one'?"
Girl (with increased emphasis): *"That one!"*
Adult: "Did you see monkeys?"
Girl (happily): "Yes!"

The answers given by young children when confronted with questions of "why" are characteristic of a generalized egocentric attitude. Such questions are not answered

15. This may be made more explicit by an example. At one moment in life, you may use a sound like "mama" in order to make your mother take you in her arms; this can be interiorized in the same way as any other sensori-motor scheme and will allow you to plan how to be taken into your mother's arms (an instance of practical causation, similar to the use of the table cloth); but only when the sound is used as a more general symbol for this all-important permanent object (normally the first object of whose permanence we are sure) will it allow you to think about your mother and not only about the specific act.

as concerned with *causes* but as regarding an anthropomorphic *aim:* "If Lake Geneva does not go as far as Berne, it's because each town has its own lake."[16]

It has been objected repeatedly to the cognitive interpretation of such statements that children answer in such unexpected ways because they understand words differently. This is certainly correct in itself. Children understand words in a way which corresponds to their answers. But they do so consistently (and in spite of the way the words are used by the surrounding adults from whom they learned them) because they are unable to grasp the usage of the grown-ups, which cannot be assimilated to their own cognitive structure. The "cognitive interpretation," the interpretation of children's sayings as evidence of their basic pattern of thought, is not only permissible but mandatory.

The "centering" character of thought can be explained through another example (in fact one of Piaget's key experiments). A child of five is shown a ball of soft clay and looks at the experimenter as he rolls it into a sausage. Asked whether the sausage contains more or less clay than the ball, three out of four children will respond that there is more, since "the sausage is longer"; the rest will claim that there is less, since "it is thinner." In both cases, the child concentrates interest on one conspicuous feature of the situation and does not take other features into account.

The irreversible character of thought can be elucidated by one of my own observations. A girl who had learned rudimentary counting by means of the number jingle (and who was thus at the threshold to the next period, cf. below) was asked as follows: "Seven birds are sitting in a tree. Then two more birds join them, but soon two birds fly away. How many are left?" The answer was simple and characteristic: "You would have to count them again." Some months later, in contrast, she was asked the same question, but dealing with "a hundred thousand million birds" (that is, a number which she could neither imagine nor remember but knew was a number), eight birds joining and eight leaving. She was not encumbered by the impossibly large number but simply answered "equally many." At this moment she could grasp the process "from above," and administer the mental process from a meta-level from where the mutual cancellation of the two changes (in other words, the reversibility of the process) could be perceived.

The absence of a meta-level in the thought of pre-operatory children is seen in another one of Piaget's question types—the one exemplified above by the question

16. (Piaget 1967: 25). Twentieth-century adult thought, of course, may still answer *some* questions of "why" as concerned with purpose. The motor-road around Roskilde is there "because cars should be led around the centre of the town" and not "because workers levelled the surface and spread concrete." But the Alps are there because of geological forces and not because they constitute such a nice skiing resort. At least in the era of secularization, adult thought tends to deanthropomorphize processes not performed by human beings.

whether there were "more girls or more children in the kindergarten." The point in the answers given by pre-operatory children is precisely the absence of a meta-level on which the total category of children can be comprehended *together with* the two distinct subcategories. In the absence of this level, singling out the girls automatically transforms the notion of 'children' into 'the remaining children', that is, 'boys'. The mental process is really an ongoing process, a chain where each step supersedes its predecessor step and makes it inaccessible to renewed treatment.[17]

The radical egocentricity, exemplified by the use of *that one* as a valid term for "monkeys," retreats during the pre-operatory phase, and from the age of four the child is able to participate in what the adult interlocutor perceives as sensible dialogue. This demonstrates that mature pre-operatory thought is a functioning and relatively adequate structure; relatively adequate functioning is indeed the very condition that a cognitive structure can be stable and thus mature. It may therefore seem misleading

---

17. Cf. Piaget 1950: 113. It is sometimes claimed that the young children produce their "absurd" answer not because of any inability to understand "correctly" but simply because they assume the psychologist to have something sensible in mind. That this is no adequate explanation follows from an observation which I made on my younger daughter when, at the age of five, she was close to the end of the preoperatory period (having already learned rudimentary counting). When asked by her sister (with whom I had just played with more complex logical problems) whether there were more girls or more children in her kindergarten, she replied "One cannot know" (this phrase we had just used in the preceding play). I then produced a model, telling her that her fingers were the children, those on the left hand the girls and those on the right hand the boys; when asked whether there were more children or more girls, she counted the children on one hand and the girls on the other—without completing the process, however, already knowing the answer she was going to give: "Equally many!"

Next I tried to let her count once more "all the children"; after she had completed the five on her right hand, I took her hand and continued with "six" on her left—and before I reached "seven" she burst into violent crying, screaming "one can't do that"; without my knowing it, my seemingly innocuous experiment had violated her world order.

As one may imagine, I did everything I could to comfort her, admitting that it could not be done according to the way she thought, but that it could in the way her older sister and I thought. I tried to explain,speaking about yellow, blue, red, and violet flowers, telling her to keep the yellow in one hand and the rest in the other. When asked, she was able to point out the yellow in one hand and "all of them" in both—and to the question where there were more she answered "All together." Returning to the question of girls and children, which I illustrated once more with closed fists in order to prevent counting, she answered "Boys and girls" (avoiding again the superordinate concept). I told her older sister that now she understood our way to think—but the younger commented with a "No!"

Two hours later, however, she could understand *nothing but* the operatory structure, and for days she went on enthusiastically posing me analogous questions about mackerel and fish, and other categories as bizarre as possible.

Neither crying nor enthusiasm would certainly have been produced if nothing but a linguistic misunderstanding had been involved.

to characterize this thinking by *what it is not (yet),* that is, not based on operations (to be explained presently). But in spite of the relatively adequate functioning of the thought of the period, its irreversibility and centering character, as well as a less radical egocentricity, remains for years, and a meta-level does not develop. As an alternative to the negative characterization *pre-operatory,* Piaget (1950: 129; 1967: 29) therefore uses the term *intuitive thought* in order to describe the typical thought of children between four and six as *what it is:* stating opinions without argument or support from facts—how should, in fact, a chain of arguments be constructible if you cannot step outside the chain of your own thinking? And what would be the use of arguments if you do not see your own thought as only one possible way to think which your interlocutor does not necessarily share?

The negative characterization *pre-operatory* is a characterization of that which typical preschool thinking is not yet but is on its way to preparing. Around the age of seven, children will know that there are more children than girls in a mixed school class, and more flowers than primroses in the garden. Similarly, they will now claim that the sausage and the clay ball contain the same quantity of clay. In the latter case, the reason normally given is typical of the acquisition of reversibility: you could roll the sausage back into its old shape. The idea of 'same' is, in fact, quite empty or at least unspecific; it will take years before the child is able to foretell that the weight of the sausage is not changed, and still more before it will predict that the volume (that is, the increase in the level of water in a glass into which the clay is immersed) is conserved. Similarly, the child who knows that flowers outnumber primroses is the garden (and who states that not all animals are birds, since "there are also snails, horses . . ." whereas all birds "certainly" are animals) will not be able to decide whether there are more birds or more animals outside the window.[18]

In Piaget's words (1973: 24), the reason for the latter failure is probably that "flowers can be gathered in bouquets. This is an easy concrete operation, whereas to go and make a bouquet of swallows becomes more complicated; it is not manipulable." Initially, the intellectual operations of the next cognitive structure are thus strictly bound up with concrete imagination. But they are what Piaget calls "operations," that is, "interiorized (or interiorizable), reversible actions (in the sense of being capable of developing in both directions and, consequently, of including the possibility of a reverse action which cancels the result of the first), and coordinated in structures . . . which present laws of composition characterizing the structure in its totality as a system." (Piaget 1973: 76; cf. 1967: 78) The explanation is beautifully

18. This delay of the unfamiliar as compared to the familiar is obviously contradicted by my observations reported on p. 245, n. 17. The difference is to be explained by the blow to which I had unknowingly exposed my daughter's world order, and illustrates the importance of affective factors even in the domain of seemingly neutral cognition.

illustrated by the "hundred thousand million birds" plus and minus eight.[19] Only when such operations have emerged can a concept of counting be constructed (younger children see no problem in having mislearned the number jingle in such a way that it ends in a loop, for instance, 1–2–3–4–5–6–7–8–9–10–11–7–8–9–10–11–7–8–...); only then is conceptual analysis possible. In general, operatory thought is felt by adults to be "logical"; actually, Piaget used the term *pre-logical* as late as 1940 instead of the technical term *pre-operatory* which he was to coin later (see Piaget 1967: 30).

Operatory thought starts in the most concrete domain, as we have seen. Gradually, larger areas are assimilated: at the typical age of ten, no doubt will remain that there are more animals than birds outside the window, nor that the weight of the clay ball does not change when it is transformed into a sausage; at the age of thirteen, most children will admit that even the volume cannot have changed. Still, the logic of operatory thought remains a *logic of the concrete.* Instead of seeing assimilation as an extension of operatory structures so as to cover increasingly abstract domains, we should indeed rather see it as an integration of increasingly wider ranges of experience into the realm of the familiar and thus concretely imaginable.[20] Purely formal operations remain inaccessible. This is illustrated by the inability of typical children below twelve to answer correctly one of the old Burt problems, an example to which Piaget returns time and again (e.g., 1950: 149): "Edith is fairer than Susan; Edith is darker than Lily; who is the darkest of the three?" The child of ten will mostly argue that both Edith and Susan are fair, while both Edith and Lily are dark. Edith must thus be in between, Lily must be darkest, and Susan is fairest. Only at (typically) twelve will the child be able to argue correctly from the purely formal sequence, for example, by inverting the first statement into "Susan is darker than Edith." From now on, the child (now, normally, an adolescent) will be able to manipulate symbols in agreement with abstract rules and without consideration of their actual meaning.

The ability to handle such purely formal problems should then demarcate the emergence of the fourth (and definitive) cognitive structure. My own investigations and observations of mathematical (that is, supposedly purely formal) reasoning, as well as the research of the last thirty years regarding the importance of writing (and,

---

19. With the qualification that the girl had been subjected to so many Piagetian problems that she was able to manipulate fairly abstract entities like the fancy number in question.

20. This follows clearly from my experience as a teacher of mathematics and physics to grown-up students of engineering or mathematics. Even many of these would make typical preoperative (centering and irreversible) errors as soon as problems dealt with entities with which they had no familiarity, and which they could not imagine concretely (for example, sound energy density).

by implication, material representation) for the mastery of abstract thought,[21] make me doubt the validity of the absolute distinction between concrete and formal thinking. The typical mathematician will immediately transform Burt's problem into (actual or imagined) writing: "E>S, E<L"; turn the first statement around into "S<E"; and combine the two into "S<E<L" (according to my repeated experience, most of them resort to pen and paper). This is no longer a sequence of formal operations but a progression of quite manifest manipulations of visually familiar and thus "concrete" entities, facilitated by an iconic symbol[22]—concreteness being, in fact, no immanent characteristic of the object but a characterization of the attitude of the knowing subject to the object (in themselves, birds are no less concrete than primroses).

Regardless of the absolute or only relative character of the transition from concrete to supposedly formal thought, the acquisition of the ability to handle mentally the endless range of problems outside the realm of direct concrete experience (be it by means of formal thought or through the construction of pseudo-concrete representations) is of course an important step; without this step, in particular, scientific knowledge could never be achieved.

SUPPLEMENTARY OBSERVATIONS

The characterization of the periods of cognitive development is a skeleton; in order to dress it in flesh and skin, one has to make a number of supplementary observations, of which I shall introduce a few—some from within the Piagetian perspective and some from the outside.

The first observation to be made extends the remark that "children of three are no less sensori-motor efficient than they were a year before." Nor are they at seven, nor at twenty-five. Sensori-motor learning continues when you learn to ride a bicycle, when you begin driving a car, and so on. Certain skills, furthermore, are learned at

---

21. Summarized in Ong 1982. If, as it seems, supposedly formal thought can only develop on the basis of writing (or some equivalent material representation), there is nothing strange in the apparent absence of formal thought from cultures without writing. Literate cultures tend to consider thought based on interiorized writing as "formal" but to see interiorization of other material representations as "concrete." The claim that "primitive man is unable to think formally" (which was current in this or similar forms in the psycho-ethnography of the earlier twentieth century) is thus nothing but a self-promoting reformulation of the quasi-tautological observation that "cultures without writing do not base their thought on writing."

22. Not only is the symbol < smaller in the end pointing to the smaller entity; the relation between < and > also corresponds to the spatial reversal in the written line which transforms one relation into the other. Most important perhaps: one < can be located within the other, as <<, in a way which appeals directly to the sensori-motor experience of putting a smaller box into a larger one and this into a still larger one.

the level of conscious thought and then absorbed as subconscious sensori-motor skills (for instance, changing gears in a car, or tying the bow knot of your shoes). To an increasing extent, the basic or primitive cognitive structures are integrated with the higher (conscious) structures and made subservient to these; as we shall see presently, however, this integration is far from complete and is no unproblematic process.

Cognitive structures were totalities integrating many schemes characterized by shared basic features and limitations. Even when broadened so as to encompass bicycle and car riding, sensori-motor cognition is thus different, and distinct, from operatory cognition. None the less, schemes may in some sense be transferred from one level or structure to another or, perhaps with a better metaphor, serve as models for the construction of analogous schemes at new levels. Such a suggestion is made occasionally by Piaget (e.g., 1967: 25f, 48). It is indeed a characteristic of operatory thought that it achieves at the level of thought the same decentration, reversibility, and composability which was achieved at the sensori-motor level during the second year of life. The process could be formulated as an assimilation of verbal and non-verbal thought to sensori-motor schemes.[23] Correspondingly, the characteristic schemes of formal thought could be seen as an assimilation of sentences to schemes used to deal with concrete objects, in as far as it is at all justified to distinguish these two levels (cf. above).

A different type of observation concerns the status of Piaget's favorite discriminative experiment: *conservation* (the clay sausage and its kin). The understanding of conservation is more or less taken by Piaget to be the essential content of operatory thought. Cross-cultural studies make this doubtful. Australian aboriginal children are indeed unable to master conservation at an age where European children do so; use of maps, however, which also requires operatory thought and only comes later to European children, comes within their range at the age when European children grasp conservation (Cole & Scribner 1974: 152f). It seems that operatory thought is a rather open-ended potentiality, the precise actualization of which depends to a

---

23. The schemes of which Piaget speaks in this connection can be characterized in a broad sense as "logical schemes": the fact that there are more flowers than primroses does not depend on the specific nature of the two categories but only on the fact that the first category encompasses the second without being itself exhausted. It might be worthwhile investigating (via the "spontaneous grammar" of young children, see, e.g., Bickerton 1983) whether the development of grammatical schemes can also be correlated with the preexisting structure of sensori-motor schemes.

There are good reasons to expect such a correlation. On one hand, the area of the cerebral cortex that is the main responsible for processing syntax is the very area that processes spatial information (commentary by Ron Wallace in Burling 1993: 43f). On the other, in almost all cases where the origin of grammatical case systems can be traced, they derive from frozen spatial metaphors (cf. Anderson 1971).

large extent on that cultural practice which brings one or the other kind of mental operation to the fore.

A final cluster of observations concerns the mutual relation between the different co-existing cognitive structures. Firstly, there is the phenomenon of *cognitive regression.* It was explained above how operatory structures only assimilate unfamiliar domains gradually, as they become familiar. Familiarity, however, is only one factor. Engineering students are more likely to fall back on intuitive thought at exams than during daily teaching. This recourse to basic cognitive structures is a fairly common consequence of emotional stress, and can be observed on many occasions: in heated discussions, when you try to find by systematic trial-and-error the code of an unknown bicycle lock and get nervous, and so on. In the history of science, repeated examples of elementary blunders committed by eminent minds approaching the borders of their understanding could be listed.

The choice of cognitive structure to apply in a given situation is indeed no conscious decision. Nor are, in fact, the schemes themselves present to conscious awareness. This is, of course, true in the case of the infant applying the ball scheme or constructing a grammatical form *wented;* but it is equally true of the daily thinking of the professional philosopher. He, of course, may notice that his conclusions follow from a particular syllogistic scheme, like the classical "All men are mortal; Socrates is a man; thus Socrates is mortal." But he needs no syllogistic scheme to know it, and he knows it immediately before correlating it with the scheme. The schemes inside which our conscious thinking takes place are themselves unconscious (cf. Piaget 1973: 31–48).

This holds in general, and normally we do not think about it. At times, however, the complex interplay between that of which we are aware and that which does not come to awareness may produce paradoxical errors and failure to grasp correctly what one is able to do correctly in practice. For an example of this "cognitive repression" I quote Piaget:

> A child is given a sling in its simplest form: a ball attached to a string which is whirled, then aimed at a goal. At first, there is no goal whatever and the child enjoys whirling the ball at the end of the string and then letting it go, noting that it flies off from his side (and in general even seeing that it flies off in the extension of the rotary direction). Next a box is placed thirty to fifty centimetres away and the child, often as early as five years old, quickly manages to reach the box by whirling the ball from his side (about nine o'clock, if we consider as clock dial the rotation surface, the box itself being placed at noon). Having done so, the child is complimented; he begins again several times and is asked where he has released the ball.
>
> A strange reaction then occurs. The youngest children claim that they released the balls exactly in front of them (about six o'clock) and that the ball left in a straight line, from six o'clock to noon (the diameter

> of the rotary circle) into the box. Others (children aged seven to eight) claim that they released the ball at noon, that is, facing the box. About the ages of nine to ten, there are often compromises: The ball is released about eleven or ten-thirty, and it is only about the age of eleven or twelve that the child replies at once that the ball left at nine o'clock, that is, tangentially and no longer facing the goal. In other words, the child soon knows how to accomplish a successful action, but years are needed before he becomes aware of this, as if some factor were opposed to this knowledge and retained in the unconscious certain movements or even certain intentional parts of successful behavior.
>
> The factor behind inhibition is easy to discover. The child represents his own action as divided into two periods: spinning the ball, then throwing it into the box; whereas without this goal he throws the moving object anywhere. But, for him, throwing to the goal supposes a perpendicular trajectory to the box, thus a release facing it. When asked to describe his action, he thus reconstructs it logically as a function of this preconceived idea and hence does not wish to see that actually he proceeded differently. Therefore he distorts and even dismisses an observation contrary to the idea he has and which alone seems right to him.
>
> (Piaget 1973: 36ff)

At the sensori-motor level, the child thus knows correctly what to do. But this correct knowledge is not brought to awareness, because it disagrees with a pre-existing conceptual scheme which reconstructs the process wrongly; one might add that even the correct awareness of the oldest children is likely to result from a reconstruction—albeit a better one—and not of immediate observation. The path leading from sensory perception, and even from intentional action, to recorded observation is far from direct.[24]

Most often, indeed, awareness only results from challenge or conflict. If a wrong reconstruction is used as the basis for the planning of further action, one may be forced to recognize that something is wrong (because the sling ends up in a wrong

---

24. This does not hold for children alone. Once Piaget's collaborator A. Papert had investigated whether children are able to tell afterwards what they did when walking on all four. The result was that the youngest provided a physically rather impossible explanation (movement "in Z," first the arms, then the legs, . . .). Somewhat older children would provide a physically possible explanation which did not agree with that they had actually done ("in N," both left limbs, both right limbs, . . .; obviously copied from ordinary walking). Only the oldest children tended to produce a correct description of what they had actually done (movement "in X") (Piaget 1976: 1–11).

Before they presented this result to an interdisciplinary symposium, Piaget and Papert had the participants walk on all four and then asked them to describe in writing what they had done. According to Piaget's account, physicists and psychologists tended to give the correct description, whereas logicians and mathematicians gave the physically possible but actually wrong description (Piaget 1973: 41).

place), and then be led to better understanding. When the activity in question is unproblematic and everything functions as expected, awareness is superfluous and need not arise. We are normally quite unaware of the precise mechanisms of walking, and we have no reason to produce awareness; walking can be integrated without such awareness even in consciously planned movement, for example according to a map. It is only when the terrain is utterly difficult, if one of our feet is severely hurt, or in similar situations, that we are forced to focus awareness on the actions involved in walking.

The moral of this observation is of extreme importance for any theory of knowledge. Since theories are themselves products of awareness and are addressed to conscious awareness, they tend quite naturally to identify knowledge with conscious knowledge. But the larger parts of our knowledge—firstly the basic schemes, but secondly even many of the actions performed within these schemes—are unconscious, and as much as possible remains unconscious as long as absence of awareness gives rise to no problems. Many activities, moreover, which are learned at the conscious level are removed from consciousness by repetition and training (training is indeed intentional removal from consciousness); changing gears and tying shoes were mentioned above as examples. This is a question of simple economy: conscious awareness can keep track of only a few processes at a time, and it does so relatively slowly; it therefore has to be reserved for these aspects of our actions which cannot be accomplished automatically. It is also easily distracted and thus more likely to commit errors than automatically performed routines.

A theory of scientific knowledge which does not take these observations into account can be blamed for neglecting the important fact that even scientific knowledge is human knowledge. It is inadequate, either by being unable to understand the economy and ease by which even scientific cognition works through the automatization of subprocedures, or by copying the incomplete awareness of the acting individuals and not recognizing the pertinence (or the very existence) of those procedures which are automated. In both cases, it also fails to explain the errors to which automatization gives rise on certain critical occasions.

THE STATUS OF SCHEMES AND CATEGORIES

These are questions to which we shall return (p. 296 onwards). At first, however, we shall have to close the discussion of individual cognition.

Kant, we remember, held certain categories to be inescapable frameworks without which we cannot know. Piaget, on his part, investigated how these categories arise, and demonstrated that at one moment in life these categories and the schemes in which they are organized were not yet part of our cognitive equipment—not only not inescapable but in fact inaccessible. At that time we gathered and organized our experience in other ways. The Kantian equipment is only a preliminary outcome and not the starting point of our knowing about the world.

But why do we end up with precisely these categories? Might suitably planned education have produced a cognitive structure not based (among other things) on the category of *permanent objects,* on the expectation that a toy which has been concealed must still *be* somewhere, and that a rabbit drawn out of the magician's top hat must have been hidden in his sleeves? Might our world have become one possessing no fixed boundaries between any *this* and any *that,* a world ever fluid and elusive?

This other world is difficult to describe, and for good reasons: it is not our world and not the world to which our concepts and our language are adapted.[25] Are then the separation of our world into objects, our logical schemes and categories simply an implicit message we could not avoid when learning our language and learning to describe our world in language?

This is certainly not completely false, as illustrated by a cross-linguistic example. Chinese does not allow the enunciation of counterfactual statements like "if printing had not been invented, then the industrial revolution could not have occurred" but only approximate equivalents translatable into "Printing has not been invented, and therefore the industrial revolution has not occurred." In a test where English-speaking subjects are asked for a conclusion from the first formulation combined with the statement "The industrial revolution has occurred," they will have little difficulty in deciding that printing has been invented. Chinese confronted with their equivalent are likely to protest that "Printing has been invented" and thus not to accept the game. Bilingual Taiwanese Chinese, moreover, tend to react "in English" when asked the test in English and "in Chinese" when asked in Chinese (Bloom 1979). Without adequate support in language (or other symbolizations) that can express them, higher logical schemes are not easily accessible.[26]

---

25. Writing from a pre-Kantian perspective which does not distinguish sharply between *our* world and *the* world, Plotinus, the founder of Neoplatonism, expresses this experience forcefully when approaching *the One* as the supreme principle of divinity:

> It is in virtue of unity that beings are beings.
>
> This is equally true of things whose existence is primal and of all that are in any degree to be numbered among beings. What could exist at all except as one thing? Deprived of unity, a thing ceases to be what it is called: no army unless as a unity: a chorus, a flock, must be one thing. Even house and ship demand unity, one house, one ship; unity gone, neither remains. . . . . Take plant and animal; the material form stands a unity; fallen from that into a litter of fragments, the things have lost their being; what was is no longer there; it is replaced by quite other things—as many others, precisely, as possess unity.
>
> (*Enneades* VI.9, trans. MacKenna 1969: 614)

26. The same influence of language may be the reason that Chinese mathematicians, when adopting Euclid's *Elements,* saw no reason to distinguish a theorem from its converse (Martzloff 1988: 105).

But most of the fundamental categories develop as practical categories before language is acquired (like the toy under the Basque beret), and others (like the girls-children scheme) may be present in language but systematically misunderstood until the moment when our cognitive structure is ready to use the linguistic structure as mature speakers do. Furthermore, basic categories like the permanent and separate object are common to all known languages, which would hardly be the case if language itself was not constrained in this domain. As far as the basic categories are concerned, language appears to be only a secondary regulatory factor, a support which stabilizes the incipient formation of individual categories and structures by lending them adequate symbolic expression.

This seems to leave us with two possibilities. The categories may be determined by our perceptual and nervous apparatus; that they develop may then reflect that this apparatus is not fully evolved when we are born but needs to go through a *process of maturation*. Or they may really correspond to the structure of the world in which we live; their development then reflects a *process of discovery*.

None of the two explanations taken alone is satisfactory. If nothing but maturation is involved, why should children in the Iranian countryside acquire operatory structures at a later age than urban children from the same country while being no less intelligent according to tests which do not require operatory thought (see p. 241 n. 10)? And why should Australian aboriginals develop operatory structures in an order which differs from that of Swiss children? On the other hand, whatever the structures in which the world around us is ordered, we would not be able to adopt them into our cognitive structure if we were not in possession of an adequate nervous system.

This can be illustrated by another reference to the category of permanent objects.[27] Frogs and other more primitive vertebrates possess no scheme for the permanent object; frogs jump at the direct perception of visual signals (for example, representing flies). A frog whose eyes have been turned around in a surgical operation will jump in a wrong direction for the rest of its life. Early mammals, however, which hunted their prey at night and had to rely on the integration of sound impressions over time in order to construct the path of the prey, developed a larger brain and a scheme for permanent objects—more precisely: a larger brain which allowed them to organize their sensual impressions as representations of permanent objects.

The larger brain with which mammals (and birds) are born has large biological costs. Birds and mammals are born immature and defenseless, and would never survive if their parents left them to themselves as crocodiles do with their offspring (mammals indeed *became* mammals as a way to take care of their litter). If the possibility to experience the outer world as consisting of permanent objects had not implied definite advantages which could balance this cost, selection pressure would

---

27. I rely for this on Jerison 1973 and *id.* 1976.

soon have eliminated the larger mammalian brain. We cannot conclude that the world *consists of* permanent objects. In fact, it does not in any absolute sense: the fox pursued by the hounds exhales and transpires, and matter which in one moment is fox is not a moment later; physics tells us that we may analyze the fox into atoms and into still smaller particles. What we *can* conclude is that the material world is constituted in a way that allows an adequate practice if we order our perceptions as representations of permanent objects. Only in this sense can we say that the world in which we live "is" itself structured in permanent objects.

Mutatis mutandis, the same will hold for other fundamental cognitive categories and schemes. We cannot claim that the world "in itself" functions according to logic and mathematics, but still that it functions in such a way that seeing it as if it functioned thus allows an appropriate practice. The evolutionary level at which such other categories and schemes have evolved will evidently differ from that where the possibility of the conserved object was discovered, and operatory thought (without which neither complex logic nor mathematics is possible) appears only to have arisen during the process of hominization.

CHAPTER 17

The Nature and Demarcation of Scientific Knowledge

Knowledge is always known by somebody, if not actually then at least before it was stored in books or other receptacles; and stored knowledge remains knowledge only to the extent that it can re-emerge as the knowledge of somebody. Therefore, all knowledge partakes somehow in the characteristics of individual knowledge.

But part of what we know is *only individual* knowledge; part of what we know is *only accidentally shared* by groups of people who happen, for example, to be witnesses of the same events; and part of our knowledge is *produced so as to be communicated and shared.*

Part of what we know, furthermore, concerns *particular occurrences*—for instance, the rainy weather in Copenhagen at this moment; another part is of a *more general character:* for instance, describing the Danish climate or the dynamics of cloud formation and movement. Some of it, finally, consists of *isolated bits,* whereas other parts are *built up as wholes* ("theories" and the like) whose single constituents only obtain their full meaning as components of these wholes.

Scientific knowledge is produced so as to be communicated and shared in stored form; it is generalizing in character; and it consists of wholes. It is, furthermore, produced by communities with a strong internal interaction by means (*inter alia*) of stored knowledge (books, scientific journals, letters). For all these reasons, scientific knowledge possesses particular qualities.[1]

1. From another point of view, it is true, these characteristics of scientific knowledge highlight general conditions for knowing, or at least for *knowing explicitly:* all expressible knowledge presupposes a language in which it can be expressed (whence also shared and stored), and language is, if anything, common property of a community. John Donne's maxim "No man is an island" holds for human knowledge no less than for the rest of our existence.

Some of these concern the relations between producers, users, and (their existence should not be forgotten) victims[2] of the knowledge in question, and the links between its production and its social and technical uses. Others (and they are the theme of this and the following chapters) concern more purely epistemological questions. For instance: Is scientific knowledge true? If it is, in what sense and under which conditions? If not, why do we then rely upon it, so often with considerable technical success? And so on.

Since scientific knowledge normally ends up as stored knowledge, traditional philosophy of science has approached these epistemological questions with regard to the stored form of knowledge. In recent decennia, however, certain workers in the field have insisted (against strong opposition from traditionalists) that genuine understanding of scientific knowledge (including understanding of the stored final

---

Any piece of knowledge, moreover, however much it may deal with particular occurrences, locates these within a framework of general concepts (cf. the observations concerning Windelband on p. 167).

2. Since these victims are none the less only indirectly visible in what follows, it may be suitable to repeat the sonnet in which Adalbert Chamisso—poet, Jacobin, scientist, and believer in scientific progress—commemorated the hundred oxen which Pythagoras is told to have sacrificed when discovering "his" theorem:

Die Wahrheit, sie besteht in Ewigkeit,
Wenn erst die blöde Welt ihr Licht erkannt;
Der Lehrsatz nach Pythagoras benannt
Gilt heute, wie er galt zu seiner Zeit.

Ein Opfer hat Pythagoras geweiht
Den Göttern, die den Lichtstrahl ihm gesandt;
Es thaten kund, geschlachtet und verbrannt,
Einhundert Ochsen seine Dankbarkeit.

Die Ochsen seit dem Tage, wenn sie wittern,
Daß eine neue Wahrheit sich enthülle,
Erheben ein unmenschliches Gebrülle;

Pythagoras erfüllt sie mit Entsetzen;
Und machtlos sich dem Licht zu widersetzen
Verschließen sie die Augen und erzittern.
(*Werke* II, 212)

(Truth endures eternally / once the inane world has perceived its light; / the theorem, named after Pythagoras / is valid today, as in his times. // A sacrifice did Pythagoras dedicate / to the Gods, who sent him the illuminating ray; / one hundred oxen, slaughtered and burnt, announced his gratitude. // The oxen, since that day, when they suspect / that a new truth may reveal itself / bring forth an insufferable roaring: // Pythagoras fills them with horror / and without power to resist the light / they close the eyes and quiver.)

phase) can only be achieved if we understand its original *emergence in process* as individual knowledge generated within a field of specific social interaction.

Chapters 18 and 19 will take up the latter approach. In the present chapter I shall introduce some fundamental concepts and terms and look at some of the established approaches and their problems.

A PSEUDO-HISTORICAL INTRODUCTION TO SOME KEY CONCEPTS

It may be profitable to have another look at the ancient anecdote told on p. 8 and to expand the commentaries made there.[3] It ran as follows:

> One day Plato the philosopher met his fellow philosopher Diogenes, who, as so often, made a teasing comment on Plato's philosophy. "Good Friend," he said, "Table and cup I see; but your Tablehood and Cuphood, Plato, I can nowhere see." "That's readily accounted for, Dear Diogenes," replied the other. "You have that which is needed to see the table and the cup: that's the eyes. But you lack what is required to grasp Tablehood and Cuphood: namely the intellect."

This story locates the two philosophers with regard to several concepts and problems which are central to established philosophy of science:

Plato is an *idealist.* That is, to him things exist primarily *as knowledge or thought,* in a mind, as concepts or *ideas* (in Greek: "shape," "appearance," "sort"), whence the term. The single material representatives of the concepts—the tables as representatives of Tablehood, the tragedy *Medea* as a representative of THE TRAGEDY as a genre—are precisely that: representatives *depicting* genuine reality.[4]

---

3. The characterization of the present section as "pseudo-historical" should be emphasized. It does not introduce the points of view of ancient philosophers in the context of their total thinking (and still less in the general context of their times—but cf. the presentation in Chapter 2); nor is it faithful to the real complexity of their ideas. It is rather a case of intellectual bank robbery, a rash exploitation of the historical material intended to procure a pedagogical introduction to a number of themes which have stayed important in the philosophy of science ever since antiquity.

4. In order to do Plato justice, we should repeat that the "Tablehood" and "Cuphood" of the anecdote are parodic distortions of Plato's real doctrine, which is concerned with "ideas" like COURAGE and THE GOOD, modelled on that TRIANGLE which is the real object of mathematical proofs even when a particular triangle has to be drawn for the proof.

Nevertheless, the parody has a point in itself, calling attention to an inner problem or even inconsistency in the Platonic and similar doctrines, already pointed out by Aristotle (*Topica* 143[b]). Some geometrical proofs concern *the triangle,* others only *the right triangle,* still others all polygons. Does this mean that there exists a particular idea of the 'Right Triangle', and which are then its relations to the idea of the 'Triangle'? Does, in a similar way, the idea of the 'Artefact' (or 'Material Object') split up into sub-ideas for tables, cups, . . . ? When

Diogenes on the other hand is a *materialist:* only tables and tragedies exist; Tablehood and TRAGEDY—the ideas—are *our* inventions. He is also a *positivist:* we must distinguish that *positive* knowledge which we get from observation of the real world from those figments of the mind which we add ourselves—interpretive frameworks, concepts, generalizations, and metaphysical entities (nobody ever saw or put hands upon the *force of gravitation,* an *animal* which was not a particular individual belonging to a particular species, *justice,* or *the tragic dimension of human existence*).

With regard to another dichotomy, Plato can be characterized as a *realist:*[5] according to him, the universals (general concepts, ideas) possess the status of something really existing. Plato is indeed an *objective* idealist. Tablehood and other ideas do not have their fundamental existence as images in our individual intellects—these are merely our only access to that higher Universal or Divine Mind where they have their real abode. Diogenes, on the other hand, is a *nominalist,* and regards the universals as nothing but *names* (puffs of the voice, as the nominalists of the late medieval universities used to say) which *we* invent and use as shorthands to sum up collections of particular experiences or objects.

> As one may perhaps guess from the expression *objective idealism,* another, "subjective" idealism exists. Much modern positivism—but in particular empiriocriticism, (cf. p. 156)—is of this breed. According to subjective idealism, every reference to an external reality is ultimately nonsense: our mind has access to and registers nothing but our *sense impressions.* These form a forever impenetrable screen between our mind and anything which may lie behind.[6]

---

taken to this consequence, the simple structure of the universe of ideas dissolves into an indefinite and virtually infinite number of nested and intertwined sub-, super- and interwoven ideas.

5. In later years, it is true, the traditional term has been taken in a somewhat different sense, grossly corresponding to materialism: the outer world exists, and our knowledge is about this (but in as far as our knowledge is formulated in general concepts, claiming that *these* exist may still come close to Platonism). To avoid ambiguities, we may then speak of Plato's attitude as an instance of *concept realism,* the doctrine according to which concepts correspond to really existing entities.

The reason for this rather confusing linguistic innovation appears to be political: Since the term *materialism* has mainly been used by Marxists since the late nineteenth century, better avoid it, whatever lack of clarity results!

6. The ultimate consequence of this principle is *solipsism* (from *solus ipse,* "oneself alone"), the tenet that one's own mind is the only thing of whose existence one may be sure—everything external, including other persons and their minds, may be nothing but dreams and imagination (however these are to be understood without being contrasted with perception of the real world). No philosopher likes being considered a solipsist, for which reason a variety of protective devices have been invented by subjective idealists. But if the same strict logic is applied to these inventions that made their inventors accept the sense impressions as an impenetrable screen, they evaporate immediately.

At this point, we have probably exhausted the philosophical contents of the anecdote. In order to complete the list of essential concepts, we will have to involve a third ancient philosopher: Plato's rebellious follower Aristotle.

Aristotle, as pointed out repeatedly in part I, was an *empiricist.* According to his view, all knowledge derives from experience and thus comes through our senses. But Aristotle was no positivist and no nominalist. He held that both the particulars (the single tables, the single tragedies) and the universals (Tablehood and TRAGEDY) exist—the latter as *essential* characteristics objectively shared by the directly observable tables and tragedies, respectively. The universals constitute the *essence* which lies behind and determines the *phenomena,* that which *appears to* and can thus be *observed by* our senses.

But Aristotle was no idealist of the Platonic ilk. He did not regard the *forms* (a term which is usually preferred when Aristotle's philosophy is discussed, cf. p. 37 n. 16) as the only really existing entities, and the phenomena as merely fleeting and ephemeral, imprecise and ultimately nonexisting representatives. Tablehood is not something existing besides and above the particular tables; Tablehood only exists as tables—the essence is only there *in* phenomena. Aristotle is thus a materialist, but he is no positivist. With regard to the dichotomy between realism and nominalism, he has been labelled a *moderate realist.* Diogenes represents a flat rejection of Plato's point of view—an antithesis, in the language of Hegelian dialectic; expressed in the same idiom, Aristotle can be regarded as integrating and going beyond both positions, producing a genuine *synthesis.*

According to Aristotelian empiricism, it is the aim of scientific investigation to determine the essence of things through observation of phenomena. The method is *induction:* Examination of many horses allows us to find that which by necessity will hold for anything that is *horse*—that is, to find the *essence* of the horse; examination of the constitutional history of many city states will lead us forward to knowledge of the essence of the state; and so on.

When induction has thus provided us with the essential truths concerning a field, we may deduce through the application of logic alone what must necessarily hold for particular phenomena without further empirical observation. If it has been shown to be part of the essence of cows to have four stomachs, we need only establish that Karoline is a cow in order to know she has four stomachs.

> Once again, *mathematics,* and in particular geometry, supplies the model for this epistemology, as it is made amply clear by the choice of illustrative examples in Aristotle's main work on the philosophy of science, the *Posterior Analytics.* From thorough investigation of the properties of geometrical figures, we establish a number of fundamental truths ("axioms") concerning ideal points, lines, and figures in the geometrical plane; from these we may deduce, among other things, that the sum of the angles of a triangle equals two right angles (that is, 180°). Once that is done, we need only establish that a given figure is a triangle, that is, contained by

three straight lines, in order to know the sum of its angles; empirical measurement is no longer necessary.

Already in antiquity it was objected to Aristotle's methodology and epistemology that induction can never be logically compelling. Even if we have dissected 10,000 cows, we can never be absolutely sure that number 10,001 possesses the same number of stomachs as the others. Couldn't Karoline belong to that fraction of a percent which has only three stomachs? Or perhaps be *the* exception? To accept four stomachs as an aspect of the essence of the cow can be nothing but an inspired guess, the validity of which is never absolutely guaranteed.[7]

This point of view is classified as *skepticism.* According to the skeptical point of view, science can never establish necessary truths concerning the essence of phenomena, only plausible truths.[8] It will be seen that skepticism, nominalism, and positivism are related philosophies: if we cannot *know* whether we have penetrated to the essence of things (skepticism), then it is close at hand to claim that our general concepts about the structure of the world are mere abbreviations and names which *we* have invented for convenience (nominalism)—and it is tempting to regard all discussion of a not directly observable essence behind observable phenomena as metaphysical rubbish contributing nothing to the real process of knowing (positivism).

---

7. Alternatively, we may of course take four stomachs as part of the *definition* of a cow. But then we shall only know whether Karoline is a cow when cutting her up. If we want to know about empirical reality, definitions do not solve our problem.

This is the point of our other Plato-Diogenes anecdote (see p. 3, n. 4): When Plato had defined *Man* as a biped and featherless animal, Diogenes plucked a fowl and brought it into the lecture room with the words, "Here is Plato's man." In consequence of which it was added, "having broad nails."

For the historical Aristotle, all this was of course no problem (cf. p. 41, n. 25). For him, a cow with three stomachs would simply be an aberration, a specimen that had not realized its form to the full. But this is a view which has been abandoned by Modern science, and therefore merely an example that ancient philosophy *has* to be simplified and distorted if it is to serve as a convenient introduction to the philosophy of *science as we understand it.*

8. One may also speak of "probable truths," as is often done. But if so, then only in a loose sense: logically seen, observation is just as unable to establish that it is "more than 90 percent sure that all cows have four stomachs" as it is to establish a *100 percent necessary* truth. Strictly speaking, talking of quantified probabilities in such cases is pure nonsense, since it tells that "in nine out of ten worlds, all cows have 4 stomachs"—there *is* only one world, as far as we can know.

This observation may seem trivial. None the less, it is often overlooked in practice. The conclusion of, for example, a medical double-blind experiment is frequently explained to be that "it is 95 percent sure that this new drug works;" but what *should* be said is that there is only a 5 percent chance that anything as suggestive as the actual outcome would occur accidentally if the drug has no effect.

According to the ancients, scientific knowledge consists (roughly speaking) in finding the *objects* which exist within this world, and in listing their properties and characteristics (in technical language, to establish an *ontology*). To them, the essences of things could thus be listed as *objects and properties.* Modern sciences, on the other hand, also look for relations, structures, and dynamical interactions. In the perspective of modern sciences, *essences* should thus involve these categories (even though they rarely use the term essence, which smacks too much of pre-Modern thinking).[9] None the less, and irrespective of terminology and the exact character of general or universal features, the old question remains whether universals have any status within reality. *Mutatis mutandis,* the concepts of objective and subjective idealism, materialism, empiricism, positivism, nominalism, and scepticism apply as much in the modern as in the Ancient world.

EMPIRICISM AND FALSIFICATIONISM

The preceding pages are not written with the intention of summing up, however superficially, the history of philosophy. They are only meant—let it be stressed once more—to serve as a pedagogical framework for the presentation of a group of concepts and views that are presupposed (implicitly or explicitly) in any discussion of the philosophy of science, and certainly also in what follows. We shall now leave pseudo-history aside, step a bit outside the confines of professional philosophy, and look at certain attitudes to the production of scientific knowledge which are widespread today even among those who take part in this production.

One of these attitudes is nothing but Good Old Empiricism, coined as a rule of conduct rather than as a stringent philosophy. Formulated in maxims, it runs as follows:

---

9. Evidently, modern sciences *also* encounter the question of existing objects or entities—in linguistics, for instance, what kind of existence to ascribe to the phoneme /t/. The status of sounds may be left to physicists, and so may also the specific *t*-sounds in *ten* and *steam* (the first of which is aspirated and the second not—cf. Robins 1971: 122). Linguists, however, who notice that the first of these sounds (t^h) cannot occur *in English* in places where the second is possible, and vice versa, speak of the two sounds as representing the same phoneme /t/. No physical analysis (whether a frequency analysis of the sound or a description of the sound production process) can reduce the two to one, since phonemic identity is language-specific—the Danish *r*-sounds in *arbejde* and *rede* (one of which is vowelized and the other not) represent the same phoneme; in Arabic, the former would be heard as representing the initial phoneme /ʿ/ of *ʿAlī*, and the second the phoneme /ġ/ used in *Baġdād.* No other science can decide for linguistics whether, and in what sense, phonemes exist—in the speakers' minds or expectations, in some objective structure of language, or as a linguists' shorthand or abstraction.

Scientific explanations are only allowed to make use of concepts and to postulate relations and structures which can be rooted in experience, observation, or experiment. Mythological explanations referring to entities with no such empirical underpinning are inadmissible: they only obstruct genuine scientific insight.

This programme contains some essential points—first of all that science should deal with reality and not be spun out of pure imagination. It also corresponds to the immediate impression which working scientists get from most of their work: we simply describe (so we feel) what we discover in our material (at times after great struggle), including the relations and structures which turn up. But it does not avoid fundamental problems.

Traditionally, the whole complex of logic and mathematics is regarded as falling outside the domain regulated by the empiricist rule. The statement, "if I know that rain makes the street humid, and that it is raining, then I can conclude that the street will be humid," appears to be unproblematically and unconditionally true—expressed in the appropriate language it will have to be accepted by a Touareg in Sahara knowing neither streets nor rain, and it stays true if *humid* is replaced by *dry* or *spherical.* It is a *tautology,* a statement which is true because of its logical structure.

That difficulty is normally solved by means of a distinction (going back to Kant) between two kinds of scientific statements: *synthetic* statements which deal with reality, and for which the empiricist claim must be upheld if they are to be meaningful; and *analytic* statements (the theorems of logic and mathematics), the truth of which follows from their own structure, but which on the other hand tell us nothing about the structure and properties of *reality.*[10]

According to this distinction, the truth of analytic statements is given *a priori,* that is, "*in advance*" (*viz.,* in advance of experience). The truth of synthetic statements only follows *a posteriori,* that is, "*after*" experience.

Kantian philosophy, it should be noted, accepts the analytic and *a priori* character of logical tautologies, the truth of which follows from definitions—"all husbands are male," if we define a "husband" as "the male part of a married couple." Likewise, it accepts the synthetic and *a posteriori* character of normal descriptive statements. But "2+2 = 4" is, according to Kant, a *synthetic a priori:* it cannot be false, whence it is *a priori;* but none the less it tells us something about reality (for example, that

---

10. Albert Einstein once summed up this distinction and its consequences in a nice aphorism: "Inasmuch as the theorems of mathematics refer to reality, they are not certain, and inasmuch as they are certain, they do not refer to reality" (1921: 1).

A remark on terminology may be useful: the English language allows a distinction between *analytic* and *analytical.* The former adjective characterizes statements whose truth follows from their structure; the second means "based on or using analysis." *Analytic* may also be used in the second sense, but the present essay takes advantage of the possibility to distinguish, and observes a similar distinction between *synthetic* and *synthetical.*

two married couples are exactly what you need for playing bridge). The same holds for all those categories which are *a priori* necessary for making experience: space, time, causality, and so on. Precisely because we cannot know reality without making use of these frameworks, knowledge about the frameworks already tells us something about *our* reality—namely, about *the only way in which we can deal with reality.*

As it will be remembered from the previous chapter, Piaget started out to find the roots of the Kantian *a priori* categories, demonstrating that they are in fact the outcome of a genetic process, and hence not properly *a priori.* Since they result from practical experience in the world (experience which is made in interaction between our biologically determined equipment and the outer world), they also seem to be synthetic though dealing with the most general properties of reality—those properties which reveal themselves in any environment in which children are brought up. Piaget's results thus appear to imply a reduction of the classification of statements into a new dichotomy: analytic *a priori,* which only encompass tautologies by definition; and synthetic *a posteriori,* which embrace not only normal descriptive sentences but even the theorems of logic and mathematics.

Certain empiricist philosophers—not least Willard Quine (1963/1951)—make the point that the meaning of words can never be fully reduced to simple definitions (what, for example, happens to the concept of 'husbands' when homosexual marriage is accepted?). This would abolish the category of analytic statements definitively. Ultimately *all* statements would have to be judged in the light of the "empiricist imperative" formulated above; according to Quine's argument, the apparent exceptions result from a too naïve understanding of logic and mathematics.

Closer investigation of the empiricist programme reveals more severe difficulties than those which can be circumvented by segregating logic and mathematics or by arguments that after all they constitute only relative exceptions.

Firstly, experience is never completely pure, that is, made before and independently of every theory; if this was not clear before the invention of the telescope and the microscope, it should at least be glaringly obvious to anybody looking at the technical equipment of a modern scientific laboratory. The cell biologist presupposes the functioning of contrast media and microscopes and thus theories belonging within chemistry and optics; the literary scholar and theatre critic watching *Medea* knows beforehand that what goes on is a play (in any other case he would be morally obliged to save Medea's children[11]); and so on. Experience and theories are elements of a network, in which certain elements can be regarded as more elementary than others and less subject to doubt; they can therefore be regarded as legitimate underpinning and background. But it is impossible to make a sharp cut between *empirical*

11. Jane Campion's film *The Piano* shows the Maoris—unfamiliar with the conventions of European theatre—doing exactly this.

knowledge (constituting a theory-free background) and *theories* derived from this background knowledge, as required by the empiricist ideal. Once again we encounter the absence of a totally firm foundation and the need to build instead on piles hammered through many layers of experience (cf. p. 232, n. 5).

Secondly, the skepticist argument remains valid. However great the amount of your experience and the number of your experiments and observations, you can never derive even the simplest theoretical generalization with absolute logical necessity. All your acquaintance with and examination of winged feathered beings will not force you to bring forth the concept of 'a bird'; reading all the books written to date and listening to all the monologues and dialogues of history will not by necessity make you discover a *deep structure* in language. Even if we forget for a moment about the impossibility of obtaining immaculate empirical knowledge, we are forced to conclude that "science" created in agreement with the empiricist letter will be restricted to listings and descriptions of singular cases, and perhaps to tables and statistics—"natural history" instead of biology. *Theory* will never emerge.

Leading in the formulation of twentieth-century empiricism was the so-called "Vienna circle" and a number of associates, who in the 1920s formulated the program of "logical empiricism" (by others often labelled neopositivism).[12] Leading in the destruction of the program was the same group, astonishing as that may seem: over decennia the school tried out strategies for a complete empirical underpinning of scientific statements, searching for a *verification criterion* by means of which precisely those statements may be singled out that can be founded upon and proved from experience, that is, *verified,* from all those which cannot be verified and which are therefore, strictly speaking, *meaningless.* Their work never produced a verification criterion which could demarcate science from nonscience precisely, but instead led to a triple negative conclusion. Firstly, *existing science* cannot be reconstructed in this way, that is, reduced to a system of single statements which directly or indirectly have their complete reference to experience; secondly, *no science* comprising theoretical statements can be constructed in agreement with the prescriptions; thirdly, *no* verification criterion can be found which distinguishes sharply between empirically meaningful and empirically empty statements.[13]

---

12. "Logical" because of the way in which it tried to get behind some of the vagueness of classical empiricism, including the "empiricist imperative" formulated in these pages: science does not build directly on experience—perceiving something and enunciating a theory belong at different levels. Experience has to be formulated in simple sentences which state what has been experienced ("protocol statements," since these are to be entered into the experimental protocol of the scientist). In agreement with the rules of logic, these statements are then to be combined into higher-level statements (generalizations, "laws," "theories," etc.).

13. The whole argument, together with the arguments against the existence of genuinely analytic statements, is summed up in Quine (1963/1951).

As an alternative to the untenable demarcation by verification, the Austro-English philosopher Karl Popper (1902–1994) has proposed his own recipe. In "naïve" formulation, and thus corresponding to the above "empiricist imperative," it can be summed up as follows:

> *We are allowed to use in our explanations whatever self-invented concepts and hypotheses we like; but we should be aware that our hypotheses are indeed nothing but hypotheses, preliminary explanatory models, and not the truth. We should therefore constantly check our hypotheses as thoroughly as we can, and we must reject them as useless as soon as they enter into conflict with our observations of reality—that is, as soon as they are "falsified."*

Popper did not invent this canon. In the main, it coincides with the idea of the "hypothetical-deductive method," which has been practiced for centuries:[14] Instead of working our way inductively from particular observations in order to find the inner regularities and laws behind phenomena (their "essence" in the classical language), we guess at a set of laws or relations and try to deduce predictions about how things will behave if our guess is correct. If the prediction turns out to be correct, the guess is regarded as strengthened or "corroborated"; if not, we try another guess. Nor is Popper's stress on the forever preliminary character even of corroborated hypotheses original; already C. S. Peirce emphasized that new counter-evidence may always turn up, and labelled this idea *fallibilism.*[15]

Popper's fundamental idea is thus less original than his own writings try to tell. Nevertheless he must be credited with spreading the gospel, and indeed with *making it a gospel,* to such an extent indeed that Peirce's original term is at times used as a synonym for Popperianism.[16] Writing in the wake of the Vienna circle (to which he was close in the late 1920s without being a genuine member), whose use of verification as the criterion of demarcation between meaningful and empty statements was presented above, he was also the first to use *falsification* as a criterion of demarcation:

14. For example, by Newton in his analysis of the planetary system, although he tried to present his results in empiricist garb.

15. Also Peirce of course has his forerunners. In the *Malleus maleficarum,* a handbook in witch hunting from 1484, it is taught that the inquisitor should never pronounce any accused innocent however much she might seem so, only declare that no evidence had so far proved her guilt (III.20, ed., trans. Summers 1971: 241). Even then, new evidence might always turn up. Nobody should be acquitted; cases should be postponed for want of proof.

16. But not always! In his discussion of Popper's methodology and aims, Lakatos (1974a: 93–132, in particular 112 and 114) makes Peirce's term cover the skeptical position that any knowledge, including knowledge of presumed facts that are supposed to falsify a theory, may equally well be mistaken. This "skeptical fallibilism" is almost as far from Popper's philosophical inclinations as can be.

Statements and theories which are compatible with every imaginable situation (and which can therefore never be falsified) have no place within science.[17]

INSTRUMENTALISM AND TRUTH

If our explanations are built on arbitrary constructions and remain forever preliminary and subject to rejection at failure, they cannot be "true" in the classical naïve sense; things which may be false tomorrow (because they may then have been falsified) cannot be true today. If theories cannot be claimed to be *true,* however, the best explanation of their role seems to be that they are *tools* for practical action. There is thus a close affinity between Popper's ideas and *instrumentalism,* the view according to which scientific theories have no truth value, that is, are neither true nor false. Since they *cannot* be judged on the scale of true and false, we should not try to do so. Scientific theories should be evaluated in the way we evaluate instruments, according to effectiveness, for instance, for prediction.

---

17. As examples of such not genuinely scientific theories, Popper (1972: 34) refers to "Marx's theory of history, Freud's psychoanalysis, and Alfred Adler's so-called `individual psychology'." He illustrates this (p. 35)

> by two very different examples of human behaviour: that of a man who pushes a child into the water with the intention of drowning it; and that of a man who sacrifices his life in an attempt to save the child. Each of these two cases can be explained with equal ease in Freudian and Adlerian terms. According to Freud the first man suffered from repression (say, of some component of his Oedipus complex), while the second man had achieved sublimation. According to Adler the first man suffered from feelings of inferiority (producing perhaps the need to prove to himself that he dared to commit some crime), and so did the second man (whose need was to prove to himself that he dared to rescue the child).

In contrast, Einstein's General Theory of Relativity would have been "simply refuted" (p. 36) if starlight had not been seen to be bent around the solar disk during the solar eclipse of 1919. According to Popper, this "risk" taken by the theory is what makes physics *scientific,* in contrast to psychoanalysis and historical materialism.

Rhetorical zeal makes Popper forget that his drowning episode is not of the same kind as the eclipse observation. When it comes to describing single aspects of events taken out of context, physics is no different from Popper's aversions. Physics too may explain that a piece of lead flies upwards (it has just been shot out from a gun) and that it falls downward (it was shot upwards a little while ago and is now falling back), or that water evaporates (the kettle stands on an electric boiler) or freezes to ice (somebody cut out the current, the window is open and the weather frosty). This oversight is rather typical of Popper's ways and may provoke the question why so sloppy a thinker is worth mentioning. The reason is threefold. Firstly, Popper is not always rhetorical and therefore not always sloppy; secondly, precisely his sloppy thinking has become extremely popular; thirdly and finally, coming to grips with Popper is a useful step in the argument that follows.

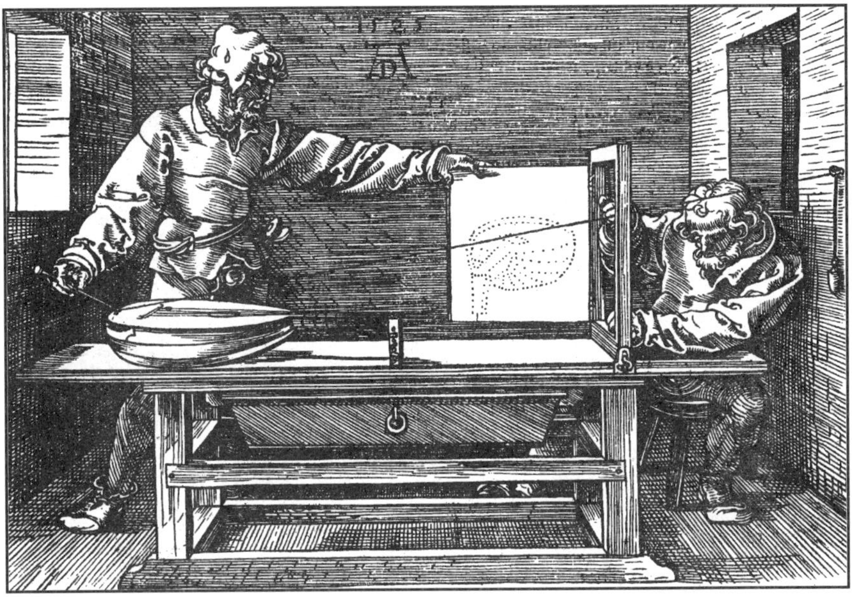

Perspective drawing of the principles of perspective drawing. From Dürer, *Unterweysung der Messung* (1525 edition).

According to instrumentalist tenets, the Copernican theory that Earth and the other planets move about the sun is thus neither more nor less true that the ancient Ptolemaic notion of a fixed Earth around which the sun, moon, planets, and stars are thought to move. Both are applicable as *models,* and our only reason to prefer Copernicus's model will be that it is simpler and therefore gives rise to less complex calculations if both models are built up with orbits corresponding to empirical observations.[18] Being no more *true* than the alternative, Copernicus's model is to be preferred for a reason that will convince any craftsman: it feels better in the hand.

Beyond the ambiguity of the notion of 'simplicity', a fundamental objection against the instrumentalist interpretation of scientific statements is this: instruments can be used for precisely that for which they have been designed; they can be used for other purposes only if their constitution reflects, or corresponds to, *features which are shared* between the intended use and the other possible uses. A screwdriver can be used for many different screws, but only because they all carry a notch; and it can only be used at all because its edge fits the notches and its rotation symmetry corresponds to the rotation by which its target is screwed in. Similarly with theories. We may claim that we judge them according to instrumental value; but we cannot

18. Strictly speaking, the Copernican model is only decisively simpler than the Ptolemaic system if we refer to Kepler's revision of the Copernican theory.

invent that value of theories freely, it is revealed (or denied them) when they are applied—"the proof of the pudding is the eating." The *applicability* of a tool is thus a consequence of its *correspondence with certain features of that reality to which it is applied*—features that themselves are not brought about by the tool. Similarly for theories regarded as tools: their *truth value* can be explained precisely as a *structural agreement or correspondence* with features of reality. Further on, this conception will be spoken of as a *materialist notion of truth.* Evidently, as also demonstrated by the example of the screwdriver, *correspondence* is something quite different from *similarity,* not to speak of *identity:* a screwdriver provided with a notch instead of an edge would be worth nothing. In general, the theory as well as the screwdriver belong to other categories than the reality described by the theory and the screw to be put in by the screwdriver. Reality consists, among other things, of atoms, birds, emotional states, and poems (according to physics, biology, psychology, and literary scholarship, respectively). Theories, on the other hand, consist of words and other symbols. Only from an idealist point of view do reality and theory belong to the same realm, both being in some way idea or concept; but then Plato's idealism postulates an absolute categorical rift between *real* reality, that is, the realm of ideas, and apparent, material everyday reality.[19]

---

19. That theories "consist of words and other symbols" points to another characteristic which they must possess beyond "structural agreement or correspondence with features of reality" if they are to be considered "true": logical consistency (or, put differently: words put together without consistency can correspond to nothing beyond themselves; they are meaningless). Much work has been done in twentieth-century formal logic to render precision to this requirement, which practical scientific workers tend to treat no less commonsensically than the idea of "correspondence with facts." The Polish-American logician Alfred Tarski, in particular, is known for having formulated a "theory of truth" determining the conditions which must be fulfilled by a formal sentence system if it is to possess this logical consistency; he is also known for having shown that attempts to determine the truth or falseness of the sentences of such a system from within the system itself lead to self-referential paradoxes of the type "this statement is false." Truth has to be ascribed from without, by a metalanguage.

Sciences are not written in formal but in technical languages which ultimately derive from common daily language. None the less, Tarski's latter observation is important for understanding the difficulty with which we are presented when we try to understand the nature of the "correspondence" between sentences (or theories) and reality. If correspondence is revealed through interaction with reality ("praxis" in a Marxist sense) functioning in the role of the metalanguage, then it cannot be discussed within the quasi-formal discourse of logical theory but only in a (genuine) metalanguage which steps outside: a metaphorical language which evokes for us *our experience* of such interaction—for example, the above screwdriver. In another metaphor: perspective drawing is a way to render three-dimensional reality in a two-dimensional plane. How you make a perspective drawing can be shown in a perspective drawing (see the figure)—but it is only because we know about three-dimensional reality and move around in it that we give the right interpretation to Dürer's woodcut, and that we see this drawing as a cube and not as a plane jigsaw:

Instrumentalism is thus right in seeing family likeness between a screwdriver and a theory, and has a good point in its subversion of the metaphysical concept of truth which ultimately presupposes an idealist stance; but it is mistaken in believing that the screwdriver and the theory are alike because they are equally arbitrary with regard to reality. Provocatively speaking, we may say that the reason for the usability of the screwdriver is that it possesses a structure which corresponds to certain essential features of the structure and function of screws—it embodies, in materialized form, part of the truth about screws and about our way to deal with them.[20]

One decisive difference remains between the screwdriver and a scientific theory. The instrumental validity of the screwdriver is static and limited; science, on the contrary, is in continual development, constantly searching for new correlations, and ever extending so as to grasp new phenomena. Kepler's Copernican cosmology is more true than Ptolemy's planetary system because it allows a unified treatment of celestial and terrestrial physics (until Kepler, the heavenly bodies were supposed to move according to other laws than those which governed movement below the sphere of the moon).[21] By saying that the reason for the usability of theories is that

---

20. We may give this a somewhat more precise formulation. The edge of the screwdriver, of course, agrees with a feature of the screw, and with what we want to do with screws. The rotational symmetry, however, beyond these, also fits *our hand* and the way we move it when driving in a screw. It may hence be claimed to correspond to the Kantian categories.

21. This extendibility is crucial if we want to formulate a truth theory which is relevant for the humanities. Claiming that your interpretation of a Platonic text is *true* because it coincides with Plato's own understanding makes no empirical sense—how do we know that it does? But interpretation of a Platonic text (or any other past text) makes use of techniques which are *also* used in the present—some of them in everyday dialogues with people with whom we share a material practice, some of them in the courtroom, where textual evidence is combined with material evidence. If the interpretive techniques which we use on Plato do not function in the communication and together with the material practice and evidence of our own age, we will have to reject them as general tools.

Extendibility and coherence are also what allow us to distinguish between historical *truth* (be it partial or preliminary) and "useful" myth. According to crude instrumentalism, the *Blut-und-Boden* "theory" and the legend that the Danish colors fell from heaven during the Danish crusade in Estonia (cf. p. 182) are as good historical truths as any, since they are socially and culturally useful (or, we might say with hindsight, *Blut-und-Boden* is false because and only because its implementation as social practice resulted in all German *Boden* being occupied by the Allied armies, not to speak of wasted German and Allied blood). As a consequence of instrumentalism, the choice between such "truths," when they enter into conflict, comes to depend on power—as does the decision whether Estonia is legitimately to be considered a Danish or a German satellite (or Russian or no satellite at all, according to still other national myths). Extendibility and coherence, on the other hand, that is, the claim that *the same techniques* should be applicable to all sources, and that the different sources at our disposal should yield a consistent picture, allows historians to dismiss the myths as, exactly, *myths*.

they reflect features of reality, we also claim that *reality carries objective features* which can be reflected by theory—"objective" in the sense that they are contained *in the object,* in that reality which is studied. The assertion that theories are better ("more true," as just said concerning Kepler) if their range can be extended implies that the objective features carried by reality are of general validity, that *reality is coherent,* that is, *potentially one.*[22] This principle is analogous to the expectation that cow number 10,001 will possess the same number of stomachs as the first 10,000, though more open-ended: it does not tell *which* kind of coherence it will be reasonable to look for; we may speak of it as "generalized induction."

The affirmation of instrumentalism, namely that there is no truth, and that we should choose our theories as it fits our aim, ends up by being inverted: No, our choice is not arbitrary and not subjective. The aim of science must be to capture (as far as it is possible in the given moment) as many of the objective features[23] in as general a form as can be done, and thus to be—*in this sense*—as true as possible. Only *then* will our knowledge be instrumental.

This is the real crux of the empiricist imperative as formulated above; this is Aristotle's old program, to find the essence behind phenomena, but stripped of the belief that any definitive essence can be found once and for all; and it is, for that very reason, the core of the dialectical-materialist understanding of the aims and possibilities of scientific insight.[24]

It may be relevant to remember in this connections that the textual criticism of Renaissance Humanists, from Petrarca to Casaubon, consisted precisely in the application of techniques used to expose forged juridical documents.

22. This is another way to approach a question dealt with in the end of chapter 16. Here it was concluded that "the material world is constituted in a way that allows an adequate practice if we order our perceptions as representations of permanent objects," and it was suggested that other fundamental cognitive categories and schemes had a similar foundation. The same kind of argument applies in the case of scientific knowledge: if the requirement of logical consistency and extensibility works (as it normally does), then this must tell something about the reality that our theories deal with.

But we might continue: the permanency of the fox was not absolute, although we might discuss its shortcomings in terms of the same principle (exhalation of air, etc.). If other fundamental schemes (including the requirement of logical consistency) are also "biologically *a posteriori,*" we have no guarantee of their absolute validity; we only know them to be a priori and hence inescapable in our actual life.

23. As we shall discuss later on, however, *our* questions to reality determine the *kind* of features that will be reflected. Only our aim of driving screws into the wall makes the edge and the symmetry of the screwdriver relevant—if we wish to use our screws as weight units, a pair of scales would be the relevant instrument.

24. A core which, it is true, was sadly misrepresented not only in the various textbook doctrines that went under this name in the former socialist countries but also in most of the dissident versions. None of them ever took seriously statements like Lenin's that "this reflection

But still it tells nothing about the ways to attain this scientific insight; let us therefore return from this excursion into the theory of truth to Popper's recommendations in the matter.

INSTRUMENTS OR MODELS?

Occasionally, Popper formulates himself as if he were an instrumentalist.[25] But his fundamental attitude is certainly different. Firstly, he says so.[26] Secondly, and more tellingly perhaps if we look for his fundamental convictions, it is made clear by his appraisal of people who do not reject a theory when (in Popper's opinion) it is falsified. His rhetoric is that of a preacher denouncing sin, not of the carpenter censuring a bungler who reaches out for the hammer when he is to put in a screw. Even if truth is only preliminary, maximal truth is set forth as a moral obligation.

[of nature in the human mind] is never simple, never immediate, never total; it is a process consisting in a series of abstractions, of being put into form, of formation of concepts or laws . . .—and these concepts *comprise* relatively and approximatively the universal laws of nature in movement and perpetual development. . . . Man cannot comprise = reflect = represent *all* of nature in its 'immediate totality', he may only approach it *perpetually* by the creation of abstractions, concepts, laws, a scientific chart of the universe, etc., etc." (*Notes to Hegel's "Logic,"* from the French translation in Lenin 1973: 171f.)

25. "The tentative solutions which animals and plants incorporate into their anatomy and their behaviour are biological analogues of theories; and vice versa: theories correspond (as do . . . honeycombs . . . and spiders' webs) to endosomatic organs and their way of functioning. Just like theories, organs and their functions are tentative adaptations to the world we live in" (Popper 1973: 145).

The framework within which this belongs is one of "evolutionary epistemology," according to which the development of knowledge is part of Darwinian evolution, that is, of adaptation to the conditions for biological survival. Popper is one of those who refer for this to Konrad Lorenz—cf. p. 199, n. 3. A similar line of thought (with no need for this authority) was indeed used above (p. 254) in the discussion of the emergence of the category of the conserved object and the possibly analogous roots of the higher schemes of logic (etc.).

26. (Popper 1972: 111–119). The main objection to the instrumentalist view of science runs as follows: Instrumental rules are tried out in the kind of situations where they are meant to be used. A theory, on the other hand, is tested by being applied "to very special cases—cases for which it yields results different from those we should have expected without that theory, or in the light of other theories. In other words we try to select for our tests those crucial cases in which we should expect the theory to fail if it is not true" (p. 112). As we see, the line of reasoning builds on the inherent *extendibility* of theories (see p. 271, n. 21 and surrounding text), *viz.* to these very special cases; it also presupposes that the test is meant to decide between alternative theories (in contradiction with Popper's normal presentation of the matter, a point to which we shall return). The whole argument is borrowed from Bacon and Descartes.

Nor is Popper's real point of view, however, systematically identical with the italicized imperative attributed to him above; this imperative he only uses for polemical purposes (which still are very visible in his writings)—the point of view he is willing to defend in a serious discussion is more sophisticated.

In order to see why we may look at the difficulties presented by the "naïve-dogmatic Popperian imperative."

Two objections were already raised against empiricism, where they were equally relevant. Firstly, observation and theory belong on different categorical levels. Therefore, *facts* cannot contradict theories; only statements (for example, about facts) can contradict other statements (for example, predictions made by theories). This was the problem which logical empiricism tried to overcome by concentrating upon the connection between "protocol statements" and theories, leaving to practicing scientists the translation of observations into protocol statements. The same could of course be done in the Popperian perspective. But this leads to the second objection. No observation is *pure,* every observation presupposes a number of general cognitive structures or theories—increasingly so in contemporary experimental science. But what precisely are we then to do when (a statement about) an observational fact contradicts our predictions? If, for example, a telescope observation of the planet Mars finds the planet in another place than predicted by the Theory of Relativity? Should we regard the Theory of Relativity as falsified and reject it? The theory of the telescope? Both? Or none?[27]

A tentative solution might be gained from the observation that the functioning of the telescope has been confirmed through many other uses, including observations of terrestrial phenomena, and that it is thus more likely than the Theory of Relativity to hold water. But "confirmation" belongs with empiricism, being in fact nothing but that "verification" which falsificationism tries to replace. The solution thus ends up with the same conclusion as that which came out of the analysis of the logical empiricists: science cannot be analyzed into single statements which are confirmed or rejected one for one; to some extent, scientific truth depends upon the totality of the scientific explanation.

---

27. This is a somewhat simplified version of a real dilemma which once presented itself to Newton and the Royal Astronomer John Flamsteed. Flamsteed did not find the moon where Newton had predicted it to be; Newton, however, was able to convince Flamsteed that he had made wrong corrections for the refraction of light in the atmosphere, and that the real position of the Moon was where (Newton's) theory would have it to be. In coming years he went on to correct Flamsteed's supposed "facts" time and again. See Lakatos 1974a: 130 n.5.

The example may also be correlated with the discussion between Galileo and his opponents when he published his new telescopic observations of hitherto unknown celestial phenomena (the lunar mountains, the satellites of Jupiter). Among the objections to Galileo was the question of how he could be so sure that the effects were not artificial creations of the telescope.

As already told, these objections hit naïve falsificationism on a par with empiricism. A final difficulty, on the other hand, comes from turning one of the objections against empiricism upside down. Empiricism could not explain the origin of theoretical concepts, since they could not be derived directly from experience. It is precisely the aim of falsificationism to make space for these unsubstantiated yet indispensable entities. But the cost is as large as the gain: falsificationism makes possible the existence of theoretical concepts by disconnecting them completely from experience. In this way, theories end up by being nothing but *computational models* which bear no more similarity to the reality they describe than the gears of a watch bear to the movements of the solar system—the only connection being that the pointers of the watch can be used to predict the position of the sun in the firmament. If the precision of the watch is insufficient, we scrap it and replace it by a different model: a digital watch containing no gears but only a quartz crystal and a printed circuit.

This is not the way real theories change. When one theory is replaced by another theory dealing with the same field, the new theory will contain some new concepts and certain new relationships. But the concepts are rarely *quite* new, nor are the relationships. As classical mechanics was replaced by the Theory of Relativity, for instance, a number of fundamental entities like time, space, and mass (the first two being Kantian *a priori* categories, we observe) had to be understood in new, more sophisticated ways than believed till then; they also turned out to be mutually connected in ways which Newton had not imagined. But they were not abolished. The pattern of which they were elements was restructured, which changed their meaning—much in the way cognitive schemes accommodate when they are integrated into new cognitive structures during individual development. As phonemes (cf. p. 263, n. 9) replaced letters as elements of linguistic analysis, this was more than a mere change of names. Yet even though there is no one-to-one correspondence between letters and phonemes, the agreement is large enough to permit us to name most phonemes after letters. It was indeed the introduction of the phoneme concept that allowed linguistics to change its focus from written to spoken language with much less abrupt changes than an unmediated reference to speech sounds would have required.[28]

Theories are thus *not* mere computational models, and predecessor theory and successor theory are more intimately related than the two watches. This falls outside the comprehension of falsificationism, which is at best able to explain the

---

28. We may also note that the decision to spell *ten* and *steam* with the same letter *t* shows the generations who introduced and adapted the writing of English (long before the emergence of any science of language) to have had an understanding of sounds as elements of language not too different from that of modern phonemic linguistics—a striking case of continuity of theoretical concepts in spite of theory change (theory *emergence*, as a matter of fact).

continuity between classical mechanics and the Theory of Relativity as a consequence of Einstein's lack of imagination.

Curiously enough, this problem is solved by empiricism with the same brutal elegance with which falsificationism solves the problem that empiricism encounters concerning the justification of general concepts: *if* theoretical concepts are after all founded in experience, then there is no reason to wonder why they undergo only relative change instead of disappearing. The two approaches to the problem of knowledge solve each other's difficulties—but in mutually unacceptable ways. They stand as thesis and antithesis, in a way which in the first instance is as barren as a marriage between Plato and Diogenes.

CHAPTER 18

A New Approach: Theories About the Scientific Process

It was already pointed out that Popper is only a "naïve falsificationist" when he needs to be so for polemical purposes. But it is the naïve Popper who is generally known; the naïve Popper is the real counterpart of empiricism; and the "philosophical" Popper is, after all, an attempt to keep together with string and tape the naïve Popper when the latter falls into too obvious pieces. For all these reasons, the naïve Popper is the more interesting of the two. For our present purpose, the philosophical Popper is mainly of interest as a step toward that "realistic" Popper which his one-time follower Imre Lakatos (1922–1974) has created through a reinterpretation of the key concepts of his theories, and toward the understanding of scientific *knowledge* as resulting from a scientific *process.*

POPPER AND LAKATOS: THEORIES OR RESEARCH PROGRAMMES?

The philosophical Popper (whom I shall call Popper_1 in the following, while Popper_0 is the naïve Popper, and Popper_2 is Lakatos's construction) differs from Popper_0 on three essential points.[1]

Firstly, Popper_1 does not take theories to be falsified by conflicts with experience, that is, by "facts." As argued above, facts and theories belong to different domains. Theories consist of statements expressed in words or other symbols, and accordingly they can only enter into logical conflict with other statements. Theories are therefore not falsified by facts but, according to Popper_1, by *statements dealing*

1. The exposition of these differences owes much to Lakatos 1974a. This article is also the source for the labels Popper_0, Popper_1, and Popper_2, the meaning of which will become clear in what follows.

with facts—"basic statements," in Popper's idiom, evidently a concept which is closely related to the "protocol statements" of logical empiricism.[2] The "theory" that "all swans are white" cannot be in logical conflict with a bird in flesh and blood; what falsifies it is the basic statement "here is a black swan."

Superficially seen, this is only a shift of the problem which does not solve it—basic statements and facts still belong to different categories—and a specification: how should a theory be in conflict with empirical facts if it was not contradicted by the enunciation of these facts in statements? But in connection with "improvement" number 3, the shift will turn out to have important consequences within all ideologically sensitive scientific domains (cf. below).

The next innovation replaces precipitate dogmatism with philosophical and historical common sense. The infant mortality of theories would be exceedingly high if every theory in conflict with (statements of) facts were to be rejected. Grossly speaking, *every* theory is contradicted by lots of facts at birth. But are we to reject Galileo's law of falling bodies because it is not obeyed by a falling withered leaf? Or a theory of prices referring to costs of production because it does not fit rare stamps?

Such rash rejections are not usual. In both cases you will have a definite feeling that the deviations from theory are due to specific circumstances, even though you may not yet be able to specify and quantify them. But you would evidently be dismayed if the speed of heavy leaden balls and the price of eggs went equally astray.

$Popper_1$ attempts to formalize this consideration by restricting the range of inconsistencies with regard to reality that will count as falsification. Evidently such a restriction cannot be specified in general. But when working on a theory, you should point out (yourself and beforehand) the specific domains where the theory should in any case hold good;[3] if it does not, you should reject it *without mercy* (Popper's rhetorical style)—and if you don't, you are dishonest (ditto). According to Popper, Galileo should thus "at his peril"[4] determine beforehand that "if my law of falling bodies does not fit within 98% the speed of a leaden ball of 20 kg falling 50 m, I will burn my manuscripts and turn to literary criticism." Theories should "stick out their neck": they should be unable to escape the hangman of contradictory experience, should he happen to pass by.

---

2. "What I call a 'basic statement' or a 'basic proposition' is a statement which can serve as a premise in an empirical falsification; in brief, a statement of a singular fact" (Popper 1972a: 43). Further on in the same book, Popper tries to construct a fence between his own concept and that of the logical empiricists, which he finds too "psychologistic" (probably because experimental psychology is the field which consistently makes use of "research protocols").

3. ". . . criteria of *refutation* have to be laid down beforehand; it must be agreed which observable situations, if actually observed, mean that the theory is refuted" (Popper 1972: 38 n.3).

4. Popper 1972: 42 n.8.

The third difference between $Popper_1$ and $Popper_0$ relates to the problem that scientific "facts" are obtained by means of methods themselves presupposing theories, as in the case of the telescope observations of the moon.[5] Even this problem Popper gets around by making it a moral obligation to choose in advance. Before testing your theories, you should also stick out your neck by deciding beforehand which theoretical fundament you accept as unproblematic and hence above criticism. Woe to the scientist who *post festum,* when his theory *has* got into trouble, starts doubting his telescope. He is, according to Popper, dishonest.[6]

Lakatos has composed an ultra-short story demonstrating the divergence between Popper's methodological requirements and the real behavior of the scientific community:[7]

> A physicist of the pre-Einsteinian era takes Newton's mechanics and his law of gravitation, (*N*), the accepted initial conditions, *I,* and calculates, with their help, the path of a newly discovered small planet, *p.* But the planet deviates from the calculated path. Does our Newtonian physicist consider that the deviation was forbidden by Newton's theory and therefore that, once established, it refutes the theory *N*? No. He suggests that there must be a hitherto unknown planet *p′* which perturbs the path of *p.* He calculates the mass, orbit etc., of this hypothetical planet and then asks an experimental astronomer to test his hypothesis. The planet *p'* is so small that even the biggest available telescopes cannot possibly observe it: the experimental astronomer applies for a research grant to build yet a bigger one.[8] In three years' time the new telescope is ready. Were the unknown planet *p'* to be discovered, it would be hailed as a new victory

---

5. Popper himself will rather point to human weakness, but the conclusion is the same: ". . . we may point out that every statement involves *interpretation in the light of theories,* and that it is therefore uncertain. This does not affect the fundamental asymmetry [between possibly falsifying observation and falsifiable theory], but it is important: most dissectors of the heart before Harvey [who discovered the blood circulation] observed the wrong things—those, which they expected to see" (Popper 1972: 41).

6. One is tempted to ask how to characterize Popper's own attitude. At the age of 17, he engendered his marvellous falsificationist epistemology, though in the naïve "$Popper_0$" version. In coming years, when he discovered the shortcomings of this program, he did not give it up. Instead he repaired it by means of the distinction between facts and "basic statements" and all the other subtleties belonging to "$Popper_1$."

But since, as we shall argue below, the prohibition of *a posteriori* criticism of the theoretical foundations of observations is ill founded, there is no serious reason to censure Popper for his failure to submit to his own rule.

7. Lakatos 1974a: 100f. As in the case of my above Mars/telescope example, Lakatos's story refers to somewhat more complex but similar real-life events.

8. If the tiny conjectural planet were out of reach even of the biggest *possible* optical telescopes, he might try some quite novel instrument (like a radiotelescope) in order to enable

> of Newtonian science. But it is not. Does our scientist abandon Newton's theory and his idea of the perturbing planet? No. He suggests that a cloud of cosmic dust hides the planet from us. He calculates the location and properties of this cloud and asks for a research grant to send up a satellite to test his calculations. Were the satellite's instruments (possibly new ones, based on little-tested theory) to record the existence of the conjectural cloud, the result would be hailed as an outstanding victory for Newtonian science. But the cloud is not found. Does our scientist abandon Newton's theory, together with the idea of the perturbing planet and the idea of the cloud which hides it? No. He suggests that there is some magnetic field in that region of the universe which disturbed the instruments of the satellite. A new satellite is sent up. Were the magnetic field to be found, Newtonians would celebrate the sensational victory. But it is not. Is this regarded as a refutation of Newtonian science? No. Either yet another ingenious auxiliary hypothesis is proposed or . . . [Lakatos's ellipsis] the whole story is buried in the dusty volumes of periodicals and the story never mentioned again.[9]

This story agrees well with what goes on within even the most exact sciences. Within the realms of social sciences and the humanities, where precise predictions are rare, and where a distinction between facts, theoretical notions, and ideological conventional wisdom is not easily established—there, as one might guess, the rules are even more rarely observed.

A follower of Popper might reply that he knows: Popper, he may claim, does not describe what scientists actually do—he proposes a program which would make science advance more rapidly and with fewer wasted efforts than it actually does *precisely because Lakatos's story is correct.*[10] Popper's rules would not ensure that no mistakes were made; but they would reduce the number of mistakes and, especially, the time that is wasted on mistakes.

---

him to "observe" it, that is, to ask Nature about it, even if only indirectly. (The new "observational" theory may itself not be properly articulated, let alone severely tested, but he would care no more than Galileo did). [Lakatos's footnote].

9. At least until a new research programme supersedes Newton's programme which happens to explain this previously recalcitrant phenomenon. In this case, the phenomenon will be unearthed and enthroned as a "crucial experiment." [Lakatos's footnote; in the next note he refers to Popper's polemics against the Freudian and Adlerian psychologies which can be made to agree with any state of the actual world, pointing out, as done in these pages on a simpler example, that the same holds for Newtonian physics].

10. This answer is not always given by Popper himself, in particular not when he lapses into Popper$_0$. The point in his discussion of psychoanalysis and Marxism versus the testing of the General Theory of Relativity in his (1972: 34ff) is precisely that the latter, representing *science,* behaves differently from the former, representing *pseudo-science.* Elsewhere, however, Popper is more clear about presenting a set of prescriptions and no description.

Our follower of Popper would be wrong. Scientific work according to Popper's rules would, like most "work according to the rules," be a synonym for a strike in disguise.

There are several reasons for that. Firstly, one may ask what happens when a theory has been falsified and therefore rejected. When planet *p'* does not show up in the telescope, should we then reject Newton's understanding of the planetary system and of mechanical physics in general? Should we stop calculating the dimensions of steel beams for bridge buildings and make them at random instead? Having rejected Newton's mechanics, we have no theory at all, since all predecessor theories have *also* been falsified. The absurdity of the claim is blatant, and it shows the Popperian notion of 'merciless rejection' to be empty jargon.

One may also—which is Lakatos's main point—observe that Popper's understanding of the nature of a *theory* is much too static and formal. A theory which is to fit Popper's prescriptions is complete and fully finished—in the style of a Tarskian formal language (cf. p. 270, n. 19); it consists of a set of formulas (verbal or symbolic) which definitively state the mutual relations between the concepts of the theory, and a set of rules which allow the translation between theory and observation, that is, allow observation to function as a metalanguage telling which statements are true and which false.

Few theories, if any, have been born in that form. It is debatable how many attained it before they died. Theories are born instead, as we shall see in more detail below, as open structures. Only through the collective work of the scientific community with the theory are its implications and potentials and its relations to other theories discovered to the full.[11] Already for this reason it is impossible to indicate when a theory is conceived at which points decisive testing should be performed.

In social and human sciences, Popper's methodology would give rise to yet another problem, which has to do with the *conventionalism* of Popper_1.

In general, conventionalism belongs to the same family as instrumentalism. Like instrumentalism it holds that one theory is no more *true* than another, competing theory—ascribing "truth values" to theories is as nonsensical as ascribing colors.

---

11. This was in fact pointed out at one moment by Popper himself (in an article from 1940): "The dogmatic attitude of sticking to a theory as long as possible is of considerable significance. Without it we would never find out what is in a theory—we should give the theory up before we had a real opportunity of finding out its strength; and in consequence no theory would ever be able to play its role of bringing order into the world, of preparing us for future events, of drawing our attention to events we should otherwise never observe" (reprint, Popper 1972: 312). And similarly: ". . . this dogmatism allows us to approach a good theory in stages, by way of approximations: if we accept defeat too easily, we may prevent ourselves from finding that we were very nearly right" (1972: 49). Yet, as observed by Lakatos, this glimpse of insight does not influence his general thinking and his magisterial preaching.

Whether we use one or the other theory for our description of reality is decided *by convention* and is no more compulsory than the convention which makes us speak of "cigarettes" and "pamphlets" and not of "cigamphlets" and "parettes." Empirical evidence may at worst force us to change the way we interpret our theory, thus modifying the "rules of translation" between observation and theoretical prediction.[12] In consequence *scientific objectivity* is nothing but agreement about a shared convention—and if you disagree with the majority of your scientific colleagues, *you* are automatically the sinner against objectivity.

This breed of conventionalism is treated with as intensive scorn by Popper as are Marxism and psychoanalysis. Conventionalism, indeed, denies the falsifiability of theories, eschewing it through a reinterpretation of the rules of translation between theory and observation. Popper sees clearly and correctly that conventionalism can function as a cloak for scientific opportunism and for facile thinking in grooves. But his own philosophy contains obvious conventionalist elements: convention and nothing but convention points out which kinds of conflict should be regarded as falsification; and convention decides which theories should be ranked as unassailable and which should be submitted to continuous attempts at falsification.[13]

In ideologically sensitive areas, that is, in particular within the social and the human sciences, even this brand of conventionalism will easily entail stagnation in ideological opportunism. What is more easily agreed upon by the majority than the set of already familiar, stereotyped ideas? Once more the objective scientist will be the one who accepts the received opinions of respectable people, and the dishonest worker the one who challenges conventional wisdom and sticks to his own scientific convictions.[14]

As stated above, Popper's one-time follower Lakatos has formulated that more accurate epistemology which in his opinion might grow out of Popper$_1$ and has baptized

---

12. All this may be more clear from an example. If you sit in a train, you will normally state that it starts moving after the doors have been closed; but you might equally well state (and at times get the momentary sensation) that you and the wagon stay at rest, and the rest of the universe starts moving. In the first case, the observation that empty bottles start rolling along the floor is explained as a consequence of the law of inertia; in the second by a changing gravitational field.

The standard example is in larger scale but built on precisely the same principle: the question whether Earth or the sun is at rest.

13. This is not kept secret by Popper, who speaks explicitly of his methodology as "methodological conventionalism."

14. Popper's requirement thus stands in curious contrast to Merton's norm of "organized scepticism" (see p. 350), which Merton illustrates by the German *dictum* "ein Professor ist ein Mensch, der anderer Meinung ist," that is, one who does *not* automatically submit to received opinions.

it $Popper_2$.[15] The central point in Lakatos's epistemology is reflected in this labelling: "Popper," in fact, does not refer to the person but to what Lakatos calls a *research program,* evidently inspired by the person; $Popper_0$, $Popper_1$, and $Popper_2$ are nothing but single stages within the development of this program.

Precisely this example may provide the easiest explanation of the concept. A research program is not, as a theory according to Popper, a static and solid entity; it is a progression through a number of static theories superseding each other. A research program thus roughly coincides with the more loose parlance of a "theory" as something which is in continuous development.

The feature which welds a sequence of theories into a research programme is the existence of a shared *hard core,* a set of notions about the entities which exist within the field dealt with by the research program. In the Popper program thus *theories* which cannot be derived from empirical observation; *falsifications* which kill theories; and some kind of facts or representatives of facts taking care of the killing. In the Newton program, material particles and forces. These entities are the tools which the theories belonging to the program apply in order to describe or explain empirical reality. In addition to this *ontology,* the hard core prescribes the kinds of explanation which should be aimed at, the methods which are to be preferred, and so on.

The hard core is "hard" in the sense that it does not bow to conflict with observational reality—it is, in another word, *irrefutable.* If we use the Popper program as an example, we see that empirical observations similar to Lakatos's short story may demonstrate that falsification *à la* $Popper_0$ and $Popper_1$ does not describe the real process of scientific development; but they cannot prove that the rejection of unsatisfactory theories may not in some way or other be understood as a "falsification." Experiments might show that the gravitational force does not depend on distance in the way Newton believed; but they could hardly prove that forces are *in general* untenable explanations.

The protection of the concepts (etc.) belonging to the core at "any price" is called by Lakatos a *negative heuristic*—that is, a guide as to what one should *avoid finding.* The core also contains a *positive heuristic*—a guidance prescribing how increasingly extended ranges of reality are to be explained, and in which order "anomalies" are to be solved. The existence and efficiency of this positive heuristic is of course the reason for the cohesion and continuation of the research program; if $Popper_0$ and $Popper_1$ had produced no interesting points, there would have been no reason to stick to the program and to a fundamental idea like falsification.

The theories which make up a research program are not only claimed to constitute a family because they are characterized by certain shared features constituting a

---

15. Rumor has it that $Popper_2$ was already in manuscripts written by the real Popper but kept unpublished—maybe because their publication might have reduced the famous $Popper_{0/1}$ to ashes.

shared hard core. As already intimated, and in the likeness of $Popper_0$, $Popper_1$, and $Popper_2$, they are also ordered in a progressing sequence, $T_1, T_2, T_3, \ldots T_N, \ldots$.

In order to introduce more fully the relations between the members of such a progression of theories, it might prove useful to look at a more substantial example than the Popper sequence. Such an example is provided by the progression of economic theories built upon the labor theory of value.

The first step—the step which originated the hard core and the program—was Adam Smith's *Wealth of Nations* from 1776. In this work, Smith formulated a theory of the price of a commodity as proportional to the working time required to produce it.[16] This doctrine was no loose postulate but argued from the competition between workers as well as between manufacturers, and thus connected with the concept of three social classes.[17] This was a radical innovation with regard to the preceding physiocratic conception, according to which only work in agriculture was productive, while all kinds of industrial transformation (for example, grain into flour and bread) were unproductive (cf. pp. 134–139, *passim*).

In his more systematic *Principles of Political Economy and Taxation* from 1817, David Ricardo took over the labor theory of value and used it among other things to explain the mutual advantage of foreign exchange in a market system. By using the concepts of competition and scarcity, he also managed to explain from the theory how the mere possession of agricultural land would allow landlords to earn money (the rent).[18] Ricardo, however, wrote in a situation where some industries were significantly more "capital intensive" than others; this difference had been less conspicuous forty years before, when Smith wrote his book. Ricardo knew that the difference had to influence prices, without being able to integrate this knowledge

---

16. Evidently, my exposition of Smith's and other economic doctrines is cut to the bare bones—simplified to that extreme where only the features which are essential for the epistemological argument stand back.

17. "Those who live by profits," that is, capitalists; "those who live by wages," that is, workers; and "those who live by rent," that is, landlords. In Smith's England, it should be remembered, landowners would normally lease their land to farmers. The latter would thus be counted as capitalists, whereas the landowners (who owned only land but no means of production) were a separate class.

18. In brief: If a country needs *N* tons of grain per year, this requires that the best *A* acres are cultivated. Some of these acres yield more than others at the same expense of labor, but the price of the grain will of course be the same. Competition will fix the price at the level corresponding to the labor costs of the poorest land, which is the highest level: if the price were lower, nobody would care to cultivate this land, which would result in shortage, famine, and raising prices; if it were higher, even poorer land could be cultivated with profit, which would lead to overabundance and falling prices. The landlords possessing the best land will thus get more from their grain than the labor costs—or they will lease to capitalist farmers who will be willing to pay the difference between their selling price and their labor costs as rent.

Keine Hexerei, nur Behändigkeit!

into the theory of value, and therefore restricted himself to the statement that the working time required to produce a given commodity would determine its price until at least 93 percent; one has spoken of Ricardo's theory as a "93 per cent labour theory of value" (Barber 1967: 85, 92 n.9). If we speak of Smith's theory as T_1, Ricardo's will be T_2.

Both T_3 and T_4 are due to Marx. T_3 solves the difficulty which arises if the labor theory of value is applied to the price of labor itself, that is, to the wages. It would seem that a labor cost determination of wages should lead to a payment for eight hours of work which would be able to buy precisely the product of eight working hours. This would leave no space for profits—in flagrant contrast with the normal state of affairs.

In Marx's writings from the late 1850s onwards (T_3) this problem is solved. Prices, according to this theory, are still determined by the working time needed to produce commodities; but the wage is not the payment for the *working time* but for the *working power*. The working time needed to produce 8 hours of working power is the time normally required to *produce* that working power, that is, to produce the commodities which the worker consumes in order to keep going for another day.[19] If the working class of a country only consumes half of its social product, we see that the time used to produce what an average worker needs to go on for another average day is produced in 4 hours. The price of *8 hours working power* equals *4 hours working time*. This leaves space for profit.

Another problem is still unsettled—*viz.* Ricardo's problem of varying capital intensities. This difficulty is only resolved in volume III of *Das Kapital*, which was published posthumously by Engels in 1894 (T_4). The breakthrough consists in a separation of the concepts of 'value' and 'price.' The value of a commodity is now *defined* as the working time normally required for its production under given technological, societal, and historical conditions. *If* prices were equal to values (after a suitable conversion of time into monetary units), capital would be invested in those sectors where a maximum of work was done (and hence a maximal profit earned) per unit of invested capital.[20] These sectors would soon be overproducing compared to actual demand, while those depending on larger investments per working hand would be underproducing. The imbalance between demand and supply would make prices fall in the overproducing sectors and make them rise in the underproducing ones. This would continue as long as profit rates differed; in the end, prices would

19. Of course averaged over life, so that the costs of procreating and feeding children is included. In this simple version, a two-class model for society (capitalist and working classes and nothing else) is presupposed.

20. Equality of prices and values would mean that the profits *per working hour* would be same in all sectors; the more capital you need in order to employ one worker, the less will be your profits *per invested £ Sterling*.

be stabilized precisely so far from values that the profit rates of all sectors were equal.[21] Ricardo's problem is solved—indeed by means of theoretical considerations borrowed from his own theory of rent. At the same occasion another difficulty dissolves: how to explain *within the labor theory of value* the incomes of banks, wholesale merchants, and *rentiers.*

After the death of Marx, only Marxist economists continued work within the framework of the labor theory of value (grossly speaking). The reason was obviously the political consequences of the doctrine, as they had been uncovered by Marx.[22] Further development was branched, as was the labor movement itself. One further development (T_5', T_6', etc.) consists of inconsistent crisis theories (Rosa Luxemburg, Ernest Mandel, and others). Another branch contains a better theory of the dynamics of economic crises (T_5, Michal Kalecki) and the solution of subsisting problems concerning the relation between value and price (T_6, Pietro Sraffa).

What follows from this whole story? First of all, that Lakatos's research programs are no description of real history but "rational reconstructions," in Lakatos's own words. Marxism is certainly more than a further elaboration of Adam Smith's research program, and one gets no genuine understanding of Marx's thought from isolated exposition of his economic analyses. Worst of all, Marxist analyses of the increasing monopolization of capitalist economies after 1870 are not easily fitted into a rational reconstruction relating everything to the Smithean starting point.[23]

At the same time, however, the story shows that Lakatos is far from being totally wrong; the rational reconstruction reflects central aspects of the historical development and can thus be claimed to be a *true* theory for the structure of scientific development in the sense explained above. Finally, the process exhibits some of the characteristics which according to Lakatos distinguish the development of research programs.

Firstly it shows that research programs may easily live with "anomalies", that is, points where they disagree with observation without being able to explain precisely why. Ricardo's 93 percent theory is in fact nothing but an attempt to gloss over an acknowledged anomaly, the influence of capital intensity on prices. The labor theory of value was not dismissed by Ricardo because he got stuck in a problem which he

---

21. In the actual world, they will of course never be completely stabilized—continuous technological development is one of several factors which cause the point of equilibrium itself to be moving, as emphasized by Marx.

22. Gustafsson (1968: 14–16) lists an array of sources which document this explicit concern. The main problem was the separation of working time and working power (T_3), which automatically entailed a concept of exploitation.

23. In justice it should be said that Lakatos did not propose the application of the research program idea to the development of the labor theory of value; but similar features would turn up in many other instances.

could not solve for the moment. Nor did the difficulty paralyze Marx, though only two theoretical breakthroughs allowed him to solve it.

Still, a research program cannot live with all its difficulties without doing *something* about some of them. If Ricardo had only been able to introduce his 93 percent restriction and had not increased the explanatory power of the program on other points (of which foreign exchange and rent were mentioned), the program would have *degenerated* (Lakatos's term), and it would have been abandoned by active research as soon as an alternative approach had been found.

Changes which are not degenerative are called *progressive problem shifts* by Lakatos. A progressive problem shift occurs when a new theory is both empirically and theoretically more powerful—if it predicts *more* than the predecessor (greater theoretical content), and if it predicts better (greater empirical content). If we forget about the 93 percent restriction, the whole sequence T_2–T_3–T_4–T_5–T_6 consists of progressive problem shifts. T_5', however, which aimed at increasing the theoretical content of the theory, was no progressive shift: on one hand, it did not increase the empirical content of the theory; on the other it was ridden by inner inconsistency. The same holds for T_6', Mandel's attempt to show by means of numerical examples how the spread of fully automatic industry would entail the collapse of capitalism.

THEORIES FALSIFIED BY THEORIES

An important feature of Lakatos's conception is his notion of falsification. A theory, as we have seen, is not falsified by an anomaly, however serious. According to Lakatos, a theory is falsified by *another theory*—by a theory with greater theoretical and empirical contents, by a theory which, so to speak, *explains why its predecessor failed to explain* specific anomalies.[24]

This is in itself a progressive problem shift, solving or dissolving no less than four of the central dilemmas presented by $Popper_1$:

Firstly, it solves the question what to do in the interlude between the falsification and resulting merciless rejection of one theory and the devising of a replacement. There *is* no such interlude, since falsification only follows from the development of a new and better theory.

Secondly, it solves a difficulty which, though not formulated explicitly above, follows from the lack of continuity of theoretical concepts through the cycle of falsification and ensuing free invention of a new theory. At best, Popper's methodology might bring forth a sequence of increasingly precise *models* of reality; but even in

24. As we remember from p. 273, n. 26, Popper suggests in one place that his testing is meant to choose between alternatives which are already at hand. This suggestion, in collision with what Popper mostly writes on falsification, already adumbrates the present idea, though without developing it in full; it is one of several points where the common sense of the real Popper has made him contradict himself and thus foreshadow $Popper_2$.

the best of cases the falsification cycle will never procure *increasingly deep theoretical insight:* every time a theory is falsified and thus rejected we replace it with something which in principle is *totally new* (like the digital watch replacing the mechanical watch). We cannot raise the question of *what* was wrong in the rejected theory; it is the model as a totality that is wrong and thus rejected. Within the framework of the materialist notion of truth (see p. 270) we may say that if the key concepts of the hard core reflect essential aspects of reality, then the research program allows an increasingly exhaustive investigation of these features, and thus an increasingly objective reflection.

Thirdly, falsification *à la* Lakatos does not invite to ideological opportunism as does Popper's methodological conventionalism. On the contrary: if a theory is only regarded as falsified by another theory which offers deeper and more precise explanations, then disagreement with superficial ideology and conventional thinking will be a less threatening argument against it.[25] Theories get greater opportunity to confront their object directly, bypassing the censorship of received opinions.

Fourthly and finally, the paradox evaporates that every observation is polluted by theoretical or proto-theoretical presuppositions. If theories are falsified by theories, this is no longer a source of logical trouble but only another expression of the tangled character of scientific (and other) knowledge—a totality which cannot be fully analyzed into mutually distinct elements, be it into the verified and thus meaningful statements of empiricism or into Popperian basic statements and static theories.

At one point, Lakatos's conception (Popper_2, as we remember) can be claimed to constitute a degeneration with respect to Popper_1 (the sophisticated real Popper). Popper's aim is to formulate a *logic of scientific research,*[26] a formalized system which can be set before the scientific community as a set of rules which it ought to obey. Lakatos's rational reconstructions preserve this aim to some extent. But Lakatos has given up the conviction that the falsification processes of the rational reconstruction (not to speak of that real history which it rationalizes) can be formulated in a way

25. Once again we may correlate with Merton's "organized skepticism": whereas Popper's prescriptions would tend to undermine this norm, Lakatos explains it and makes it a methodological necessity.

26. This is the best English translation of the original German title *Logik der Forschung* from 1936 of Popper 1972, and describes Popper's intentions adequately. The actual English title (*The Logic of Scientific Discovery*) may be better for advertisement purposes but misses the point completely: the only part of the research process which according to Popper should be completely free and subjectively creative and not subject to any logic is precisely the phase of *discovery,* the invention of a new theory. Formalization and strict rules belong with the *control process,* the compulsory stubborn attempts at the life of the assumed discovery.

which complies in full with the requirement of formalization.[27] Who is able to balance the degenerative versus the progressive elements in the shift from Smith to Ricardo? Everybody, of course—but very precisely to balance, not to state the definitive and indubitable decision. In many instances, moreover, it is also only through extensive and meticulous research that one is able to decide whether a theory possesses greater empirical content than a competitor; it may even be that agreement with observation improves on some points and decreases on others. The decision can no longer be reached by an impersonal and objective judge; it is attained instead in an informal arena, *viz.* the scientific community "voting with its feet." The choice between theories is the sum of individual choices made by individual workers deciding which theory they are going to make the foundation for their own further work. Paraphrasing the jibe against Ricardo, Lakatos's theory can be characterized as a "93 percent logic."

Rather than speaking against Lakatos's epistemology, however, this observation tells something about the concept of 'degeneration': degeneration need not be a defect, even though this is the obvious moralistic implication of the term. It may just as well be a rejection of empirically degenerative aberrations contained in earlier theories. In this vein we may look at $Popper_0$ and $Popper_1$ as aberrations within a research program starting informally from a hypothetical-deductive understanding of scientific method, and at Lakatos/$Popper_2$ as an alternative and more fruitful development from the same roots. In any case, theoretical degeneration need not be a development for the worse—as is well known by military planners, a tactical retreat from unwisely occupied positions may be the only way to avoid imminent defeat and the best preparation for further offensives.

Two final points should be addressed before we leave the presentation of Lakatos's ideas. Both may, like the discussion of degeneration, be approached through the example offered by the theory itself.

According to Lakatos, the normal situation within a scientific discipline is the existence of several competing research programs. The philosophy of science offers a striking example of this: during a protracted period, logical empiricism and Popperianism were indeed both pursued (along with other programs). Similar examples can be found in many other disciplines; still others, on the other hand, seem to be dominated by one program at a specific moment.

Here it is important to remember that falsification takes place *within* research programmes. One research program cannot (if we follow Lakatos) be falsified by

27. Conversely, of course, Lakatos also gives up the belief in the completely unfettered process of invention: as long as innovation takes place within the same research program, it is guided by the positive heuristic and restricted by the negative heuristic, both of which are constituents of the hard core of the program. But even if guided and restricted, and thus no act of pure subjectivity, innovation has not become formalized.

another program, since its hard core is irrefutable. Research programs are not falsified, they are given up when they degenerate and when alternative choices are at hand.[28]

To some extent, the process by which one research program displaces another thus constitutes a parallel to the falsification within a research program. Yet the parallel is imperfect: a new research program does not necessarily get the upper hand because its empirical content is larger than that of the program which it supersedes; it may be preferred because it explains a specific anomaly which has resisted the predecessor so stubbornly that the whole validity of this program has come to be doubted. This is what happened within chemistry in the late eighteenth century, when a program explaining combustion as the absorption of a new chemical element *oxygen* replaced a predecessor theory which explained combustion as the liberation of an igneous substance called *phlogiston.* Among other things, the phlogiston theory had explained the colors of many chemical substances. Within the framework of the oxygen theory, these colours became inexplicable. None the less, the oxygen theory was preferred, because the areas where it possessed increased empirical content were considered more important.[29]

From Lakatos's point of view, this is a decisive difference. All things considered, however, it seems to amount to no more than a difference of degree. True enough, the abandonment of one program for another cannot be described as a formalized process—it results from a process of balancing and "voting with the feet." But precisely the same was, though to a lesser extent, the case when we considered the falsification process within a program.

The last feature of Lakatos's epistemology is that it is *reflexive,* that is, able to describe itself. Although it is nowhere said, exactly this must be the coquettish point in Lakatos's use of the term $Popper_2$ as a label for his own approach. In so far as it

---

28. In rare cases, programs have even been considered degenerating beyond hope and have been given up by practitioners notwithstanding the absence of alternative programs. In such cases the whole discipline has been abandoned by the scientific world and considered a pseudo-science, and an earlier belief in the results of the program is declared superstitious or at least illusive. One example of this process is *phrenology,* the nineteenth-century science about the alleged relation between people's character and the form of their skull (cf. p. 160). The rejection of astrology by astronomers in the seventeenth century may be looked at in the same perspective.

29. One of these areas was the specific weights of chemical substances: *if* weight was to be conserved in chemical processes, phlogiston had to have changing and sometimes negative specific weight. In the earlier part of the eighteenth century, weight measurements seemed to deliver no consistent picture, for which reason weight could hardly be considered a key parameter; with the increased experimental precision of the second half of the century, however, the outcome of the experiments became consistent. See Partington 1961: II–III, *passim,* in particular III, 606ff.

claims to be a description of the actual process of knowing and of its conditions, that is, to be itself *knowledge* about that process, reflexivity must of course be required from any epistemology.[30] Yet far from all epistemologies are in fact reflexive. As already hinted at, Popper's own rules would have forced him to give up his ideas as exhaustively falsified if he had followed them. Empiricist philosophy is no better off; logical empiricism, in particular, would probably be forced to see its own statements as ultimately meaningless if it applied its own verification standard. Similar autodestructive conclusions will be reached in the cases of instrumentalism and conventionalism.

Still, Lakatos's theory is not the only reflexive offer on the epistemology market. The epistemology of dialectical materialism has the same characteristic.[31] Moreover, full reflexivity is only attained by Lakatos if he is given a materialist interpretation, through which a truth value can be ascribed to a research program and its appurtenant theories *in spite of* the metaphysical and irrefutable character of its hard core.

THE LIMITS OF FORMALIZATION

Lakatos's epistemology is able to grasp essential features of the development of scientific knowledge. But it is not able to grasp *all* essential features—no theory is. And it has not solved all the problems to which it directs attention.

Like Popper and the logical empiricists, Lakatos still regards science as a formalized system: a theory consists of unambiguous statements dealing with concepts and their mutual relations, and of "rules of translation" telling how the predictions of

30. Evidently, reflexivity is in itself no proof of the adequacy of an epistemology. As pointed out on p. 270, n. 19, the truth of a system cannot be proved by self-reference. But as a minimum, the self-references contained in or implied by a system that pretends to be true must be required to be of the type "this statement is true" and not variants of the so-called *liar's paradox,* "the present statement is false."

31. But not the average textbook doctrines that went under that name, cf. p. 272, n. 24; nor *a fortiori* the kind of vulgar Marxism that claims consciousness to be nothing but reflection of the socioeconomic circumstances under which it is produced and rejects the relevance of any discussion of it in terms of truth value. If thinking in general should be understood on a par with Pavlov's conditioned reflexes, why should the status of vulgar Marxism itself be different? Similar conclusions will be reached in the case of other unrestricted sociologisms, as also when we look at B. F. Skinner's behaviorist epistemology or other deterministic theories—indeed for all epistemologies which deprive knowledge of the possibility of being a *true* description or reflection of the real world (cf. p. 203).

In general, *any* epistemology claiming possession of a complete and exhaustive explanation of the nature of knowledge can only be reflexive if it is able to explain the existence of complete and exhaustive knowledge. No middle road appears to exist between Platonism and epistemologies which are as open as that of Lakatos.

theory may be compared with empirical observation;—and science, on its part, consists of theories.

This view Lakatos shares with Popper, and for that matter with logical empiricism. To be sure, the formal understanding of the system of scientific knowledge does not imply that Popper and the logical empiricists (nor, *a fortiori,* Lakatos) have not discovered the importance of ideas without formalizable foundation for the development of knowledge. Quite the contrary, the logical empiricists distinguished sharply between the "context of discovery" and the "context of justification"; they were fully aware of the possibility that the context of discovery of ideas may be far removed from empirical verification, involving intuition, religious and metaphysical ideas, and so on. What they asked for was that an idea, once proposed, in the context of justification could be "verified" empirically. Popper, on his part, made a cardinal virtue of what logical empiricism had felt forced to accept: new theories should be freely devised, any attempt to make rules for this process would be inimical to science. Only in the moment when the theory *has* been formulated does the merciless effort to falsify set in. The whole model looks as if it were inspired by traditional liberalistic ideology: anybody should be allowed to settle down as a shoemaker or as a railway baron, free of state regulation and control; but in the moment when he has started his business, he should be subjected to the objective verdict of the market, which kills unsound undertakings without mercy. Like attempts to keep a falsified theory alive, efforts to keep unsound businesses afloat through public intervention will only do damage to the common good.

Lakatos is less of a (vulgar) liberalist than Popper in his epistemology. The concept of a 'research program', of progression within the program, and particularly the idea of a positive heuristic, describe the process of invention as less than fully arbitrary and as somewhat open to theoretical comprehension—as taking place within a certain pattern. But the origin of research programs and of their hard cores is still regarded as inaccessible to formal analysis and hence left outside the area to be considered by the philosophy of science. This does not invalidate the rest of Lakatos's analysis, but it remains a serious restriction that the theory leaves out of consideration the life-giving moment of the development process as unexplainable; what would we think of a theory of ecological metabolism which explains that animals live from each other and ultimately from plants but disregards the photosynthetic process through which plants secure the energy supply for the total ecosystem?

In other respects too, Lakatos's search for formalized structure (and his desistence from describing what cannot be sufficiently formalized) creates more problems than it solves. The distinction between falsifiable theories and the irrefutable hard core of a research program is surely meaningful. Yet the two levels can not be regarded as absolutely distinct; in the case of the labor theory of value, for instance, we must probably see the separation of price and value (T_4) and perhaps even more the separation of working time and working power (T_3) as so radical reinterpretations of the foundations of the program that its hard core is changed. If we look at the shift from

$Popper_1$ to $Popper_2$, we must also acknowledge that the new concept of falsification is so far removed from its predecessor that even here the hard core is affected—no wonder that Popper rejects $Popper_2$.

Finally, the absolute separateness of research programs is dubious. It is true that different research programs build on different sets of cardinal concepts, and no complete translation from one program into the other is possible (cf. the relation between empiricism and naïve falsificationism). But they are still connected via their reflection of features of the same reality (if we presuppose a materialist view) and often of the same features (even though the problem for empiricism and naïve falsificationism seemed to be that each of them was formulated with regard to features which were inaccessible to the competing program). Thereby the possibility emerges that the concepts of one program may be explained at least with some approximation in terms of the core and theories of the other; perhaps one program may even develop to the point where it is able to explain the accomplishments and the failures of the other—which was the criterion for falsification *within* a program.

Apart from being built upon the basic premise of a materialist view (*viz.* that knowledge reflects features of a reality existing independently of the knowledge in question), this conclusion corresponds to experience borrowed from the history of a variety of sciences.

If we look at the confrontation between the phlogiston and oxygen theories, the former was mere nonsense as seen from the stance of early oxygen theory, and its triumphs nothing but lucky accidents. In some instances, in fact, the "substance" phlogiston "turned out" (thus oxygen theory) to be identical with carbon or hydrogen; in others it represented the absence of oxygen. But the development of the concept of "degrees of oxidation" in the later nineteenth century provided an explanation of what was common for carbon and the absence of oxygen. It thus became clear which features of reality (as seen by the mature oxygen theory) were reflected in the doctrine of phlogiston. Phlogiston theory, which had originally been *abandoned,* could now be seen as *falsified* in Lakatos's sense.

Corresponding examples can be found everywhere. Most obvious is perhaps the relation between Ptolemaic and Copernican planetary astronomy. If we accept the Copernican system (or one of its later variants) it is easy to calculate how planetary orbits behave as seen from the Earth, and hence to see how the Ptolemaic model manages to account with relative precision for the position of planets on the celestial vault. But even the more ambiguous field of social sciences offers some instances—as many, indeed, as can be expected in a domain where woolly conceptualizations and cross-fertilizations often make it difficult to speak of distinct research programs.

A striking example is provided by the relation between the labor theory program and that "neo-classical" or "marginalist" program which replaced it within academic economics after c. 1870, when the former program had become politically unacceptable (cf. above, p. 286, n. 22).

> The neoclassicists started out from concepts and problems which were explicitly incompatible with the labor theory of value. Asking for a theory which was equally valid for rare stamps and for eggs, it had to take its starting point in consumers' preferences and not in the costs of the producer.[32] But gradually the neoclassicists were forced to approach the questions that had occupied Smith and Ricardo, *viz.* the global economic process of society. At that moment they had to develop a concept of the *price of production* which determined the long-term price level of products which (like eggs) could be produced in any quantity.[33] This price of production turns out to be explained by arguments that follow the fundamental structure of the discussion in *Das Kapital* III of the relation between value and price (only published some years later). Marx is hence able to explain Marshall, just as Copernicus/Newton is able to explain Ptolemy.[34]

Lakatos, we may say, sees the development of a scientific discipline as consisting of a number of parallel lines, each representing a research program, competing for the favor of the scholars of the field and terminating at the moment when favors fail. A more realistic view, on the other hand, would have to look at the lines as partly interconnected. Lakatos's idealization is correct in so far as the connections between research programs are weaker than the connections between theories belonging within the same program; but an understanding which aims at getting beyond Lakatos's formalization should start by recognizing the existence of interconnections.[35]

We may add that even the lines themselves (the single programs) possess an inner structure. Branchings are common, and it is not always possible (as it was in the case of the labor theory of value) to distinguish between a sound trunk and vain aberrations (we may think of the splitting between the Malinowski and the Radcliffe-Brown variants of functionalism, cf. p. 209, n. 15). The solution of single problems (concerned, for example, with specific anomalies) may be the occasion for the emergence of specific subprograms within the same global research program. At times such subprograms may be absorbed into the main program when a satisfactory

---

32. William Jevons, *The Theory of Political Economy*, published 1871.

33. Alfred Marshall, *Principles of Economics*, 1890.

34. In both cases, on the other hand, the reverse explanation (Ptolemy of Newton, Marshall of Marx) turns out to be impossible, for the simple reason that Newton's and Marx's theories include dynamic explanations which fall outside the scope of their competitors.

Marx, on the other hand, is not able to explain John Maynard Keynes's theory of the economic cycle; this is only done by Kalecki (T_5)—see Robinson & Eatwell 1973: 48ff.

35. This is no point of pure philosophy but carries an important message for practical scientific work: You should never dismiss the reflections and theoretical results achieved by another school with the argument that they belong within another research program and that they are therefore irrelevant for you. *Dialogue is possible* and often the crucial condition that you may progress along your own road—not least within the human and social sciences.

solution to their specific problems has been found; at times they may provide the starting point for a new discipline or subdiscipline.

KUHN: PARADIGMS AND FINGER EXERCISES

Chronologically and historically, Lakatos's concept of 'research programs' is an attempt to describe from a Popperian point of view an approach to the problem of scientific knowledge which in many respects constituted a radical break with established ideas. Making Kuhn—the originator of the new approach—appear as a commentary to and an elaboration of Lakatos's ideas is thus a pedagogical trick which inverts the real historical process and connection. None the less, the trick may be useful.

Thomas Kuhn (1922–1996) was a former physicist turned historian of science and no philosopher. This is clearly reflected in his book *The Structure of Scientific Revolutions* (1970; [1]1962), in which his ideas were first presented in print. It does not, like Popper's presumed "logic" of research, attempt to prescribe rules which are supposed to guarantee more steady scientific progress. Instead, Kuhn's first aspiration is to find structure and coherence in the baffling imbroglio of the history of the sciences (Kuhn had worked on the Copernican Revolution and been perplexed by how much "old" thinking he found in the new approach). His second aim (which need not be secondary) is to understand why this structure is able to produce *knowledge,* and to show how it may indeed be adequate and perhaps even necessary for the production of scientific knowledge, regardless of its conflict with time-honored ideas about the nature of good science.[36]

The central concepts in Kuhn's understanding of scientific development are *the 'paradigm,' 'normal science,'* and *the 'scientific revolution.'* Normal science is science whose development is governed by a paradigm, and a scientific revolution is the replacement of one paradigm by another. The *paradigm* itself is thus an adequate starting point.

The term *paradigm* is borrowed from traditional language teaching and is another name for the *exemplar.* An exemplar or paradigm is a word which is used to train a conjugation scheme—as in Latin *amo, amas, amat, amamus, amatis, amant,*

---

36. The second question was formulated in the title of Kuhn's contribution to a symposium on "Scientific Change" held in Oxford in 1961 (published as Kuhn 1963): "The Function of Dogma in Scientific Research." Some years later, Kuhn (1974: 237) formulated his double approach as follows: "The structure of my argument is simple and, I think, unexceptionable: scientists behave in the following ways; those modes of behaviour have (here theory enters) the following essential functions; in the absence of an alternate mode *that would serve similar functions,* scientists should behave essentially as they do if their concern is to improve scientific knowledge."

or in German *ich liebe, du liebst, er liebt,* . . . . Other words belonging to the same category (a-stem verbs and weakly conjugated verbs, respectively) will then have to be conjugated "in the same way," in a way which is understood quite as much through subconscious training as from explicit rules. The point of using the paradigm instead of the abstract sequence of endings *-o, -as, -at, -amus, -atis, -ant* is precisely this subconscious way of functioning. If you had only learned the latter system you would have to switch from speaking to analytical thinking each time you were to use a verbal form. The paradigm, on the other hand, functions much in the same way as the subconscious sensori-motor schemes described by Piaget[37]—or it may serve at least as the starting point for the construction of a subconscious scheme.[38]

The key point in Kuhn's approach to scientific activity is that it is a creative and active practice, as is the use of language. You learn to use German verbs through reading and speaking German and through the training of paradigms, not from the mere reading of grammatical rules; you learn to ride a bicycle by trying and not merely through explanations of the role of the handlebars for maintaining balance (I still remember the explanations I got at the age of five; they would have sent me headlong into the pavement had I ever tried to follow them); you learn to play the piano through finger exercises, transposition exercises, and training, not from mere explication of the musical notation of the correspondence between single notes and keys and of the major/minor system. In a similar way, you learn to work inside, for instance, the Newtonian research program by using its theories and by observing their use, not from a mere abstract exposition of "Newton's three laws" and the law of gravitation or of the "hard core" of the program with its appurtenant negative and positive heuristic; you learn to perform a structuralist analysis of a literary work by doing and by following analytical work, not from mere exegesis of the principles of structuralism.

We learn not *merely* from theoretical and abstract exposition, though evidently *also* in this way. Scientific work does not stop at being skill and knack, it is also a conscious activity. Researchers are not sensori-motor infants but analytically minded adults who integrate the schemes of their cognitive unconscious as tools for conscious operatory thought. The essential point—a point which is neglected by both Popper and Lakatos and indeed by almost all philosophers of science—is that scientific activity *also* contains an essential element of skill.

---

37. Cf. also what is said on p. 240 regarding grammatical schemes.

38. It should be observed that in structuralist linguistics the term *paradigm* is used in a way which differs fundamentally from Kuhn's: in the rudimentary sentence structure "subject—verb," the phrases "a dog," "the bird," and "Susan" are part of *one* paradigm, the set of words or phrases which may serve as subject, and from which precisely *one* element is to be chosen; the phrases "runs," "dies," and "is born" belong to another paradigm.

This use of the term is derived from its meaning in traditional grammar, too; even from the sequence *ich liebe, du liebst,* . . . , *one* element is to be chosen when a sentence is to be constructed.

From where, then, do scientific workers get their skill? In former times, before the systematic training of future research workers in universities, by reading THE BOOK—that decisive book which had moulded the basic understanding of their discipline. Astronomers read Newton's *Principia;* before this seminal work was published, they read Kepler or Copernicus;[39] and before Copernicus was accepted, they had read Ptolemy's *Almagest.* The *Principia,* Kepler's *Astronomia nova,* Copernicus's *De revolutionibus,* and the *Almagest* functioned, each in its time, as those exemplars through which astronomers were trained to see the planetary movements, consciously and subconsciously, as astronomers could be expected to see them, and to analyze the problems of their field as currently done.

These books thus functioned in a way analogous to that of the paradigms of language training, which explains the origin of Kuhn's central term. As it often happens, however, the actual meaning of the term came to differ from the etymological origin—in fact this occurred already in Kuhn's own book. The paradigm concept, if it had referred to the role of such books alone, would have described an earlier stage in the development of the sciences and nothing more. In modern times, natural scientists are trained by means of *textbooks* and prepared exercises, not by following immediately in the footsteps of the founding fathers of their field; they will only be confronted with original research papers at a relatively late stage of their education, and rarely at all with the classics.[40] In the humanities, early confrontation with research literature is customary; but one will rarely (neither at present nor in the past) find a field to be defined by one book to the same extent as physics was once defined by Newton's *Principia* and economics by *The Wealth of Nations.*

Even though Kuhn does use the term *paradigm* to denote the pivotal book which once defined a field, he therefore mostly uses the term in as somewhat different sense, *viz.* about that *collective attitude,* that *collective intuition,* those *shared techniques* and that "tacit knowledge"[41] which natural scientists once got by working

39. With one historically important exception: after the Galileo trial, Jesuit astronomers read Tycho Brahe, who allowed Earth to remain quiet. Even after the publication of the *Principia,* they were supposed to do so. However, since they undertook to translate and publish explanations of the Newtonian system, we may presume that they used it just as much as other astronomers at least for training purposes, irrespective of conceivable metaphysical reserve.

40. Quite a few biologists, of course, will read (passages from) Darwin's *Origin of Species,* some physicists may take a look at Galileo's *Discorsi,* and many economists may study some chapters from Smith's *Wealth of Nations.* But these classics are so removed from what goes on now in the respective fields that they can have no genuine training function; having them on your bookshelves and having looked into them is rather a way to affirm your professional identity.

41. The concept of 'tacit knowledge,' which has been amply used in explanations of the Kuhnian view, was created by the philosopher-chemist Michael Polanyi. The insights which (with an extra delay of ten to fifteen years) gained wide currency after the appearance of Kuhn's book were thus not totally unprecedented—Polanyi is not the only precursor.

their way meticulously through THE BOOK, and which is now acquired in other ways. In a postscript to the second edition of his book, Kuhn even proposes to reserve the term *paradigm* for the shared "constellation of beliefs, values, techniques and so on" (1970: 175; cf. 191), and to label the "exemplary past achievements" instead as *exemplars* for the sake of clarity.

The paradigm in the sense of a "constellation of beliefs . . ." is a totality, and those constituents which can be brought to light through analysis will only direct and govern scientific work *because* they are parts of an integrated whole. Recognizing this restricted value of analysis, however, should not prevent us from having a look at the constitution of the whole.

One element of the paradigm may be familiarity with an exemplar, a fundamental work or a group of central works. In certain cases this exemplar need not belong to the discipline itself; thus, the works of the anthropologist Claude Lévi-Strauss (*La Pensée sauvage,* etc.) and the linguist Ferdinand de Saussure (*Cours de linguistique générale*) have played the role of exemplars for structuralist currents within many disciplines.

More important than the exemplar itself, however, is what you learn from it. The contribution of the exemplar to the paradigm may be found on several different levels. From Newton's *Principia,* for example, you may learn about the actual movement of physical bodies influenced by forces. In general, you learn that the forces acting upon one body originate with another body, and that the acceleration of the body multiplied by its mass equals the total force acting upon it; you learn mathematical techniques, and you learn that these techniques are the means by which you compute the movement of bodies. You learn a precise, "Euclidean" deductive construction of your line of argument, and thus that physics may (and, implicitly, *should*) be constructed as a rigorous deductive progression of propositions and calculations. You learn that physical theory should relate to and explain phenomena, and you learn how to relate theory to phenomena.

From Adam Smith's *Wealth of Nations* (which functioned as an exemplar in classical British political economy) you also learn on several levels at a time. You learn to divide the population of a country into social classes according to the source (not the size) of their income. You learn that the relevant sources are *wages* derived from work; *profits* derived from the possession of means of production (capital); and *rent,* derived from the possession of scarce resources like land; for which reason the classes are working class, capitalist class, and landed proprietors (cf. p. 284, n. 17). You learn about competition and its effects, and about the formation of monopolies and about their consequences. You learn about quantity of work as the factor which determines prices within a market economy. You learn that economic analysis presupposes social statistics and historical considerations. You learn a specific way to analyse and to argue.

What comes from work with an exemplar may thus be summed up as follows:

— You learn about the kinds of entities which constitute the world of the discipline: physical bodies, forces, . . . / kinds of income, social classes, . . . (in philosophical jargon: an ontology).

— You learn which types of explanations belong legitimately within the discipline, that is, which explanations should be used by a physicist and an economist, respectively. Implicitly, you also learn which kinds of explanation should be avoided (the moving intelligences of celestial bodies and just prices, respectively, to mention kinds of explanations used before Newton and Smith).

— You learn about a number of techniques which can be used to attack the problems occurring within the discipline, and you learn how to use them.

— You are provided with a total idea of what the world (of the discipline) looks like, a global perspective on (the pertinent) reality.[42]

The paradigm is thus related to the "hard core" of a Lakatosian research program (no wonder, since the hard core is just Lakatos's explanation of the paradigm concept from a Popperian perspective). But there are important differences. The Kuhnian paradigm is not as precisely delimited and formalizable as Lakatos presumes his hard core to be. A "total idea" and a "global perspective" cannot be summed up exhaustively in well-defined propositions. Learning "how to use" the techniques of a discipline is the acquisition of a skill; skills one may speak about, but a skill in itself is not something which can be enunciated (as can a theory or an ontological presupposition). Presupposing his hard core to be clearly expressible, Lakatos can imagine that a scientist may reject one research program and start working upon another by a fully conscious choice. The idea of the paradigm as containing a prominent factor of training and skill, on the contrary, implies this shift to involve more of a new learning process and less of a free instantaneous choice (the choice actually involved is the choice to start learning anew, to assimilate a new perspective which *is not yet yours* and thus is not fully understood in the moment you choose). Ultimately, the paradigm shift is not an individual affair but rather a process affecting the whole disciplinary community; "a new scientific truth does not triumph by convincing its opponents and making them see the light, but rather because its opponents eventually die, and a new generation grows up that is familiar with it," as Kuhn (1970: 151) quotes Max Planck (the physicist who took the first step in the development of quantum mechanics in 1900). The paradigm involves elements of collective intuition, and intuition, as we all know, cannot be changed by deliberate choice or majority vote.

---

42. In the postscript to the second edition of *The Structure* . . . (1970: 187ff), Kuhn introduces some more precisely defined constituents of the paradigm; but since these are geared specifically to the paradigms of physical sciences, they need not concern us here.

Becoming familiar with an exemplar is not the only way you learn to work within a paradigm. It is even questionable whether you learn it in full in that way. The kind of knowledge which is contained in the exemplar may contribute to the collective intuition; yet it is mainly through *working as* a physicist or an economist while using the exemplar as a navigation mark that you make the exemplar paradigmatically productive. This is why the contemporary training of natural scientists (and, to a large extent, economists, sociologists, linguists, etc.) can be successfully effected without exposition to exemplars but by means of textbooks and appurtenant exercises, the gist of which is that the exercise is to be performed as presupposed within the paradigm; and this is why many fields of human science can transmit their paradigms through exposing students to select pieces of current research literature combined with independent work.

THE STRUCTURE OF SCIENTIFIC DEVELOPMENT

Kuhn's primary aim was never to describe the socialization of future research workers. It was to understand how scientific fields develop. But once the insight was gained that the dynamics of scientific development could not be understood unless the moment of production of this knowledge was taken into account (cf. the initial passage of Chapter 17, p. 257), the socialization of workers turned out to be pivotal. The distinctive character of *scientific* knowledge must then depend, among other things, on the particular way workers within a field see and deal with this field, and hence on the process that makes them see it thus.

When a new field becomes the object of systematic ("scientific") investigation for the first time, there *is* as yet no such particular way to see it and deal with it. Those who approach it do so from common sense understanding of its character and common sense definition of its contents and delimitation,[43] *and* from a general intention to understand it "scientifically." As examples of such "pre-paradigmatic" (proto-)sciences one may take "women's studies" from around 1970 and the study of electrical phenomena between 1600 and 1750.

In a preparadigmatic science, the approaches are multiple and uncoordinated. The results obtained by one worker will normally build on presuppositions and refer to concepts which are not shared by others, and others will therefore have difficulties in assimilating them. Instead they will tend to be neglected and eventually

---

43. With the reservation that they will often have been trained as scholars within other fields. Their "common sense" is thus the common sense of the general scholarly community as tainted by their specific training within particular fields. As a friend of mine once asserted about a former physicist who had gone into peace research and from there into sociology, where she had met him as a teacher: "*A* is a physicist; he will never be anything but a physicist."

forgotten, possibly to be rediscovered twenty or forty years later.[44] Borrowing the Piagetian idiom, we may say that the workers in the field possess no common cognitive structure which is fit to integrate unexpected results and to keep them available for further use and elaboration by the community as a whole. Preparadigmatic sciences are not *cumulative*—at best, single schools with a certain inner coherence (as found, for instance, in women's studies from the mid-seventies onward) exhibit cumulative characteristics.

It may happen, however, that a particular contribution or a specific school obtains such convincing results that other workers of the field accept its approach as an exemplar, and try to emulate it (the precise nature of the contribution is irrelevant, as long as it only convinces and is able to transmit *some* relatively coherent approach to the field). This breakthrough may start the development of a genuine paradigm, and as long as this paradigm serves, the field is in a phase of *normal science.*

During such a phase, work is directed at *expanding the paradigm:* to understand more and more of reality through the paradigm. One may speak of applying the theory to new areas, or of developing new theory for new areas on the basis of the paradigm; the latter formulation may be preferable, since the expansion to new areas may require the addition of new concepts and presuppositions, as well as an *articulation* of the paradigm with regard to its original content by which it is made more precise, explicit, and conceptually clear.

This articulation is the other aspect of what goes on during the normal science phase. Clarification of concepts and increasing adaptation of the paradigm to that reality which is studied may result as secondary effects of the expansion of the paradigm—if you apply the outlook generated within women's studies to the situation of sexual minorities or suppressed racial groups, then you get a new perspective even on the original core of your field, and you get new skill in dealing with it (cf. below, p. 372). Similarly if you apply Newton's laws to the flow of water through pipes, or the principles of structural phonology to analysis of kinship structures. But articulation may also follow from conscious efforts to get better insight into the foundation of earlier results.

Much work within normal science is concerned with the solution of "puzzles" (Kuhn's term). The metaphor refers to such everyday phenomena as riddles, crossword puzzles, and chess problems. In all of these we know that a solution exists; it is only up to our ingenuity to find it. The same thing characterizes normal science:

---

44. This is, for instance, what happened to Vladimir Propp's analysis of the invariable morphology of (Russian) fairy tales from 1928: it only became influential in the 1950s, when Lévi-Strauss and others had established the structuralist paradigm—within which, by the way, the implications of Propp's findings were interpreted in a way that differed decisively from Propp's original "diffusionist" understanding of the matter. Cf. the prefaces and introductions to Propp 1968 and 1984.

since the paradigm "knows" what the world (of the discipline) consists of and which types of relations hold good between its constituents, all problems within this world *must* (if we are to believe the implicit claim of our paradigm) be solvable; the question which remains open is whether *I* am smart enough to find it. If a problem resists my efforts, my colleagues in the field will conclude that I was not—after all, they know from proper experience that the paradigm is fully efficient for all relevant purposes. Only if others fail like I did will the problem cease to be a mere puzzle and become an *anomaly* which challenges the paradigm.

The appearance of an anomaly may lead to focusing work on precisely this stubborn problem (cf. Lakatos's story as told above), and then perhaps to a solution; or it may remain unsolved, and if the paradigm remains productive in other respects the anomaly may then be encapsulated while work goes on under the paradigm without being disturbed.[45]

The puzzles which are taken up during a phase of normal science are not selected at random. In combination, the global view which the paradigm gives of the constitution of its object and the array of results obtained until a given moment will suggest which problems are *now* most urgent and most likely to be solved.[46] This explains the recurrent phenomenon of *simultaneous discovery:* extremely often, the same essential discovery is made by workers who have no direct connection and indeed nothing in common beyond a shared paradigm; this, however, is enough to make them take up the same problem at approximately the same time, and provides them with sufficient knowledge about what should be expected to make them see an actual outcome as noteworthy.

Preparadigmatic science was not cumulative. Normal science is. Stubbornly, it sweeps up everything which the paradigm is able to interpret, shelving anomalies encountered in one direction if it is still successful in others. Eventually, however, anomalies accumulate and tend to turn up at all essential points. The efficiency of the paradigm for puzzle solution shrinks, and during the effort to explain one or another anomaly, the paradigm is articulated in increasingly divergent fashions. Eventually, without having been given up, the paradigm may exist in almost as many

---

45. This happened to the discovery that the perihelion of the planet Mercury (the point where it comes closest to the Sun) rotates ("precesses") in a way which cannot be explained by Newtonian mechanics. The anomalous precession was discovered in the 1840s, and everything suggested that it *should be* explainable. It was not, and was thus shelved—and was ultimately solved by the General Theory of Relativity some seventy years later (Lévy 1973: 278).

46. This process is seen *en miniature* each time a professor allots thesis topics to doctoral students. The teacher is expected to know in advance which questions are now solvable: it would be irresponsible to make students run on a track leading nowhere, but equally irresponsible to make them repeat what has already been done. The teacher is, so to speak, supposed to be the paradigm in person.

versions as there are active workers within the field; ultimately, this will of course undermine its credibility (and obliterate its character of *shared* beliefs, etc.). The field will end up in a state of *crisis,* where doubt about the efficiency of the paradigm grows into general distrust of its previous accomplishments: If phlogiston theory runs into paradoxes when trying to account for weight, are we then really entitled to believe in its explanations of colors? Are these not likely to be spurious and only accidentally in agreement with observations? Willingness spreads to engage in quite new approaches, varied proposals come up, and for a while different schools may exist alongside each other. In many ways, the situation is similar to that of the preparadigmatic phase. Only when one approach has proved its ability to solve precisely those problems which had become central during the crisis period (and only if competitors are unable to do it as satisfactorily) will this approach come to serve as the starting point for a new paradigm, inaugurating a new phase of cumulative normal science.

The shift from one paradigm to the next constitutes a *scientific revolution,* which is characterized by sharp rupture. Taking the Copernican Revolution as an example, Kuhn (1970: 149f) suggests that we consider

> the men who called Copernicus mad because he proclaimed that the earth moved. They were not either just wrong or quite wrong. Part of what they meant by "earth" was fixed position. Their earth, at least, could not be moved. Correspondingly, Copernicus's innovation was not simply to move the earth. Rather, it was a whole new way of regarding the problems of physics and astronomy, one that necessarily changed the meaning of both "earth" and "motion." Without those changes the concept of a moving earth was mad.

The content of a concept is only partially to be derived from empirical observations (cf. what was said above in the discussion of the problems of empiricism); in part it depends on the total theoretical structure in which it partakes and the practice inside which it serves[47]—"our statements about the external world face the tribunal of sense experience not individually but only as a corporate body" (Quine 1963/1951). *Mutatis mutandis,* the observations concerning the changing meaning of terms must therefore hold for all paradigm shifts. The discourses before and after a change of paradigm (or across a paradigmatic border) are "incommensurable." A conference may "bring people to talk to each other who would never read each other's papers";[48]

---

47. In the sciences, this practice is in part constituted by the research process, in part by teaching and applications. For the astronomers of the later sixteenth century, astronomy teaching in universities (which *had* to be traditional) and the computation of planetary positions to be used in court astrology (which by necessity asked for these positions as seen from Earth) were no less weighty than astronomical research (see Westman 1980).

48. Mogens Trolle Larsen, formulated at the dinner table the last evening of the symposium "Relations between the Near East, the Mediterranean World, and Europe—3rd to 1st millennium BC," Århus 1980. The participants were mostly archaeologists falling into two groups:

but it was my definite impression on the occasion where this was formulated that they did not understand each other too well.

Many of Kuhn's early critics (and quite a few superficial followers in later years) have taken the claim for incommensurability to imply that no communication and no rational argumentation is possible across the paradigmatic border. This is evidently a wrong conclusion, built among other things on an absolutistic concept of rationality, and it was never intended by Kuhn.[49] Breakdown of communication is *partial.* This suffices to exclude unambiguous *proof* that one part is right and the other is wrong; but it does not prevent critical, rational discussion, where appeal can be made to those cognitive structures which are shared across the border;[50] what

---

those oriented toward social anthropology and the use of statistical analysis of the distribution of finds; and those for whom "the only facts are artefacts," that is, for whom archaeology should make no theorizing about societies and their structure and interaction but simply dig and meticulously describe the finds and their stratification.

During the symposium, an exasperated member of the former group had commented upon the attitude of the latter with the phrase "Oh yes, the only acts are artefacts!"; the immediate answer was a candid "Yes, of course."

49. In his postscript to the second edition of his *Structure* . . . , Kuhn (1970: 198ff) takes up in some detail the problem of incommensurability and the misunderstandings to which his original statements had led.

50. The *only partial* breakdown of communication distinguishes Kuhn's analysis of the paradigm shift from two apparently related lines of thought: Wittgenstein's notion of 'language games' and Foucault's "archaeology of knowledge." Wittgenstein's analysis (1968: §11 onwards) comes close to Kuhn's in pointing out that a language game is connected to and rooted in a *particular* practice; but it leaves no space for description of the process by which one "paradigmatic language game" develops into another (for good reasons, since this is only a characteristic of certain "games," like the paradigms of scientific disciplines and—with some modification—artistic "schools"), and it leaves aside how the practice underlying the language game is itself to some extent a result of the game.

Foucault's analysis produces a much coarser grid than Wittgenstein's multiple coexisting language games. He speaks (1966: 13f) about two prominent discontinuities in the Western *epistēmē*, one around the middle of the seventeenth century (which we may connect to the reception of Descartes and Galilei) and one in the early nineteenth century (the epoch, among other things, of Comtean positivism). But he is even more explicit than Wittgenstein in his statement that seeming continuities within single sciences over one of these watersheds (for example, between the "general grammar" of the middle of the seventeenth century and modern linguistics) are nothing but surface effects; in Foucault's view, Carl von Linné's biology has much more in common with seventeenth-century general grammar than with George Cuvier's comparative anatomy or Darwin's theory of evolution. This may be true ("much more" is an elusive concept in the absence of a yardstick), but only in a perspective which concentrates on other aspects of the disciplines in question than their relation to their object, and which *eo ipso* (as also stated by Foucault) excludes any idea of cognitive progress through the shift, and indeed any critical communication. Maliciously one might maintain that (this early) Foucault is only right (in this respect) under the perspective where it does not matter

it precludes is the "absolutist" rationality of the strict proof, not the rationality of dialogue, in the terms introduced on p. 231. The situation bears some similarity to the description of the same situation in two different languages possessing nonisomorphic conceptual structures.[51]

The analogies between Kuhn's and Lakatos's formulations are evident. As already stated, the *paradigm* corresponds to the *research program, normal science* to *work within a research program.* Still, differences are no less conspicuous.[52] One of them turns up if we look for the analogue of Lakatos's "hard core." A paradigm possesses no hard core, that is, no sharp distinction between the absolutely inviolable and that which can be freely reinterpreted and changed in order to obtain better agreement with observations. All levels of a paradigm may be affected by articulation.

In spite of articulation, however, the main task which normal science sets itself is the solution of puzzles, where the paradigm not only "ensures" (that is, assures) that solution is possible but also mostly tells what the approximate outcome should be (in experimental or other empirical research), or how an explanation will have to look (in theoretical investigations). If things turn out in a totally unexpected way, and if they cannot be explained even with hindsight to agree with what could be expected (that is, if they constitute an anomaly), the results will often be neglected (as told above). Rebellious thought is rare, "dogmatism" prevails.

This agrees badly with common-sense ideals concerning the character of science and the behavior of scientists (not to speak of Popper's rhetoric). It is also at variance with the way scientists experience their own work: the efforts to grasp things in new ways, the struggle to get around apparently impossible obstacles, and the eventual success by means of a sudden deep insight—these are felt to be the predominant features. How come?

---

whether what he says (in this and any other respect) is right, only that the way in which he says it reflects a particular French intellectual style and the make-up of the French book market; one need not be a follower of Foucault to find this to be a distorting and reductive perspective.

51. Cf. the relation between the conceptual clusters 'knowledge/cognition' and 'Wissen/Erkenntnis/Erkenntnisvermögen'. *Cognition* encompasses only little of what is covered by *Erkenntnis* and most (all?) of what is meant by *Erkenntnisvermögen,* and *knowledge* correspondingly more than *Wissen.* This is one among several linguistic reasons (nonlinguistic reasons can be found) that epistemology looks different in English and German; still, translations *can* be made that convey most of a German message to an English-speaking public.

52. Analogies and differences taken together illustrate the partial *yet only partial* breakdown of communication between incommensurable paradigms. Where Lakatos will see a research program and look for its hard core, its positive and negative heuristic, etc., all of which can be put on paper, the Kuhnian will look for interpretations and collective understandings and intuitions, which can only be described with approximation, and which together constitute a seamless whole.

The latter problem may be postponed for a while. But the first, "how science can make progress if it is so rigid and dogmatic," should be approached now.

For one thing, the "dogmatism" of normal science does *not* imply that the exemplar (or the textbook) is regarded as holy writ that *cannot* be wrong. Firstly, the very principle of cumulative science is to use preceding knowledge (including the exemplar) in order to succeed where predecessors (including again the exemplar) have failed. The attitude is nicely summed up in the statement that we are like "dwarfs perched on the shoulders of giants [seeing] more and farther than our predecessors, not because we have keener vision or greater height, but because we are lifted up and borne aloft on their gigantic stature" (see p. 64, n. 19). Secondly, what is learned from the exemplar is not specific sacrosanct results but a general and open-ended way of thinking (in Aristotelian jargon: not the content but the *form* of the exemplar is important); nothing prevents workers from using this thinking to correct concrete mistakes committed in the exemplar.

But normal science is functional not only because it allows the errors of an original accomplishment to be corrected; it ensures that the carrying capacity of the paradigm is tried out to the full, and that it is not rejected at the encounter of the first apparent anomaly or the first change of fashion. It regulates and structures systematic examination of the field which it covers (as opposed to what takes place in the preparadigmatic phase, where work is unsystematic, unstructured, and largely ineffective); finally, and for the same reasons, the paradigm is a most efficient instrument for bringing forth and establishing the anomalies that eventually make it break down.

The latter point is contained in an aphorism which Engels formulated long before Kuhn: "In chemistry, only a century's work according to the phlogistic theory supplied the material which allowed Lavoisier to discover in the oxygen that Priestley had produced the real counterpart of the imaginary phlogiston substance, and thus to throw over the whole phlogiston theory."[53] Approbation of the "dogmatism" of normal science is thus no conservative attitude and no endorsement of static thinking; it is associated with a view according to which scientific progress comes from that "essential tension" of which Kuhn speaks in the title of another book, and not from gratuitous rhetoric à la Popper (cf. the quotations on p. 295, n. 36 and on p. 334).

COLLECTIVE AND INDIVIDUAL KNOWLEDGE

The question may be shifted somewhat. A theory of scientific development is a theory about the production of (collective) knowledge. There is thus nothing strange if scientific development exhibits features which are similar to the characteristics of other

---

53. *Dialektik der Natur*, MEW 20, 335f. Another "Kuhnian" point is also contained in the passage: Joseph Priestley produced oxygen, but understood it within the framework of the phlogiston theory as "phlogiston-free air"; only Lavoisier "invented" oxygen, thus engendering what developed into a new paradigm.

types of cognition. Science is indeed not the only domain where stability appears as a prerequisite for development and change.

Let us first have a brief look at *art*—postponing to a later chapter the intricate question of which kind of knowing (if any) is involved in art. A comparison between scientists working within a paradigm and artists belonging to the same "school" or tendency ("impressionists," "serial composers," "absurd drama") is close at hand. Even here it is obvious that working out the possibilities of one school is one of the factors which make innovation and even rupture possible.[54]

Another parallel may be followed further and more precisely. If we replace Kuhn's "paradigm" with "scheme," his "expansion of the paradigm" with "assimilation to the scheme," and the "articulation of the paradigm" with "accommodation," we shall get the gross structure of Piaget's epistemology. Even here, as we know, the child is only able to search for and gain knowledge because it possesses a cognitive structure organizing the search and the transformation of sense impressions into comprehensible *experience.* Only because this structure exists and is relatively stable can it create the conditions for its own replacement by a higher structure.

In individual cognitive development, the cognitive structure is evidently individual, even though, for example, the over-all character of one individual's preoperatory cognitive structure is very similar to that of another individual. But in that process of knowing collectively which is the essence of the scientific endeavour, the cognitive structure must by necessity be shared. At the same time, its development is not regulated and monitored by a pre-existing language and set of concepts, a pre-established stock of relevant everyday experience, and an already unfolded lifeworld (although, evidently, all of these are there and contribute to the formation of the scientist's mind). The collective cognitive structure must be brought forth by the scientific community itself—and this is precisely what is done through the establishment of the paradigm, through the common reading of and work from exemplars, educational "finger exercises," and so on.

It is an important feature of Piaget's epistemology that the cognitive structure is not constituted by conscious and explicit knowledge; it belongs at the level of the cognitive unconscious. That this is so is an empirical fact as good as any that can be established by psychology. If we oppose Kuhn's and Lakatos's approaches, we will remember one of the important contrasts to be that Kuhn supposes the paradigm to consist much less of formulated theory and explicit statements than Lakatos assumes regarding his "hard core." According to all we know about general human cognition, Kuhn's view is thus more empirically plausible than the alternative; similarly, the parallel suggests that those structures which both see as mandatory for scientific

54. Looking with historical hindsight at Paul Cézanne's or Anna Ancher's paintings, one will easily see cubism working itself toward the surface; mindful listening to Richard Strauß's early operas or even to Schönberg's still Late Romantic *Gurre-Lieder* may make one understand why Schönberg came to feel the need for his dodecaphonic technique.

development result to a considerable extent from socialization and training, and not exclusively from conscious choice. On the other hand, scientific knowledge in stored form presents itself in explicit and relatively unambiguous statements.[55] Important elements of the paradigm or the hard core must therefore consist of clearly expressible statements. Part—but only part—of the paradigm consists of "tacit knowledge."

From the discussion of the relation between sensori-motor schemes and operatory thought (Chapter 16) it will be remembered that whole structures (for instance, "riding a bicycle") may be made subservient to conscious operatory thought ("in order to go downtown I shall have to get the bus at the bus terminal; to the bus terminal I may ride on my bicycle") without being themselves brought to awareness, that is, without requiring conscious reflection regarding details (the problem of balance, coordination of the feet, etc.). The way in which scientific work integrates subfunctions assimilated during professional upbringing (laboratory technique; the way a literary scholar reads a novel, etc.) appears to be more than a vague analogue of this aspect of general human cognition and problem solving.

Above, we touched on the conflict between Kuhn's description of normal science and the participants' own experience of the situation: even within normal science the worker will often feel his activity to be a continuous struggle with the material—a struggle that is only brought to success by means of new ideas. The apparent paradox may be elucidated through this parallel between scientific and general cognition and the displacement of routines from the focus of awareness. According to Piaget's epistemology, we remember, *every* act of knowing is at the same time assimilative and accommodative. Every new item of knowledge which is assimilated to a scheme alters this scheme, at least by extending its scope—cf. *Aha-Erlebnisse* of the kind "My God! This thing that has puzzled me for so long is really nothing but . . ." and "Oh, *that* is how it is to be understood."

Scientific processes of knowing carry the same Janus face. The historian of sociology who looks at the first applications of functionalist explanations within the sociology of science will tend to see the assimilative aspect of the event: "here, the sociology of science simply expands the familiar anthropological paradigm so as to cover a new range of phenomena—typical normal science." As a practising sociologist you will see things differently. The application of the paradigm, of everything

---

55. Some ambiguity remains. Reading research papers from an unfamiliar discipline is difficult, not only because you have not read the textbooks but also because you have not been brought up in a way which makes you understand the hints, connotations, and implicit arguments contained in the texts. Reversely, writing out from your own discipline may be difficult not only because you have to present simplified versions of that textbook knowledge which your readers do not possess but also because you have to bring your own implicit knowledge to awareness; if you do not, you will neither be able to explain it explicitly nor to communicate with the readers' implicit knowledge by means of hints, connotations, and metaphors.

which is self-evident, of all your tacit knowledge, will not be in focus. You may be aware of this aspect of the matter, precisely as you know that you use your computer and your typing skill when putting your results in writing. But why bother about such peripheral and trivial matters if you are to tell the crux of your endeavor? The crux is clearly the thing which was *difficult,* that which did not run smoothly, the *new* ideas and the reinterpretations of familiar concepts which were necessary before a theory explaining the functioning of religious rituals and kinship structures in tribal societies could be used to explain patterns of scientific communication.[56]

Such extensions of our scientific knowledge which are *only* assimilative, which require no cognitive initiative whatsoever but only trite routine, will normally be regarded as mere *applications* of science (to be distinguished from "applied science," cf. below, p. 359, n. 5) and not as genuine scientific activity. One reason that many ideologues of science (not least Popperians) reacted so strongly against the concept of normal science will probably have been that Kuhn, through his emphasis on assimilation as an important aspect of the process, appeared to equate most scientific activity with what they see as a rather boring and unimaginative routine.[57]

TWO KINDS OF "LOGIC"

The parallel between Kuhn's specific and Piaget's general epistemology may carry universal implications. The similarity between the development of collective scientific knowledge and individual knowledge suggests the double (assimilative/accommodative) constitution of both to correspond to a necessary characteristic of human knowledge. Seen in this light, Kuhn's theory turns out to be a "logic" for the development of scientific knowledge as a *social, productive* process.[58] What Lakatos has

---

56. Another aphorism may highlight the matter: "In normal science, 95 percent of everything is routine; during a scientific revolution, only 90 percent is routine" (Donald T. Campbell, at the "Symposium on Evolutionary Epistemology," Ghent 1984).

57. Thus John Watkins, in an insipid panegyric of Popper's genius (1974: 32): "The careful drawing up of a horoscope, or of an astrological calendar, fits Kuhn's idea of Normal Research rather nicely." Or Popper, in an otherwise much more interesting essay (1974: 53): "The 'normal' scientist, as described by Kuhn, has been badly taught. He has been taught in a dogmatic spirit: he is a victim of indoctrination. He . . . has become what may be called an *applied scientist,* in contradistinction to what I should call a *pure scientist*" (Popper's emphasis).

58. And still, in fact, a "rational reconstruction" and no faithful rendition of actual historical processes. Twenty years ago, much fun was made of the fact that the word *paradigm* did not turn up at all in Kuhn's book (1978) about Max Planck's first steps toward quantum theory. Without reason, I would say: it is quite legitimate to derive an overall structural "logic" from the shimmering of real historical processes—but this logic should not necessarily be used to redraw in black lines the contours of the single historical process, thus eliminating the shimmering from view.

formulated is (in agreement with his Popperian starting point) rather a logic for the development of (stored) knowledge viewed as an abstract process, in which the productive mediating role of the working scientific community between one stage of stored knowledge and the next is regarded as immaterial.

At the same time (and in the same moment), Lakatos understands "knowledge" in a way which excludes creativity and fantasy from the domain which epistemological theory can legitimately investigate. Kuhn, on the other hand, who regards the production of knowledge as carried out by actual human beings, opens up the possibility that the creative and the systematic aspects of the process of knowing may be regarded together—even though he makes no remarkable suggestions himself in this direction, and does not want to do so.

OBJECTIONS AND FURTHER MEDITATIONS

The primary purpose of these pages is not to present a survey of the opinions of Piaget, Popper, Kuhn, and others, nor to investigate systematically what may be the shortcomings of their theories.[59] It is to offer a general (though neither complete nor encyclopedic) view of the characteristics of scientific knowledge and of the social process through which it is established; presentation as well as critical discussion of the theories are subservient to this aim.

Even for the purpose of establishing a general view, however, it is worthwhile to consider some of the problems left open by Kuhn or even called into existence by his work.

A problem of the latter type presents itself when Kuhn's arguments about the efficiency of normal science are used for science policy purposes (in a way Kuhn never proposed himself, it must be stressed). It is mentioned first because it justifies some of the dismay called forth by *The Structure of Scientific Revolutions.*

The scientific understanding of a problem area (say, the failures of education) does not really progress as long as it stays in the preparadigmatic phase where many approaches compete—thus certain policy-makers' (so far sensible) reading of Kuhn. If, however, this area has become socially important or politically hot, then we'd better know something scientifically about it in order to implement a sensible policy—thus managerial rationality since 1945 (more about this in chapter 21). Alas, research about the area (educational studies) is so obviously preparadigmatic

59. Nor is it therefore important to follow the later development of (for example) Kuhn's thinking; let me just mention that his latest attempt to clarify the notions of incommensurability versus comparability of theories referring to different paradigms (1992) gives up the most innovative insights of his early work—the emphasis on the role of skills and on values and norms rather than rules (below, p. 334) as the factors that regulate scientific practice. Instead, he returns to a model inspired by neoliberalist economic theory, a model which presupposes full formalization and calculability of the advantage of one theory over the other.

(or at least nonparadigmatic) that we cannot expect genuine progress to take place within a reasonable time horizon; then *let us do something about it,* and declare one of the approaches to be the paradigm, and channel all research monies accordingly—thus the conclusion.

This line of argument may seem attractive to bureaucrats, whether professional officials or academic members of advisory bodies. It should be obvious from the above, however, that the reasoning is highly fallacious: paradigms indeed acquire their status by *deserving it,* by convincing workers in the field of their efficiency; and they lose it (and *should* lose it, irrespective of the preferences of grant-giving authorities) when no longer convincing. The underlying epistemology is conventionalism and not Kuhnian dialectic: if all theoretical approaches are equally valid, then bureaucrats can be allowed without risk to choose, and there can be no serious objection to their choosing the one which seems most immediately promising or which agrees best with their preconceived ideas.[60]

Updated versions of conventionalism in Kuhnian disguise are not too rare, but not of urgent interest in the actual context.[61] We shall thus leave further discussion of this matter aside, and turn to issues deriving from what Kuhn actually says.

Firstly, we may focus on a question which has been emphasized much more in the discussion than it really deserves, but which has the merit to suggest further reflection: during phases of Kuhnian normal science, disciplines are as a rule dominated

---

60. This was already the underlying idea when conventionalism was formulated by the Catholic philosopher of science Pierre Duhem in the early years of the twentieth century. According to conventionalism, indeed, Cardinal Bellarmine had been quite right when allowing Galileo to discuss the Copernican cosmology *as a hypothesis* but not to set it forth as indubitable truth—and Galileo had proved himself to be a mediocre philosopher when he did not understand the wisdom in this directive.

61. To mention only one example, Barry Barnes (1979) presents Kuhn straightforwardly and with great sympathy as a "conservative thinker," without noticing—if we use the same political metaphor—that Kuhn would praise the conservatism of the *Ancien Régime* for being able to produce such radical novelty as the French Revolution, and point out that British Liberalism—the analogue of empiricism, both indeed having received their most famous philosophical formulation from John Locke—never gave rise to such real innovation.

Genuine conservative thinking is rather the (unacknowledged) consequence of certain recent currents: postmodernism and (much) deconstructivism. On the surface it may not look so: that everybody is entitled to think as he pleases since no point of view is better than any other was never a prominent conservative stance. But the core of these programs is the denial of the critical potentialities of human reason. As in the case of conventionalism (with which the postmodernist persuasions have much in common), the flat rejection of reason (and not just of particular limited rationalities) leaves power and inertia as the only generally effective criteria for decision, and lends them that legitimacy which the Enlightenment project had denied them. This is where "conservatism" creeps in.

by a single paradigm; Lakatosian disciplines, on their part, are normally split between discordant and competing research programs. The view appears to depend heavily on the eyes?

In any case, the view depends critically on the direction in which you look, on the way you describe it, and on the sense you give to your words—*in casu* to the ambiguous term *discipline.* The hard core of a research program, we remember, encompasses among other things a distinctive view of reality and hence a specific demarcation of the discipline. Competing research programs within (what university administrations and Lakatos see as) the same discipline may therefore define themselves so divergently that it makes better sense to speak about *competing disciplines* dominated each by its own paradigm or program.[62] Much depends on the question whether disciplines are to be defined institutionally (for example, from appurtenance to specific university departments) or cognitively.

But much also depend on the choice of prototype disciplines. Kuhn tends to choose his from the physical sciences: astronomy, physics, chemistry. Lakatos, on his part, looks more often to softer sciences (cf. my illustrations through economics or philosophy of science) or to phases in the development of physical sciences where Kuhn would see a crisis or a preparadigmatic field of research.

All in all, the paradox dissolves into disagreement about concepts (a typical case of "incommensurability") and about the delimitation of the "typical." It points, however, to a much more fundamental question: How much, and how many, fall under a specific paradigm?

As an example, I shall take my own situation in 1969, at the moment when I finished my master thesis in high-energy physics. I was associated with a subdiscipline comprising at most around 100 publishing participants (fifty may be a better estimate). Everybody within this circle followed closely[63] everything that was done under a paradigm that had been born in 1967 and was still under articulation (it never matured before it was superseded), and we had our own quite distinct methods and our own argot. Evidently, other methods and techniques we shared with other high-energy physicists, under what can be described as an open-ended paradigm ten to fifteen years old. A common trunk we shared with physicists in general, for

62. Cf. the two kinds of archaeologists described on p. 303, n. 48. People who "would never read each other's papers" are not really members of the same discipline, even though they may have their positions in the same university department.

That two disciplines compete about understanding the same section of reality does not imply that they have to be understood as two variants of the same discipline—cf. the unending discussion between psychiatrists and psychologists about who has the better understanding and who dispenses the correct treatment of the same patients. As discussed below (chapter 21), disciplines only form through *specific approaches* to reality.

63. At one moment the head of my working group, the late Koba Ziro, presented a publication as "several weeks old."

instance quantum mechanics (brought to maturity between 1926 and 1935) and the theory of relativity (1905–1912).

Kuhn speaks about "the" paradigm. If so, how much belonged to the paradigm under which I worked, and how many were my companions?

One may observe (Kuhn does so) that quantum mechanics *when seen as a paradigm* (that is, as a practice to which you are socialized, not as a body of formulas or theory) is not the same thing in chemistry, solid-state physics, and high-energy physics. The way you have *learned to work* with quantum mechanics differs from one field to the other. Quantum mechanics is a constituent of all three paradigms, but not the same quantum mechanics.[64]

The consequence of this point of view is that the paradigm characterizes the small unit with its 100–200 publishing participants. Normal science phases hence become relatively short, and revolutions rather frequent—but also rather moderate. A revolution in the small unit to which you belong does not imply that you have to learn everything anew: those who replaced "my" paradigm with the first version of string theory could continue to practice quantum mechanics much as they had done before.

Since even much of that knowledge and many of those professional skills which are conserved through a revolution of the "local" paradigm share the characteristics of a paradigm (resulting as they do from socialization, training, and practice), it may be reasonable to modify the idea that the scientist is submitted to a single paradigm, and to look at him as a member of several circles—some broader and others narrower, some intersecting. Each circle corresponds to a cognitive pattern of paradigmatic character, and single workers may well experience a revolution in their local circle without being for that reason forced to rethink and relearn everything in their professional practice. Such things only happen when the larger circles are struck by a revolution. No scientifically living branch of physics (if we stick to that example) was practiced in the same way after the maturation of quantum physics and relativity theory as before 1900; even the structure of subdisciplines was thoroughly revised, many specialties disappeared and many more emerged. Similarly, the persuasion that *weight* was more important than color in chemistry (cf. p. 290) depended on changes in the larger circles.

Whether one wants to reserve the term *paradigm* for that which is shared by the small unit or one accepts the notion of paradigmatic circles or levels may be a matter of taste. Whatever our choice, we should remember that the complete cognitive structure of the individual scientist is, precisely, *a structure* or network of mutually dependent elements and is not exhaustively described as "a paradigm" cast in a

---

64. This I can confirm from personal experience: the quantum mechanics I taught to students of chemical engineering dealt largely with concepts and techniques I had scarcely heard about when studying physics. To be able to teach it adequately, I had to work hard.

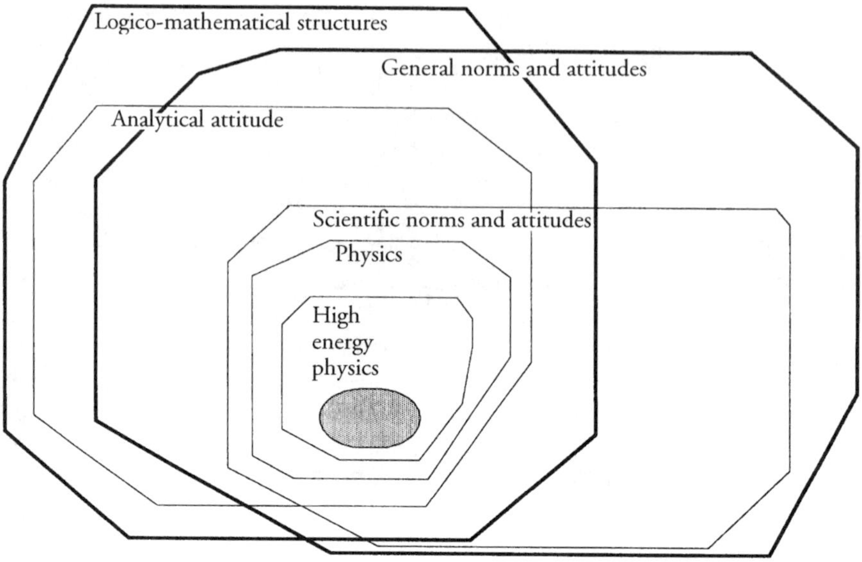

The embedding of a high-energy physicist's paradigm. The shaded area corresponds to the paradigm of his small unit.

single piece (cf. the diagram on this page, which may seem complex but is actually utterly simplified and only meant to be suggestive). Some of its constituents are shared with everybody else who has attained fully mature operatory thought; some—not least an analytical attitude—are shared by scientists from other disciplines; some may be shared by members of certain disciplines but not by the members of others (we remember how physicists and psychologists fell into one group in a Piagetian experiment and mathematicians and logicians into another, see p. 251, n. 24). Some elements of a physicist's paradigm he will share with other physicists since Galileo: I recall my own awe when reading Galileo's *Discorsi* as a young student—here spoke an eminent *physicist,* however much his theories have been buried under repeated paradigm replacement. Many, of course, he will only have in common with contemporaries or with other members of his subdiscipline (the Kuhnian paradigm *stricto sensu*).

Beyond these elements, the diagram refers to "general" and "scientific norms and attitudes."[65] As will be argued below (chapter 20), it is impossible to separate normative or moral attitudes completely from the cognitive structure—as a matter of fact,

65. These, we may note in passing, encompass (but are not exhausted by) what makes it possible for Foucault to see the multitude of sciences of a certain epoch as representatives of a single *epistēmē*—cf. p. 304, n. 50.

much of the paradigm (or the hard core) is of normative or quasi-moral type, prescribing what *should* be done and looked for.

The composite nature of the single scientist's cognitive structure provides us with a scheme inside which *creativity* can be accounted for. If you encounter a problem within your discipline—be it one requiring the "5 percent creativity" of normal science or the "10 percent creativity" of a scientific revolution—this discipline does not constitute your sole cognitive resource. You will indeed have been trained in many practices beyond your scientific specialty and have acquired a wide array of skills and patterns of thought through these processes; in many cases, these provide you with models which you may transfer to the solution of your scientific problem and use in combination with each other and with what you know from your paradigm.

> Scientists rarely tell the sources for their ideas except when these sources seem "honorable," that is, scientifically relevant. As a consequence, I shall have to illustrate this by having recourse once again to an example from my own experience:
>
> In the early 1980s I began working on the corpus of mathematical cuneiform "algebra" texts from the early second millennium BCE. I was soon led to a complete reinterpretation of almost every term and technique. If more than a dozen people around the world had been active in the field, this could have been characterized as a (local) revolution; in the wider contexts of the history of mathematics or Assyriology, it looks more like the assimilation of the field to an anti-anachronistic paradigm which looks at Greek geometry or Babylonian laws not as incomplete forerunners of *our* thinking but as expressions of the culture within which they were created.
>
> But neither Assyriology nor the anti-anachronistic ideal provided me with the tools which allowed me to reinterpret the Babylonian texts. One aspect of my method was instead *structural semantics;* that, however, I did not know at the time, and my improvised method was actually inspired from my supervision of student projects analyzing literature structurally (even this I did not think about in the process, but it could be seen with hindsight). Another aspect was *hermeneutic reading*—yet not taken over from what I knew about hermeneutics, but an application of the way I had once used to read wrong answers to mathematical exercises closely in order to discover the underlying reasoning and thus be able to make the errors pedagogically fruitful. At least one visualization of a procedure I might have taken from many places, but I happened to borrow it from my half-forgotten particle physics via a different use I had made of it in an analysis of ancient Egyptian mathematics.
>
> However frivolous these inspirations may seem, they can all be traced unambiguously to various kinds of professional (and hence intensive) experience of paradigmatic character. The anti-anachronistic drive, on its part, did not come from any allegiance to a historicist program, but rather (as far as I can see—but at such points introspection becomes suspect, as any psychoanalyst can tell) from deep-rooted personal norms

and attitudes also reflected in the way I once read my students' mathematics exercises.

Innovation and creativity cannot be reduced to mere heaping of such accidental extra-disciplinary inspirations, and extra-disciplinary inspirations change when they are brought together and applied in a new context (my hermeneutic readings of Babylonian texts were certainly more analytical than those I made intuitively as a school teacher). But the anecdote will hopefully show that the fine structure of the paradigm provides us with a framework within which the creative process can be discussed and thus, to some extent, *understood.*

CHAPTER 19

TRUTH, CAUSALITY, AND OBJECTIVITY

From the making of scientific knowledge we shall now return to three classical issues: the problem of *truth,* or correspondence between theoretical statements and reality (where we shall have to connect a number of discussions and reflections from the previous chapters); the nature of *causality;* and the question of *objectivity* versus *subjectivity.*

TRUTH

As all empiricists, the logical empiricists worked from the implicit premise that the correspondence between the single observed fact and the statement of that fact was straightforward.[1] For them, as for the whole philosophical current descending from them, the "theory of truth" is concerned with how truth values can be ascribed consistently to the sentences of a formal language (cf. p. 270, n. 19), and not with the question of how a simple statement can "agree with facts."

The position of Popper_0 is similar: the difference between the two approaches hinges not on any disagreement concerning the "naïve" correspondence theory concerning elementary facts but on the explanation of the way from single statements

---

1. That any philosopher would do so after Kant may seem astonishing: whatever one thinks of Kant's solution, he should recognize the existence of a problem. If we can only know by imposing *our* categories on the world, then simple correspondence appears to be excluded.

Yet the explanation need not be neglect of Kant's insight (even though the logical empiricists wanted to render earlier philosophy superfluous rather than continuing it). It could also be that they took Kant fully to the letter: categories which we cannot help applying in an invariable form that we cannot influence may be counted as a part of that reality which it is the aim of science to describe—"observed facts" are precisely *observed* facts. As we have seen (p. 223), such a solution had already been proposed by d'Alembert decades before Kant formulated the problem.

of facts to theory. Popper$_1$, as well as the later phases of logical empiricism, came to admit that "naïve" correspondence does not hold water, but none of them succeeded in creating a credible "critical" substitute. The consequence drawn by the logical empiricists instead verged toward skepticism, whereas Popper's later writings lean toward an unacknowledged instrumentalism formulated as "evolutionary epistemology" (cf. p. 273, n. 25).

In classical philosophical terms, the consequences drawn by Kuhn from *his* view of scientific development would also have to be characterized as skeptical. He argues strongly that scientific progress is real, but in the sense that "later scientific theories are better than earlier ones for solving puzzles in the often quite different environments to which they are applied" (1970: 206). But scientific theories do not come closer and closer to truth through successive revolutions, as common sense would have it. Scientific progress, like Darwinian evolution, is instead to be understood as a steady movement "*from* primitive beginnings but *toward* no goal" (1970: 172).

At closer inspection, the truth which Kuhn cannot discover as the goal toward which scientific development moves turns out to be *an ontology*, a "match, that is, between the entities with which the theory populates nature and what is 'really there'" (1970: 206). So far Kuhn is indubitably right. Newton introduced *forces* into nature, and Einstein's General Theory of Relativity abolished them again; if Newton's move was one toward greater ontological truth, and forces are "really there," then Einstein's move was in the wrong direction. If Lavoisier's explanation of combustion as absorption of oxygen constituted ontological progress, then the acceptance of the phlogiston paradigm had been a mere error, irrespective of its actual successes (including its role in Priestley's production of oxygen).

In the case of competing paradigms approaching the same subject matter in different ways, we are no better off. Word classes, for example, may be defined from morphology and meaning or from syntactic function (cf. also p. 119). If we choose to interpret, for instance, the form *gone* in "Peter is gone," as a conjugated form of the verb *go*, and if we follow this principle throughout, we get one type of insight into the structure of language; if we choose to interpret it as an adjective because it is parallel to *clever* in "Peter is clever," we get other insights.[2] If one of the two approaches is ontologically correct, then the insights gained from the alternative approach are spurious and not to be relied upon. If we insist on applying both approaches at a time, word classes loose their character of clearly delimited entities, and thus their ontological full civil rights.

Certainly, situations exist where one approach turns out to be mistaken and the other not (or not clearly) so. In other cases, however, later developments show that

2. To be sure, the example is simplified into the extreme—but not significantly more than the habitual references to planetary systems and combustion theories in philosophy of science. See Diderichsen 1971: 20ff for an exposition of the problem in the context of a descriptive grammar.

none of the two was quite mistaken. Above (p. 293), the interpretation of the phlogiston theory in terms of "degrees of oxidation" was mentioned. In linguistics, we may refer to the relation between the neogrammarian theory of language development (which referred to laws of phonetic change) and Saussure's early structuralist description of language as it looks at one particular moment. In Saussure's view (1972: 193–195), the two approaches are incompatible (cf. p. 170, n. 7); when discussing development he works within the neogrammarian paradigm (when needed correcting specific laws—thus 1972: 201). A couple of decades later, however, structural linguists of a new generation (in particular Roman Jakobson) reinterpreted the sound shift laws as resulting from structural constraints, in a way Saussure had adumbrated but without realizing it. In both (and many similar) cases we may conclude that the later integration of the two approaches into a more mature theory has shown both of them to be in some way true—which requires that we formulate a concept of truth where this can be said meaningfully.

As a matter of fact this concept was already formulated on p. 270, where the truth value of a theory was interpreted as a "*structural agreement or correspondence* with features of reality"; on the same occasion it was observed that "*correspondence* is something quite different from *similarity,* not to speak of *identity.*" Such structural agreement is most easily identified when we have to do with formalized sciences (physics, theoretical economics, transformational grammar, etc.), where mathematics and diagrams describe the interconnection between the elements of the theory with some precision; but Darwin's notion of 'natural selection'—which sees the natural biological process as a parallel to the artificial selection by which breeders bring forth better varieties of cattle or crops—demonstrates that even discursively organized theory can be analyzed along the lines of such structural agreement.

This materialist notion of truth is not ontological—or at least not necessarily ontological. Certainly, if we suppose that a certain entity does exist (be it with qualifications, like the fox on p. 255), then the best theories will be those which contain it. In this case we may speak of "ontological existence."

We may surmise the element *oxygen* to be such an entity. During the two centuries that have passed since its discovery, we have come to know increasingly more *about* oxygen: that it consists of atoms composed in a specific way; that three different isotopes of oxygen exist; that it has a particular spectrum; and so on—and even how oxygen can change into other elements through nuclear processes. We have come to know answers about oxygen to which neither Priestley nor Lavoisier suspected the mere possibility of the questions; we have not, however, succeeded in getting rid of oxygen as a particular way in which matter can be organized.

Phlogiston had no similar ontological existence, however real the regularities of chemical processes which were explained through the assumed existence of this substance. When taken to be a substance, it turned out to resemble the elf maiden: one sees her, one hears her enticing proposals; but when he tries to grasp her, her back is hollow, and his arm catches nothing but thin air. Newtonian forces are similar; they only manifest themselves through the regularities which they formalize, and one can get into no other contact with them. Word classes may fall somewhere in between: a delimitation made on the basis of inflection turns out to have semantic and syntactical implications (if only approximate). Social classes defined from income sources as done by Adam Smith may also be located at an intermediate position, since individuals who belong to the same class are likely also to share much of their general life experience and their material and spiritual culture. The labor value of goods (or the customer preferences of the competing marginalist economic theories) and the Gothic style in architecture, on the other hand, are probably no be better off than the Newtonian forces; few of the basic entities to which social and human sciences refer seem to be. If we expect truth to be ontological truth, we do not need Kuhn to tell us that few of the theories about human social and cultural life are *true.*

But why should truth be ontological? No sensible scholar would claim that the most adequate (or "true") description of Shakespeare's works is a dictionary listing his vocabulary, and no biologist would be satisfied by taxonomy alone. The "truth" of the screwdriver did not consist in the isolated correspondence between its edge and the notch of the screw, but in that combination of edge and rotational symmetry which corresponded to the entire constitution of the screw and allowed us practical interaction with it (namely to put it into the wall). Taken in isolation, the "ontological" truth of the edge is even meaningless, since only the practical interaction turns the edge into something which corresponds to a notch—screws have no edges, they are constituted (and used) in a way that makes the application of an edge appropriate. The truth of a theory, we may repeat once more, consists in its "*structural agreement or correspondence* with features of reality" as revealed in practical interaction with the object (even when interpretation is concerned, cf. p. 271, n. 21).[3] In a pun which is allowed by the English language, the theory is *true to* rather than *true about* the reality it deals with.

---

3. Experimental science, it is true, may aim at controlling whether a certain theoretical entity can be interacted with in new ways, so to speak testing whether the elf maid whom you see and hear and whose hair you smell can also impress the sense of touch. It was precisely because they could be contacted through several channels (semantics, inflection, syntax, or economy, living conditions, and culture) that word classes and social classes were held above to be more ontologically real than phlogiston. But since physical science provides the scale on which degrees of real existence are conventionally measured, it is worth remembering that quantum mechanics (though ridden with paradoxes as it still is when used to describe more than isolated experiments) dissolves the ontological existence of physical matter: if an electron has to be something specific, it is a particle in some experimental settings and a wave in others—maiden to the eyes and thin air to the sense of touch. A consistent description (relative as this consistency is) can only be reached at the cost of giving up its ontological separateness—and thus ultimately by relativizing the validity of the category of the (conserved, separate) object more radically than we were forced to do by the breathing and perspiring fox.

Then, however, the basic categories which form the framework of our cognition become *true*—cf. the discussion of the practice allowed by the category of the permanent object on p. 255; even the principles of induction and of "generalized induction" (cf. p. 272) turn out to be "true" in the sense that the world is "constituted in a way that allows an adequate practice" if we presuppose them as first guesses. Similarly, the progress within a paradigm becomes progress toward greater truth, not a mere accumulation of solved puzzles: expansion of the paradigm means that more features of reality are accounted for coherently, whereas articulation implies greater precision in the structural agreement. Even the replacement of one paradigm by another is a progress toward greater truth, at least if the replacement follows the Kuhnian pattern

expansion—> accumulation of anomalies—> crisis—>
convincing new interpretation achieving paradigmatic status[4]

(cf. the discussion (in terms of "research programmes") on p. 293.)

> In formal as well as Aristotelian logic, a meaningful statement is either true or false; the everyday idiom of "not quite true" and "almost true" has no place here. Of two conflicting statements, furthermore, at most one can be true. Neither seems to apply any longer if we allow, for instance, both the phlogiston and the oxygen theory to be materially true, however much one is held to be "more true" than the other. As long as we move within the same conceptual framework (within which the problem of noncommensurability does not present itself), it is no more difficult to speak of the sentence which is *less true* than the other as *false* than it is to say in traditional logic that the statement, "Peter is a boy, John is a boy, and Joan is a girl" is *true*, while the statement "Peter is a boy, John is a boy, and Joan is a boy," is *false* and not just "67 percent true." If the two statements belong within incommensurable conceptual frameworks, things are less simple: even if we regard the early oxygen theory as more true than the phlogiston theory, the statement "carbon and hydrogen have nothing in common beyond being elements" cannot be declared "so true" that the phlogiston theory identification of the two under specific circumstances is completely false; only when a common framework (in this case the developed oxygen theory) ripens is it possible to decide—in the actual case that the phlogiston theory had a point.

In the end, the Lakatosian-Kuhnian criticism (or even rejection) of the naïve correspondence theory of truth thus unfolds (in materialist interpretation) as a

---

4. The situation is different if an ideologically inconvenient paradigm is replaced by something more convenient—as was the case in the neoclassical revolution in economics, cf. p. 286, n. 22 and p. 293. In such cases, progress toward greater truth is evidently not assured. Nor is, however, the Kuhnian progress "from primitive beginnings"—Jevons is certainly more primitive than Ricardo.

genuine *critique.* While skeptical postures in the manner of Pilate ("What is truth"—John 18:38) are often meant as a mere way to wash one's hand (cf. Matt. 27:24), the critique tells us that truth, though never final nor absolute, is *not arbitrary* nor to be decided from fancy.[5] Not every point of view is as good as any other.

CAUSALITY

The concept of causality goes back to two types of immediate or daily-life experience. One is acquired already in the sensori-motor period, in the form of that "practical category of causation" which makes you pull a table cloth in order to get hold of a flower vase which your mother has tried to put outside your range (cf. p. 242). When your cognitive structure develops, the category enters awareness. In this mature form, it is the foundation of any planned action aiming toward a specified end, and thus of every technology—physical, chemical, medical, psychological, or social. Action according to this scheme falls into separate phases: first you conceive a strategy; then you start pulling the table cloth; after a short while, the vase falls over the edge of the table onto the floor, and you get hold of it (maybe in some unforeseen state, but this problem is common to all strategic planning and not our actual concern).

The other kind of proto-causal immediate experience only enters our life with pre-operatory thought: it is the question "why?" To the question "Why is the floor wet and covered with broken glass?" you may answer, "because I wanted to get hold of the flowers"; alternatively you may report that the vase fell over the edge of the table (and your brother may sneak on you and tell who was responsible for pulling it). The latter explanations are what we are used to calling "causal," dealing with various aspects of the process behind the broken glass; the former is "teleological," an explanation of the purpose which you had in mind when pulling.

The sciences also ask and answer *whys.* In general we feel that theoretical sciences are characterized precisely by posing and trying to settle such questions for their own sake, whereas applied sciences translate practical aims into questions and translate the answers back into strategies. In the humanities ("theoretical" as well as "applied"), the following question types will be familiar:

---

5. Francis Bacon, in the opening passage of his essay "On Truth," also drew on St. John for a comment on those who claim truth to be a mere whim:

> What *is Truth;* said jesting *Pilate;* And would not stay for an Answer. Certainly there be, that delight in Giddinesse; And count it a Bondage, to fix a Beleefe; Affecting Free-will in Thinking, as well as in Acting. And though the sects of Philosophers of that Kinde be gone, yet there remaine certain discoursing Wits, which are of the same veines, though there be not so much Bloud in them, as was in those of the Ancients.
>
> (Bacon 1937: 5).

— Why will so many people use time and money on reading the yellow press?

— Why will so few people read Thomas Mann?

— Why did jazz develop among the Blacks in New Orleans?

— Why is the normal form of the Oedipus complex absent from the Melanesians of the Trobriand Islands?

— What made the Roman Empire collapse?

— What were the reasons that made research a central activity for the nineteenth-century university?

All six questions may be understood causally. The first two (which ask about the actions of people) may also be given teleological answers (for example, "because the majority sees no point in playing highbrow," or "because the yellow press arouses one's feeling of being alive, while Thomas Mann's prose is so complex that you fall asleep").

In sciences which do not deal with the conscious decisions of human actors, only causal answers to the question "why" are normally accepted nowadays. Biologists do not believe that the giraffe got its long neck "in order to" be able to eat leaves from trees; instead, what eighteenth-century theologians would see as examples of God's design is explained as the adaptation to a specific ecological niche through Darwinian selection pressure.

The different meanings of the *why* have been used by some philosophers to delimit the humanities—in particular the followers of (the streamlined image of) Dilthey. Causal answers to the above questions may be quite legitimate, according to this view; but the sciences which provide them (the sociology of literature and art, psychoanalytically oriented anthropology, economic history, etc.) belong outside the humanities (more precisely: the *Geisteswissenschaften*). *Explanation* is causal; *understanding,* the purpose of the humanities, may well investigate how reading the yellow press or Thomas Mann affects one's cultural status. But this knowledge is only relevant inasmuch as it is also known by the potential readers—what *we* know is only relevant for understanding the actions and opinions of people who *also* know it.

Others have claimed that the only way the humanities can pretend to the status of *sciences* is if they allow nothing but causal explanations. Explaining reading habits through the delusive motives people give for their actions (or, still worse, by inventing motives and claiming that these are the motives of the actors) is no better than referring to the keen desire of the giraffe for green leaves. Often this latter stance is coupled to a professed positivist view and to the claim that causation has to be understood according to Hume's definition and not through the multiple causation proposed by Aristotle. Since the positivist understanding of scientific knowledge is not without problems, and in view of the importance of the *why*s for every scientific practice, an investigation of the characteristics of scientific knowledge will have to probe this claim.

Hume's view is set forth in his *Enquiries Concerning Human Understanding.* Causation is no necessary connection between one event ("the cause") and another ("the effect") (section VII,ii,59, ed. Selby-Bigge 1975: 74f). It is nothing but *an expectation on the part of the observer* produced by habit. When we have seen innumerable times that a billiard ball starts rolling when hit by another, then we *expect* that billiard ball *A*, when hit another time by billiard ball *B*, will start rolling as usual. This, and nothing more, is meant when we say that being hit by *B causes A* to start rolling. Moreover, thus Hume, the concept of causation requires that the effect come *after* the cause.

To this a modern Aristotelian will object that there are many answers to the question *why A* moves as it does. Being hit is evidently one; but if the balls had consisted of soft clay, the outcome would have been different; so it would if *A* and *B* had not been spherical, or if *A* had been located at the very edge of a table not provided with a cushion. A complete answer to the question *why* will thus involve *efficient causes* (the hitting); *material causes* (ivory, not clay; the surface of the cloth); and *formal causes* (the laws of semi-elastic impact and of sliding/rolling, as well as the geometrical forms involved). If we want to understand what goes on, we will also have to notice that somebody plays billiards and wants *B* to move (perhaps as it does, perhaps otherwise), ultimately wishing to win the game and to gain the stake; both of these are *final causes*.[6]

The modern "positivist" view of causation may presuppose the naïve correspondence theory of truth for the observation of single events and hence replace Hume's habit of mind by a regularity of nature; or it may accept the criticism of naïve correspondence and stick to the subjective expectation. In both cases it identifies *the cause* as an *efficient* cause, an event which is invariably followed by another event. It is generally held that this kind of causal thinking is the one about which the physical sciences speak, and therefore the one which should be emulated by social and human sciences.[7]

The premise is blatantly wrong (and the conclusion thus no conclusion but a claim that must be assessed independently). Firstly, a physical description (in case, by Newtonian mechanics) of the billiard game will involve all the aspects listed under the Aristotelian explanation: physical configuration and shape, masses, friction, elasticity. The events "hitting" and "starting to roll" are at best *aspects* of *moments* of a complex process without any privileged status. Actually they are even less: they do not exist as events. When *B* touches *A* both will be gradually compressed, and

---

6. The medieval scholastic tradition and later anti-Aristotelianism have spoken in the singular of *the* efficient, *the* material, *the* formal, and *the* final cause. Yet according to Aristotle's point of view, "the modes of causation are many," even though they can be grouped in classes according to their character (*Physica* 195^{a}28; trans. Hardie & Gaye 1930).

7. Many of those who do not accept the relevance of Humean causality in social and human science, on their part, claim that it was the one which prevailed in Newtonian physics, and that it has been left behind by modern physics. Why—they ask—should social and human sciences imitate a model which has shown itself to be erroneous in the physical sciences?

increasing compression will be accompanied (not followed in time) by increasing mutual repulsion. This repulsion will accelerate *A* and decelerate *B,* and after a short while *A* runs faster than *B.* After another short while the two balls therefore separate, and we see *A* first sliding and then rolling along alone.

In this description of the physical process, as we see, there is no event "hitting" preceding another event "starting to move." Both are processes, and indeed *the same process.* In the idealization where both balls are absolutely hard, the processes contract to a momentary event, it is true; but then to *the same* moment. This is nothing specific for billiard balls; every description of classical physics has the same property. So has every description according to relativistic physics and quantum theory. The reason that Newtonian forces have to be given up is precisely the unacceptability of delayed causation. Hence relativistic physics will not speak of one electric charge q_1 acting at another q_2, which it could only do with a delay corresponding to the velocity of light; instead, q_1 is told to produce a local electromagnetic field; this fields propagates, and acts on q_2.

Hume was indeed quite right when connecting his "events" to mental habit: his "hitting" and "rolling" are not moments of the process described by physical science; they are moments of the awareness of the observer. *He* sees and hears *B* approaching and hitting *A,* and afterwards he notices that *A* has started rolling.

Hume's scheme thus does not lay bare the underlying structure of the physical description and is certainly no critique of Newtonian causation.[8] It is a formalization of the sensori-motor scheme of practical causation and of strategic action in general: *First,* I get the idea of how to get hold of the vase and start drawing the table cloth; *afterwards,* the vase falls to the floor. Here, as in the Humean explanation, everything in the configuration is taken for granted: the table, the table cloth on which the vase stands, and so on. Formally, of course, Hume has left aside the anthropomorphic notion of decision and planning. Fundamentally, however, the emphasis on temporal separation shows that the "effect" is an *end result* and thus *an aim*—no step in an ongoing process is ever the end unless you define it to be because you are not interested in what comes afterwards. Similarly, the privilege of *the hitting* over the other aspects of the process is that it corresponds to *an action which you may undertake intentionally.* The table and the elasticity of the balls are given; as a player you push.

Yet Humean causality, even if a formalization of strategic action, is not relevant for technological thinking. What I need to know (intuitively, in this case) in order to win the billiard game is the degree of elasticity of the impact, the way friction transforms sliding into rolling movement, and so on. The knowledge which really serves technological planning is indeed "Newtonian" or "Aristotelian," not "Humean": what you do is to interfere with or determine specific features of a process (acting as

8. That Hume attempted such a critique but did not produce one was already pointed by Kant in the *Critik der reinen Vernunft* (B19; *in Werke* II, 59).

an "efficient cause")—yet you can only determine the ultimate outcome if you understand how the features on which you act interact with and are conditioned by other (formal, material, and structural) features.

Humean causality is hence neither relevant for theoretical natural science nor for technological thinking. The question remains whether it can be given any meaning within the humanities.

The answer is easily seen to be negative. We cannot notice it to be a regularity that all Roman Empires collapse, since there was only one. Nor can we test the hypothesis that relatively few people read Thomas Mann because he is difficult by seeing if more people will read authors who are in all respects like Thomas Mann except that they are easy to read. As everybody knows who has enjoyed *Joseph und seine Brüder,* an author whose prose is easily read would be different from Mann on almost all accounts.

What one *can* do is to look in general at the *Collapse of Complex Societies* (the title of a book published a few years ago—Tainter 1988), or to investigate the public of a variety of authors, thereby finding similarities and divergences in possible "cause" and possible "effect." This is often believed to be the closest one can get to (Humean) causality in the humanities.

It may well be the closest on can get—but still it has nothing at all to do with the Humean concept. In the moment we single out a specific class of societies as "complex" or single out an array of features by which authors can be characterized, we have already introduced a screen of theoretical or at least pretheoretical thinking between the events and the way in which we interpret them—in a more radical sense than that in which even simple observations are by necessity tainted by theoretical presuppositions.

The only kind of causality which is meaningful within the humanities (and the social sciences, for that matter) is the one which also makes sense in the physical sciences: *correlation with, or explanation in terms of a theoretical framework*—though evidently a theoretical framework which is not the one of the physical sciences, and more often an open framework than a finite and formalized set of statements.[9] A less shocking formulation of the claim is that *causality is never an extra component of the scientific explanation of phenomena beyond theory, neither in the natural nor in the social or human sciences.* It is (at best) *a way to formulate the theoretical explanation which singles out one element* as the most interesting and considers other elements and features as a background with which the element in focus interacts.

9. The historian Fernand Braudel, who was interested in global and long-lived structures but knew of course about the effects of particular human actions (dynastic marriages, wars, etc.), tried to specify a less open framework by distinguishing long-, medium-, and short-term determination in history—cf. [Kinser 1981]. What he achieved, however, was conceptual clarification and no formalizable structure.

That causality tells us nothing beyond theory may be a dismal conclusion for those who accept that theories are fallible and therefore want to base their science or their technology on more firm foundations. As we have seen, however, this aim is not attained by attempting to adopt a Humean idea of causality, and giving up illusions is no real loss. Moreover, as we have also seen, the inescapable "fallibility" of theories only means that no theory can pose as absolute truth. Inasmuch as a theory can be claimed to be *true* in a materialist sense, as argued above, causal explanations in the sense of correlation of phenomena with theory are *also true.*

When one element of a theoretical explanation is singled out as a cause, it may be so because we want to use our knowledge in some kind of technology: that which it is of interest to consider as a cause is what we can influence or determine, that which is technologically manipulable; those features of the situation which we can do nothing about are then understood as a background, the conditions to which our action is subject. But even purely theoretical investigations may speak of and discuss causation. In such cases, the choice of the cause is obviously rather free, and may depend on which features of the situation we find most interesting, or which aspects of the process we want to scrutinize more closely while relegating others to the background.

One of the more grotesque (or, if you are in that mood, tragic) facets of scientific life is the jealous and hostile passion with which scholars discuss in such situations which is "the" cause. Was the tuberculosis which killed several of my grand-uncles *caused by* bacteria or by their living conditions? These competing causalities are evidently complementary aspects of *the same* theoretical explanation: the disease cannot occur unless one is infected; but infection is much more likely to result in disease when one is badly fed and his lodging never dry. This example may be outdated except when one or the other sanitary policy is advertised (that is, when it comes to technological choices[10]), precisely because a single theory now explains how the two explanations concur. If the two derive from different theoretical approaches (which, like the early structural and the neogrammarian approach to linguistics may both be "parts of the truth," even though their mutual relation is not, or not yet, elucidated), similar discussions are still seriously meant. In the humanities, where theories *stricto sensu* are rare and open-ended frameworks the rule, discussion may easily degenerate into calumny or nondiscussion because both parts regard the other as mistaken beyond hope of salvation by argument.[11]

---

10. Cf. the case discussed in Leavitt 1992, in particular note 1.

11. This was indeed the situation in my own discipline, the history of science, until a few decades ago. As late as the early 1980s the discussion between "externalism" (the explanation of scientific development as "caused" by general cultural, social, technological, and economic factors) and "internalism" ("causation" by the inner dynamics of the sciences) was still regarded by most members of the field as meaningful.

Sticking stubbornly to your own paradigm may well be an efficient way to make the sciences progress, as held by Kuhn: as an advocate in court, you should do what can be done to find arguments in favor of your client. But like the advocate, you should also listen to the other part and acknowledge that the final verdict may not be totally in your favor. Dismissing that possibility, and denying *a priori* that approaches which differ from one's own may be legitimate, is no more productive in science than in court, and no less preposterous.

OBJECTIVITY, SUBJECTIVITY, AND PARTICULARISM

If the choice of the element to be singled out as a cause is "rather free," does that mean that causal descriptions are subjective and not objective? In general, is *scientific knowledge* subjective or objective? This question is more trendy than the question of *truth,* maybe because it is easier to assert that the other part in a discussion is subjective than to prove that he is wrong.

Is science objective? Certainly not if we understand objectivity as *coincidence* with the object. In this sense, nobody can carry an objective picture of a house in the brain or the mind, since the house is larger than the skull and the mind contains no bricks. Theories are always abstractions that strip reality of some features and describe others—stripping for instance the house of materiality and heaviness, retaining only the mathematical structures of shape, material strength and gravitation. As formulated by a student of mine when I objected to an explanation he had given at an examination that it was incomplete:[12] *Any complete model of reality is—by definition—reality.*

Criticizing a scientific description for failing objectivity in this sense is thus either absurd or foul play. Science, like any other kind of knowledge, is *by necessity subjective.* Even the way a house appears on a photograph depends on the optics of the lens and on whether the film is black-and-white or in colors. What science sees also depends on the conceptual structure through which it looks: Priestley made oxygen but saw phlogiston-free air.

But science, as knowledge in general, is also subjective in a sense which corresponds to the dependency of the picture on the place from where it was taken and the dependency of an answer upon the question. Knowledge never comes from passive reception of whatever passes before one's indifferent eyes; it is always the result of an active practice—as a minimum, of selective attention, but often of much more intentional operation. In science, knowledge comes from experiment, investigation, critical reflection, and so on. Inasmuch as our practice vis-à-vis the reality we want to know about is not in itself fully contained in that reality, whatever knowledge we earn is by necessity subjective and not determined by the object alone.

---

12. Mark Madden, Spring 1990.

This much is common sense.[13] It should have been evident at least since Kant, and it does not illuminate the relation between objectivity and subjectivity very much. An interesting quality of the Piagetian and Kuhnian epistemology is that it allows us to discern a sense in which the subjective aspect of knowing may in itself be more or less objective.

In order to see how, we shall first return to the discussion of the status of the schemes and categories of our general cognitive structure (see p. 252ff). When we conclude, for instance, that "the material world is constituted in a way that allows an adequate practice if we order our perceptions as representations of permanent objects," then we have concluded something about the material world—namely about its response to a particular practice. This predictable response is a property of the material reality in which we live, that is, of the object of our knowing; the category of the permanent object, however much it is a constituent of our subjective cognitive equipment, is *also objective.*

The basic categories and logical schemes, however, though arising through cognitive development, are end points. Once my daughter had acquired the scheme according to which there has to be more children than girls in a mixed kindergarten (see p. 245, n. 17) she could understand *nothing but* that, and once we have organized our way to experience in permanent objects, we cannot avoid doing so. It is left to further empirical studies to find the degree of permanency of actual objects, and to construct, for instance, a theory which allows us to discuss in which sense salt is conserved when dissolved in water. Irrespective of their genesis through biological evolution and individual development (which make them synthetic, that is,

---

13. It ought to be, at least; from actual scientific discourse one would not believe that it were, which may be a result of the perplexingly confused use of the word. In the introduction to one of several works on the genesis and development of the notion of 'objectivity' since the nineteenth century, Lorraine Daston (1992: 597f) points out that

> Current usage allows us to apply the word as an approximate synonym for the empirical (or, more narrowly, the factual); for the scientific, in the sense of public, empirically reliable knowledge; for impartiality-unto-self-effacement and the cold-blooded restraint of the emotions; for the rational, in the sense of compelling assent from all rational minds, be they lodged in human, Martian, or angelic bodies; and for the "really real," that is to say, objects in themselves independent of all minds except, perhaps, that of God.

It may be a Sisyphean labor to extract and assert a sensible concept from this "thick layering of oddly matched meanings." But since discussions of objectivity versus subjectivity are anyhow pervasive, it can still be useful: Sisyphos, although he never manages to put his stone to rest at the top of the mountain, at least makes sure that it does not stay at rest in the bottom of the valley. Dishwashing, no less Sisyphean and to be repeated every day, is indispensable if the kitchen is to stay viable.

informative about reality), the basic categories and schemes remain *synthetic a priori* with regard to the ongoing process of knowing; they thus belong to a cognitive species other than the *synthetic a posteriori,* the actual outcome of our observations. Once the synthetic a priori have arisen, it is no longer possible to distinguish in their domain between *more* and *less objective* subjectivity.

This distinction is only pertinent when it comes to discussing the synthetic a posteriori, in particular scientific knowledge. To see this, we may compare the Kuhnian cycle with the hermeneutic circle—I quote the explanation given on p. 183:

> At our first approach to a foreign text (in the wide sense, that is, to any spoken, written, sculptured, painted, or similar expression of meaning), we interpret it in agreement with our own presuppositions and prejudices, which are in fact our only access to the universe of meanings. But if the foreign text does not fit our interpretive expectations on all points (which it rarely does), and if we investigate the points of noncompatibility seriously, we will be led to revise our prejudices—so to speak, by "listening." The revision will enable us to understand the presuppositions of the foreign mind (or understand them better) and hence even to understand ourselves from the foreign point of view. Understanding the other leads us to better insight into our own expectations to universes of meaning, and hence allows us to approach the foreign text (and other foreign texts) with better prejudices.

Some features of this structure are certainly different from what we see in the Kuhnian cycle of normal and revolutionary phases. The relation between the scientist and the object of a science is less symmetric than that between the interpreting mind and the interpreted mind.[14] The "better prejudices" with which we approach the object after a change of paradigm do not come from understanding ourselves from the point of view of the object. But they remain *objectively better,* that is, they reflect the *features of the object* more precisely or at a deeper level, according to the arguments on pp. 293 and 321. The progress of a scientific discipline through scientific revolutions (when these follow the Kuhnian ideal scheme), which was spoken of

---

14. Anthony Giddens (1976: 146ff), when discussing the similarities between the two circles, characterizes the process of interpreting a foreign text or social world as a double hermeneutic:

> Sociology, unlike natural science, stands in a subject-subject relation to its "field of study," not a subject-object relation; it deals with a pre-interpreted world, in which the meanings developed by active subjects actually enter into the actual constitution or production of that world; the construction of social theory thus involves a double hermeneutic that has no parallel elsewhere . . . .

What this hermeneutics looks for is thus an understanding of (for instance) Rousseau's *Émile* and its impact which builds on *how Rousseau and his contemporaries understood* the(ir) world.

above as "progress toward greater truth," can also be understood as *progress toward greater objectivity of its subjectivity.*

So far so good. Yet "subjectivity" is not only used about the inescapability of knowing only in response to specific questions and in terms of a particular conceptual framework. These two kinds of subjectivity were assimilated (in reverse order) to the making of a photograph on a particular film by means of a camera with a particular lens and from a specific perspective. Given these conditions (together with the shutter speed, etc.), what will appear on the picture is determined by the object.

Or at least it should be. If it is not, the picture has been retouched. Retouching corresponds to those kinds of scientific knowing which are affected by factors other than the object, the kind of question (purportedly) asked, the instruments used, and the conceptual framework which is referred to or which is shared by the community of workers in the field.[15] This kind of subjectivity is better spoken of as *particularism.* To some degree it cannot be avoided: you always have motives beyond those which you confess to yourself or reveal to your psychoanalyst, even when it comes to knowing. But since (by definition) private distortions cannot be controlled by others, they detract from the value of knowledge to the same extent as they are present. Scientific knowledge is shared, or should at least be shareable; but knowledge expressed in a code which cannot be deciphered by others cannot be shared. Particularism, though to some degree inescapable, should hence be minimized through critical discussion.

---

15. It may be a profitable aside to point out that the seemingly innocuous phrase "which is referred to or which is shared" hides a serious dilemma. Many social scientists, noticing that social thought is always involved in conflict and cannot avoid taking sides at least by asking the questions of one party, have held that this inescapable subjectivity should be brought under control by each social scientist telling his side, his employer, and his sympathies explicitly and honestly. This stance (which is in particular associated with Gunnar Myrdal) may seem attractive, and certainly has a point. But apart from the naïveté of the expectation that people (and social scientists are also people) should tell honestly when their sympathies and aims contradict those of their employer, Myrdal's cure against unbridled subjectivity suffers from the same weakness as Lakatos's formalized conception of his "hard core" (cf. p. 307): too much of the framework that is shared by a scientific community consists of tacit knowledge, and it will not be revealed if one asks workers for their political position; the paradigm under which they have been trained may well be built around questions and concepts asked and formulated from a quite different position.

Participating in a community which shares a paradigm, including the tacit knowledge which goes with it, is therefore just as important as honesty if subjectivities are to be kept as objective as possible.

CHAPTER 20

The Role of Norms

In 1965, a symposium was held in Oxford, at which, among others, Popper, Lakatos, and Kuhn were present. Its topic was a discussion of Kuhn's *Structure of Scientific Revolutions.*[1] As Popper and Lakatos saw things, the question was whether Kuhn was right in replacing the(ir) logic of research with a description of the social psychology of scientists. Kuhn, on the other hand, asked whether Popper and Lakatos had really brought forth a *logic,* and answered that their actual output was an *ideology.*

Much may speak in favor of Kuhn's reply. Still, both Popper_1 (and even Popper_0) and Lakatos/Popper_2 are too close to aspects of the scientific production process to be dismissed as *nothing but* ideologues (in the vulgar sense). Also Kuhn's own work, as we have seen above (p. 309), results in a kind of logic for the social production of scientific knowledge. Even though there may be many similarities between *science* and *organized crime* (to paraphrase a discussion running through the same conference proceedings; both are indeed social activities perpetrated by relatively closed and highly specialized communities), there are also noteworthy differences. We can therefore only come to understand the nature of scientific knowledge if, so to speak, we grasp how the process of science, *in spite of its similarities with organized crime,* can be described approximately in terms of a "logic" through which it manages to produce some kind of reliable knowledge.

Traditional hagiography explained the specific character of science by the exemplary character of scientists. Science is reliable because scientists are eminently reliable; science is objective because scientists are heroically objective and dedicated to their cause; and so on.

---

1. The revised contributions from the symposium were published in Lakatos & Musgrave 1974. Many of them were cited above.

Bankers will probably not agree that scientists are significantly more reliable than average people when it comes to their use of a checking account. If they are reliable within their professional work, it must be explained in other terms than through general moral perfection. This will lead us into an investigation of the relation between scientific practice and norms—and, in general, between knowledge and morality. As it will turn out, the analysis will provide us with important insights, both as far as the nature of morality is concerned and regarding the scientific process. Since the discussion has to move back and forth between these levels, it may appear rather intricate.

LOGIC AND NORMS

A "logic" is a fixed pattern. If the development of scientific knowledge in social process follows a specific logic (or just follows it to some extent), we must presume that this social process is itself governed by a set of general "laws" or regularities.

Two questions can then be asked. Firstly, whether the logic of scientific development is exact and formalizable, or rather to be described as a "dialectical logic of development." Secondly, which *kinds* of social regularities are involved, and how they succeed in bringing about a logic.

Kuhn's offer is not formalizable, and is in fact dialectical. In the closing passage of his article on "The Function of Dogma in Scientific Research," he points out that

> scientists are *trained* to operate as puzzle-solvers from established rules, but they are also *taught* to regard themselves as explorers and inventors who know no rules except those dictated by nature itself. The result is an acquired tension, partly within the individual and partly within the community, between professional skills on the one hand and professional ideology on the other. Almost certainly that tension and the ability to sustain it are important to science's success.
>
> (Kuhn 1963: 368f).

Elsewhere he insists that the whole process only functions because what he speaks about here as "established rules" are indeed not explicit and unambiguous rules but shared norms or values which individual workers interpret differently:

> . . . individual variability in the application of shared values may serve functions essential to science. The points at which values must be applied are invariably also those at which risks must be taken. Most anomalies are solved by normal means; most proposals for new theories do prove to be wrong. If all members of a community responded to each anomaly as a source of crisis or embraced each new theory advanced by a colleague, science would cease. If, on the other hand, no one reacted to anomalies or to brand-new theories in high-risk ways, there would be few or no revolutions. In matters like these the resort to shared values

may be the community's way of distributing risk and assuring the long-term success of its enterprise[2]

The Kuhnian framework, however, is not the only place where we have encountered the need for a dialectical understanding and for application of the concept of norms. An obvious dialectical tension manifested itself in the contrast between the empiricist imperative,

Scientific explanations are only allowed to make use of concepts and to postulate relations and structures which can be rooted in experience, observation, or experiment. Mythological explanations referring to entities with no such empirical underpinning are inadmissible: they only obstruct genuine scientific insight.

and its falsificationist counterpart,

We are allowed to use in our explanations whatever self-invented concepts and hypotheses we like; but we should be aware that our hypotheses are indeed nothing but hypotheses, preliminary explanatory models, and not the truth. We should therefore constantly check our hypotheses as thoroughly as we can, and we must reject them as useless as soon as they enter into conflict with our observations of reality—that is, as soon as they are "falsified."

As they stand, the two rules of scientific conduct solved some of each other's problems, as we remember—but in mutually unacceptable ways. None of them, however, could be rejected as plainly irrelevant. They stand in much the same relation as these two passages from Deuteronomy:

Thou shalt not kill. (5:17)

and

But of the cities of these people, which the Lord thy God doth give thee for an inheritance, thou shall save alive nothing that breatheth. Namely, the Hittites, the Amorites, the Canaanites . . . . (20:16f)

These rules, like the empiricist and falsificationist maxims, express a *moral dilemma,* the admissibility of which is perhaps the most important distinctive characteristic of a system of (moral) norms or values as opposed to a set of binding juridical rules or a theory.

---

2. Kuhn 1970: 186. When this was formulated, the idea that a system may function better if its components are allowed a margin of unpredictability (and the notion that real systems function that way) was highly untraditional (for which reason Kuhn resorts to the metaphor of "risk distribution"). During the last decade, of course, the fuzzy metaphor of "chaos theory" has popularized the idea.

One of the things we demand from a theory is that it should be free of inner contradiction (to be sure, the existence of a recognized contradiction is no reason for automatic and immediate rejection, as exemplified in the history of quantum physics; but it is at least an anomaly which one should try to solve): if the same theoretical system predicts on one hand that a bridge we try to build will stand and on the other that it will fall down, we should obviously try to determine what is actually going to happen, or at least to find out why our theory cannot tell. One aim of theory construction (though not the only one, nor necessarily its actual scope) is that it should be fit to serve instrumentally rational action (Weber's *Zweckrationalität*), which it cannot do if producing contradictory predictions.

The same demand for internal consistency we make to juridical laws; for this reason, the existence of capital punishment presupposes a clear distinction between that sort of homicide which is murder, and hence to be punished, and the executioner's work for which he gets his salary. No juridical system could live undisturbed by a clear contradiction like the one exemplified by Mosaic law; when a contradiction occurs in real life (that is, not only as the outcome of a thought experiment), the judge or some other instance of authority has to decide which norm is primary; if the contradiction presents itself recurrently, the legal system has to be adjusted.

EXPLANATIONS OF MORALITY

Norm systems do live with contradictions, whether we like it or not; that they do so is one main reason why different social actors choose differently—cf. Kuhn as quoted on p. 334 regarding the "individual variability in the application of shared values" (another obvious reason, certainly no less important, is that norms may be contradicted and effectively blocked by our drives and desires;[3] often the two are intertwined in a way which in the actual case makes it difficult to distinguish moral dilemma from pretext). To understand *why* norm systems have to live with contradictions, we may take a brief look at the nature and origin of norms and morality.[4]

One very influential explanation (which is historically coupled to empiricist philosophy, and which shares its matter-of-fact attitude and its bent toward atomistic

3. ". . . the good that I would I do not; but the evil which I would not, that I do"—thus St. Paul (Rom. 7:19).

4. It will be seen that I treat morality (or ethics) and norms as one thing. This is not done by everybody. Certain authors would see *norms* as *that which people think is right,* whereas *morality* is concerned with *what is right* in itself. Others would reserve the term *morality* for serious matters and use only *norms* to denote, for example, norms concerning good manners. (The existence of cultures where a person's unwillingness to apologize for pushing somebody involuntarily is a sufficient reason to kill him in a duel suggest the latter distinction not to be cross-culturally valid).

analysis) is *utilitarianism:* behavior is morally good if it is useful, that is, if it promotes general human happiness.

Three problems at least inhere in this understanding of morality. Firstly, it reduces morality to instrumental rationality with an undefined aim: what, indeed, is *general human happiness* if not itself a moral issue? Secondly, it makes no sense of the experience of the moral dilemma: if two alternative actions are both prescribed, that one is obviously best which according to a cost-benefit analysis is *most* useful. Thirdly, it presupposes that the consequences of a course of action are finite[5]; if not, cost-benefit analysis is impossible.

Equally influential is Kant's approach through the distinction between the *hypothetical* and the *categorical* imperative in *Critik der practischen Vernunft* (cf. p. 142): my present action is morally good if, when generalized into a rule, it is of absolute ("categorical") validity.[6] Instrumentally rational action ("under the hypothesis that you want to achieve *X* you should do *Y*"—Kant's hypothetical imperative), on the other hand, is thereby neither morally right nor wrong.[7] From a Kantian point of view, it

---

5. Or rather that the *actual value of their sum* is finite, in the same way as the present value of £1 to be paid each year from now to eternity is £21 at an interest rate of 5 percent per year, that is, if consequences which our children and neighbors have to bear are less important than those which hit ourselves, and those which hit our grandchildren and our neighbors' neighbors count even less.

This is, of course, precisely the presupposition of every cost-benefit analysis: consequences which are beyond my horizon *do not exist;* consequences within my horizon are only counted to the extent that those who suffer them count for me.

6. Such a generalization can of course be performed at different levels. For most of Kant's contemporaries, it would be obvious that the execution of certain criminals was morally right—not because of the generalization "you should kill other people" but, for instance, because "you should protect society against the damage which could be wrought by incorrigible criminals."

7. It is often assumed that the norms governing scientific work—be it the rules implied by the paradigm or the empiricist or falsificationist imperatives—are merely hypothetical imperatives and not to be reckoned to the category of morality: if you wish science to progress rapidly, you should follow Popper's directives; if you want to trace the deeper levels of an author's thought, you have to read him in the original language; if you want to be sure of the results of your chemical analysis, make sure that your test tubes are clean; and so on. As discussed above in the case of Popper's prescriptions, and as it already follows from the presence of *dilemmas,* the assumption is wrong; hypothetical imperatives, like utilitarian rules in general, do not admit the existence of dilemmas but only doubts as to which course of action is really best. In principle, nothing prevents such doubts from being solved; a dilemma, on the other hand, is unsolvable as it stands, and may (at best) be resolved from a higher vantage point. For these and other reasons (in part to be discussed in the following), the attempt to regard the norms directing scientific work as *nothing but* technical rules is doomed to fail. The Kantian distinction does not allow us to keep the philosophy of science aloof from the imbroglio of moral philosophy.

is hence not the action in itself which is judged morally right or wrong; the moral judgement can only deal with the action in its context of justification and intention.

In any case, it will be seen, there is no more place for the moral dilemma here than inside the framework of utilitarianism (with a reservation to which we shall return). One, at least, of two rules in mutual conflict cannot be of general validity; one, at least, is hence no moral rule.

Under the impact of Kantianism, utilitarianism has been split conceptually into *act utilitarianism* (the classical stance, which judges acts individually, from their particular consequences) and *rule utilitarianism,* which does not ask whether single acts, but only whether *rules of conduct* are conducive to maximal human happiness. The difference between the two positions is larger than it may seem at first. Firstly, they may judge the same action differently: if I need the money and you already got all you want, act utilitarianism may find it justified if I omit to pay my debt to you. But rule utilitarianism would see that the rule "you need not pay your debts" would be the end of lending, and thus damage those in temporary economic distress (a real-life instance of the same divergence is cited below on p. 350, n. 30).

Secondly, rule utilitarianism, while fitting our immediate feeling that morality is concerned with rules and not with the expediency of single acts, misses the "positivist" simplicity which constitutes the merit of act utilitarianism to such an extent that it is dubious whether it still deserves the utilitarian label. It can never be given within a particular action under *which* general rule it should fall (cf. n. 6): is "hanging *X*" an instance of "killing a fellow human being," of "annihilating an enemy," of "executing a war criminal," or of "celebrating our victory by liquidating a war criminal belonging to the enemy's side"? Classification of acts under rules already presupposes some kind of moral theory or proto-theory telling the pattern of possible rules. No more than in descriptive science is there any smooth road from "positive fact" to generalization.[8]

Rule utilitarianism, if it is to be believed, has no more space for moral dilemmas than Kantianism. *If* it is at all possible to calculate the utility of a rule (which is certainly no simpler than calculating the utility of an action), then one can also calculate which of two conflicting rules is "more useful" and thus primary. It is only through its theoretical shortcoming—namely because the ascription of an action to a rule is itself ambiguous—that a specific action may present a dilemma. This weakness (or force through weakness) it shares with Kantianism.

---

8. That this is so was indeed pointed out by Kant in his third Critique, *Critik der Urtheilskraft* (A VII, B VII; A XXIIIf, B XXVf; A XXIV, B XXVI): Only *the faculty of judgement,* the very same which allows us to judge art and which cannot be reduced to strict proof, allows us to decide whether or not a particular act falls under a specific rule, and, in general, to perceive *the particular* as a case of *the general* (be it rule, or principle, or [natural or moral] law). And only this faculty allows us to "find" the general from the particular (or construct it, or whatever we do; changing our terminology does not change the problem).

It is thus not possible to derive actual morality with its dilemmas from utilitarianism of either one or the other kind; but from a small set of socially accepted core values (inner solidarity within the Twelve Tribes of Israel, combined with the right to conquer and subdue, if we think of the Deuteronomy example), we may derive a complex of norms which looks so similar to the standards of real morality that the utilitarian idea can hardly be totally wrong.[9] Kant, on the other hand, is certainly right when emphasizing that moral norms are characterized by claiming general validity. The kind of utilitarianism which offers the best prospects as a "rational reconstruction" (or retracing) of the emergence of morality is thus indubitably rule utilitarianism, its inherent difficulties notwithstanding.

MORALITY, LANGUAGE, AND SOCIAL PRACTICE

The phenomenon of the dilemma may advance our understanding by another step. Norms should be seen as social phenomena. Human beings living together in a society socialize each other (in particular, of course, their children) and make sense of their existence by describing their situation and actions in language.[10] Patterns of action which serve shared interests will be formulated in general terms, *as norms*— and patterns which go against shared interests will be forbidden, again in general

---

9. Evidently, the "socially accepted core values" also belong to the category of morals. The immediate scope of utilitarianism is thus only to reproduce the process by which a few core values (in the traditional formulation summed up as "general human happiness") unfold as a complete moral system.

If a similar argument was to be used in order to explain the emergence of the core values, it would probably have to involve something like Habermas' "universal pragmatics" (see below, n. 10) as molded into a less universal shape by the necessarily restricted horizon within which any human society lives, together with other aspects of human biological nature ("happiness" is certainly a cultural variable, but only in exceptional cases reconcilable with starvation or extreme physical distress). A complete philosophy of morality would then, furthermore, have to take into account that the core values, *qua* participants in the general system of moral norms, are themselves reshaped by the processes through which this system unfolds.

10. Habermas, in his "universal pragmatics," derives certain basic aspects of morality not from the contents but from the very presence of this communicative situation: if communicative messages could not be *presumed* to be true, they would not be messages (even the lies of commercial advertisement only function because their character of communicative messages make us accept their overt or tacit implications at one level of consciousness, in spite of what we may know at other levels). In Montaigne's classical words (*Essais* I.ix, ed. Thibaudet & Rat 1962: 37), "lying is a condemnable vice, since we are only human and only relate to each other by means of the word." Similarly, the mutual character of communication entails at least rudimentary human equality.

The pattern of Habermas's argument is obviously (though tacitly) borrowed from Merton's notion of "institutional imperatives," to which we shall return below (p. 348ff). At present, however, we are concerned with other aspects of the communicative practice.

normative terms.[11] These normative prescriptions and proscriptions will be *practical knowledge,* knowledge about *how one should act.* To the extent that norms are made explicit, there will be a certain amount of bargaining between them at higher communicative levels, for instance through their integration into religious systems, through the "thought experiments" of myth, drama, and literature (cf. p. 385), and through their interaction with legal systems. The outcome will be total structures which are categorically different from the structures of practical action and which possess a certain degree of autonomy.

In these structures, as already when rules of conduct are enunciated in language, the general formulations of norms will be *more general* than those patterns of action and common interest which they formulated; conflicts which never occur in the living existence of society will arise if the norms are taken to the letter and not in agreement with that social practice which they formulated in the first place (even though bargaining and restructuration tend to clear the most obvious conflicts away). In this way, dilemmas may arise, at least as virtual or philosophical problems—as Wittgenstein (1968: §38) tells, "philosophical problems arise when *language goes on holiday.*" In its real life, a culture will possess a large amount of "tacit knowledge" concerning the range within which norms should be applied (the ancient Israelites will have had no difficulty in recognizing *whom* "thou shalt not kill" and whom to exterminate; nor have the fighters of countless Christian armies, nor the host of eminently Christian judges). None the less, because conflicting interests may interpret the range of shared norms differently and tacit knowledge is necessarily blurred, the theoretical dilemmas may also turn up in practice.

No social situation, moreover, is completely stable. Norms, on the other hand, once they have come into being, become embedded in language, religious ritual, myths, literature, and so on, and are thereby provided with a fair amount of temporal inertia. For both reasons, and in particular through the two in combination, norms may end up governing behavior far beyond the range of experience from which they grew originally; and what were once merely potential conflicts between norms may thereby suddenly be actuated in possible behavior.[12]

---

11. Indeed, formulation in language *cannot avoid* being in general terms: general terms constitute the backbone of any language, proper names are by necessity peripheral and could never serve to tell anything on their own.

Already our use of language (for example, the concept of 'lending') thus provides us with an *a priori* pattern of possible rules with regard to which we interpret actions, that is, with a first version of the prototheoretical foundation that rule utilitarianism presupposes without recognizing it.

12. The actuation of conflicts because of changing social (here technological) realities is exemplified by one of the central norms in traditional medical ethics: the patient should be kept alive as long as possible. New medical technologies make "as long as possible" indefinitely

Change has been a condition of human existence as long as society has been human and communicative. The distinction between the two sources for dilemmas is thus merely analytical. It is not possible to point to a stage of innocence where only theoretical dilemmas existed. Human societies, as long as they have been in possession of moral norms, have always been troubled by moral dilemmas.

But not only total societies: shared experience, specific patterns of discourse, conflicting interests, and change over time also affect single institutions and communities *within* society at large. These will therefore develop a specific ethos of their own, which will share many of the characteristics of the normative structure of general culture.[13]

This brings us back to the creation of scientific knowledge. The empiricist and falsificationist rules of conduct are, in fact, nothing but community-specific norms, generalizing the experience and pragmatic rules of conduct of working scientists, and brought into mutual conflict when formulated and taken to the letter ("on holiday") by philosophers. The shared anomaly of the two imperatives—the absence of purely observational knowledge which can serve verification or falsification—on its part, is of the same kind as the dilemma presented by indefinitely life-saving medical technologies. If not created, it has come to the fore through the development of sophisticated instrumentation, from the invention of the telescope and the microscope onward.

It is thus not only in Kuhn's interpretation that the "logic" of the scientific process will have to be nonformalizable and dialectical. The underlying pattern (or "logic") of any social process governed to an appreciable extent by norms or values has to be so.

KNOWLEDGE, NORMS AND IDEOLOGY

Should we then conclude (as Kuhn tends to do) that Popper's and Lakatos's attempts at formalized nondialectical methodologies are descriptively irrelevant ideologies? Or, more brutally, that they have to be descriptively irrelevant *qua* ideologies?

---

long, while transforming the concept of 'alive' in a way which changes the meaning of the norm through and through.

13. This observation suggests another perspective on the concept of the 'living philosophy' (cf. p. 217). Neither the nascent bourgeoisie of the eighteenth century nor the working class of the later nineteenth century were homogeneous entities; in both cases, already the situation of the single member was complex, and all the more ridden by contradictions and conflicts were the interests of the group as a whole. However, when recast as worldview or "philosophy" ("ideology" in a sense to which we shall return forthwith), the generalization involved in the process brought forth structures from which most manifest contradictions had been removed or suppressed; as we have seen, later developments were to expose a number of latent inconsistencies.

The question may be approached via a discussion of the concept of 'ideology' in the light of the theory of morality. Ideologies may be understood as systems which possess both normative and descriptive aspects, but which cannot be fully analyzed into merely normative and merely descriptive elements—systems which, in the same breath, tell how and what the world *is* and how one should (therefore) behave, which merge theoretical and practical knowledge. Whether the Popperian and Lakatosian methodologies can be descriptively relevant is thus a specific case of a general question: whether the normative aspect of an ideology by necessity invalidates its descriptive aspect (or by necessity invalidates it completely)—and then, in the second instance, since the normative aspect was implied by the description, nullifies even this?

Utilitarianism of either kind explains norms as instrumentally rational prescriptions (with a tacitly presupposed aim of undefined "happiness"). Since instrumental rationality cannot exist without knowledge about the connection between our actions and their ensuing consequences, utilitarianism presupposes that its norms consist of translated knowledge. Quite as much cannot be said if norms are generalizing reconstructions of patterns of action which *on their part* are adequate within a specific and more restricted sociocultural horizon. None the less, they generalize from what was implicit knowledge about what action is adequate within this restricted domain, given the shared interests of the participants in the sociocultural system as understood by these. In terms of the materialist notion of truth introduced in chapters 17 (p. 270) and 19, they therefore contain a core of *truth* reflecting features of this social situation—a core which may be larger or smaller, depending upon the degree of generalization and on the kind of reconstruction which has taken place.[14] Norm systems are thus *already ideologies,* possessing an aspect of implicit descriptive knowledge; they are "proto-knowledge."[15,16] Even presumedly

---

14. Reconstructions aiming at inner consistency and utilitarian explainability may, like the search for coherence of scientific knowledge, increase the truth value of a norm system. Reconstructions aiming at agreement with mythical frameworks need not do so, unless these same frameworks also carry embedded practical knowledge about the same life situation.

15. An implication of this that cannot be pursued systematically in the context of this text is the possibility that the cognitive aspect of a norm system may then, like any piece of knowledge, be *wrong, mistaken* or (if we refer to the materialist notion of truth that is presupposed in the argument) *less true* than is possible.

So much remains of utilitarianism, even when norm systems are seen as reconstructions of generalizations, that erroneous cognitive presuppositions (inasmuch as they can be traced) can be used to judge the norm system. This is the reason, or at least part of the reason, that value rationality as defined on p. 140, n. 17 is a possible concept.

16. Another implication—even this one only to be outlined—continues the line of thought from n. 13, casting light on essential aspects of the political process since the French Revolution. As long as parties see themselves as organizing group interests directly, a pragmatic

descriptive knowledge, on the other hand, presupposes a framework, that is, a cognitive structure implicitly prescribing the way in which the subject matter *should* be understood; since the selection of the interesting features of a global situation corresponds to a particular practice (remember Australian aboriginal children developing map understanding well before conservation, in contrast to European children—see above, p. 249), a prescription of *how to know* also involves an implicit delineation of a range of *possible actions*. Both from this point of view and through its prescription of how to know, it involves *practical knowledge*. Knowledge systems, like norm systems, are *already ideologies* and thus "proto-morality." No absolute distinction between the two (or three) domains can be made; differences are of degree although—it must be stressed—the span from one extreme of the spectrum to the other is immense.[17]

---

compromise can be accepted without difficulty, as illustrated by Montesquieu's argument for accepting a House of Lords (cf. p. 131)—in a Confucian saying quoted by Mao Ze-dong, one should not lift up a stone in order to drop it on his own feet. Once parties become based on interests reconstructed as ideologies ("dividing according to convictions, not interests," in the words of a Danish agrarian Liberal politician from the late nineteenth century), any compromise with opposing interests (similarly reconstructed as ideology) appears to both participants as a compromise of *truth* with *falsehood,* and therefore morally abject until the interests that call for the compromise (for instance, common political or biological survival) have been digested and assimilated into the ideology; this is what happened when the Communist parties opted for the "Popular Front" instead of the suicidal "class against class" strategy in the mid-1930s, and when the democratic Right made a corresponding shift of emphasis in its conception of democracy a few years later.

Social life cannot exist without the category of morality; nor can political life without a reconstruction of interests as ideologies or convictions. What the above reflections suggest is that an unsteady world requires constant integrative rethinking of convictions—producing morally compelling insights into the conditions for our shared shaky life on this planet that are *more true* than the competing and mutually exclusive ideologies they are going to replace, if it is not going to become even more shaky. Dismissing the traditional "great tales" and replacing them by "small tales" or uncommitted "lifestyles" corresponding to even more myopic interests—the postmodernist program—instead of rethinking them leaves us in a world where TV magnates and brute force have not only the power but also the undisputable right to make the final decision.

17. The nonseparability of description and prescription, it is true, contradicts the implications which Hume drew from a famous observation made in the *Treatise of Human Understanding* (III(i)1, ed. Mossner 1969: 521), and which have been widely accepted within philosophy since then: there can be no logical derivation leading from sentences built around *is* or *is not* to sentences built around *ought* or *ought not.* Since the former sentences are descriptive and the latter prescriptive, Hume's observation seems (and is generally taken) to imply that knowledge and norms are not only separable but actually separate.

The arguments that knowledge is protonormative and norm systems protocognitive does not invalidate Hume's logical observation; but they do go against what he concludes from it. What they say is that both norm systems and seemingly neutral descriptions share the character of the question "When did you stop beating your wife?": they make presuppositions which

It is, no doubt, legitimate to characterize Popper's and Lakatos's views as ideologies for scientists; but that does not prove in itself that they have no descriptive point. In Lakatos's case this was argued quite extensively above. As far as Popper is concerned, the question is rather, *which* point? Is he right when declaring his methodological prescriptions to be purely utilitarian norms, to guide how science will progress most rapidly? Or are they misleading in this respect, which would mean that the purported cognitive (and hence also the normative) contents of the ideology is *wrong* (in the sense suggested on p. 342, n. 15)? That the latter possibility is to be preferred was argued in some depth above. But Popperianism also has another level of cognitive content with normative implications—*viz.* that science progresses because it follows Popper's precepts (cf. p. 268, n. 17), which implies that scientists have to be respected as "objective beasts" by the surrounding society, and in particular have to be more highly respected than Marxists and psychoanalysts and their kin. This norm certainly *is* useful for the scientific community—but mainly so with regard to its similarities with the world of organized crime.[18]

---

are not stated in the sentence itself, but without which the formulation of the sentence becomes meaningless.

It can be argued that the very distinction between "is-" and "ought-sentences" is of the same character. The understanding of the descriptive statement presupposes our familiar "naïve" correspondence theory of truth, and that meaning is reducible to a finite definition (with no further implications); the understanding of the prescriptive statement builds tacitly on a no less naïve understanding of the freewill problem. A prescriptive sentence "You ought to do *X*" presupposes, at least as understood in Hume's argument, that "you" are in possession of a Free Will and that you are hence able to decide sovereignly to do or not do *X*. But the relation between free will and determination is more complex than this. The naïve conception of the Free Will renders the momentary feeling of deciding freely (for example) to shout at your neighbor because the gangster really deserves it. But thinking back tomorrow at today's rage you may conclude that you overreacted on Mr. Jones's jesting provocations *because* you had slept badly; if it is your husband who shouts, you may think so already in the moment when it happens. What one decides to do is thus not as independent of what *is* as he feels in the moment. But further, reversely: what *is* and influences one's way to act encompasses not only lack of sleep and actions performed by his neighbor but also norm systems—norms which Mr. Jones has transgressed as well as norms which allow one to scold the scoundrel or which constrain one more than he can bear.

From the statement that "Mr. Jones has stopped beating his wife" we may conclude that Mr. Jones is or has been married, since the statement would be meaningless without this presupposition. From statements building upon an indefinite array of presuppositions, we cannot decide *a priori* which conclusions can and which cannot be made; in particular, if an "is-statement" has normative presuppositions (as "Mr. Jones is a criminal," where the notion of a criminal is defined legally and may encompass norms about what should be done to criminals), then Hume's argument fails even on this account.

18. Members of the Sicilian Mafia speak of themselves, and want to be spoken of, as *uomini d'onore,* as "men of honor."

VALUE RELATIVISM AND VALUE NIHILISM

Another look at the categories of the general philosophy of morality will permit us to understand better the discussions about scientific development.

It was asserted above that morality is able to live with dilemmas. This is true only up to a point. Firstly, of course, the reconstruction of incoherent rule systems through bargaining and various kinds of thought experiments has the function (and, as far as many literary thought experiments are concerned, the deliberate aim[19]) to expose and thereby to solve, to surmount, or to reconcile them. Secondly, awareness of the existence of inconsistencies in the moral system may (in particular in secularized and enlightenment periods, when the Solomonic wisdom of religious authorities is questioned) lead to *value relativism:* norms there must be, but we choose them freely—*man* is the measure of all things, as once formulated by Protagoras (and held by utilitarianism). Or the conclusion may be *value nihilism:* "If God does not exist, then everything is permitted" (Ivan Karamazov), or "*Good* is only what is good for the strongest," as Plato expressed the ultimate consequence of Protagoras's relativism in *Gorgias* and the *Republic.*

Both value relativism and value nihilism can be found within the philosophy of science. Lakatos's whole methodology of "research programs" is, indeed, a relativist reaction to the breakdown of Popper's claim that an absolute methodology *could,* and *should,* be applied if we want to know in spite of the fallible character of all knowledge: the community of experienced scientists is, in fact, the "measure of all things" within their science; it is they who decide which methods are to be accepted. Another runaway Popperian has taken the nihilist standpoint: Paul Feyerabend, whose discovery that Popper's proclaimed "rational method" does not work made him publish a book with the title *Against method,* in which it is claimed that "there is only *one* principle that can be defended under *all* circumstances and in *all* stages of human development. It is the principle: *anything goes.*"[20]

---

19. Think, for instance, of Sophocles's *Antigone,* which confronts the norms of the city state and of political society (represented by Creon) with those of ancestral morality and human love (represented by Antigone and her *fiancé* Haimon).

20. Feyerabend 1975: 28. Feyerabend himself declared the philosophy to be *anarchist* in his subtitle. Many currents, it is true, can be found within anarchism. Still, Feyerabend's principle reminds most of all of that which later anarchists have held to be malicious lies about Mikhail Bakunin. The political quotation in the chapter leading forward to the principle is from Robespierre: "Virtue without terror is ineffective," which *may* be meant as a claim that universal standards like those proposed by Popper will necessarily lead to mass murder, but may just as well have any other meaning: the "epistemological anarchist," Feyerabend's *alter ego,* opposes "ideas such as 'Truth', 'Reason', 'Justice', 'Love'" because they are "Absolute," even though he may well find it convenient to pay lip-service to them "as if he believed in them" (p. 189). Feyerabend appears to have been an anarchist because it sounds so nice, and in spite of his confession (p. 20) to detest most anarchism and most anarchists past and present

Value nihilism is a tempting inference from the discovery that absolute norms have a hollow ring. It sounds like another version of the Socratic principle that "the only thing which I know is that I know nothing." But the probable outcome of practiced value nihilism is not Ivan Karamazov's gentle desperation but Rodion Raskolnikoff. Outside literature, and in a somewhat less drastic illustration: in local democracy it is unavoidable that those who decide know some of those whom they decide about. An absolute prevention of local favoritism and similar corrupt behaviors is therefore only possible if local democracy (and local government altogether) is abolished. The nihilist supporter of local autonomy will therefore have to drop the prohibition of favoritism—any norm which cannot be upheld absolutely cannot be upheld at all. The result, of course, will be that vaguely endemic corrupt manners become epidemic.

Norm systems, indeed, are not only reconstructed reflections of adequate patterns of behavior. They are first of all what we usually take them to be: *regulators* of behavior (and only thereby, of course, reflections of adequacy). Conflicting norms—in this case the norm of democratic government as close to those concerned as possible, and the norm of decent behavior and of equal opportunities irrespective of kin, friendship, and protection—are so too. They cannot be absolute prescriptions,[21] but through acts of balancing (affected, among other things, by socialization and by tacit knowledge about the scope of each norm) they may still put certain limits to behavior in a situation which is strained by contradictory claims and interests. Irrespective of the anomalies which were discussed above, this also holds for methodological norms like the empiricist and the falsificationist imperatives—Feyerabend and other nihilists notwithstanding.

INSTITUTIONAL IMPERATIVES

Terms like *prescription, methodology,* and *rule* are often used in the vicinity of Popper, Lakatos, and Kuhn (and in many other quarters of the philosophy of science). Principles from the philosophy of morality and norms should therefore be applicable to the process of scientific development and work—as they were indeed applied in the preceding pages. As a matter of fact, however, this approach is far from traditional. The aspect of the sciences which is traditionally discussed through the concept of norms is the *sociology of science.*

The seminal (indeed paradigmatic) work which launched this norm-based sociology of science was an article by Robert K. Merton from 1942 on "science

---

because of their seriousness and their lack of respect for human lives. In language and in the interest of philosophical provocation, too, *anything goes.*

21. For this we do not need norms in conflict. The validity of rules in practice can never be more absolute than the linking of single acts to particular rules.

and democratic structure."[22] Concerning the "ethos of science," Merton explains that

> The institutional goal of science is the extension of certified knowledge. The technical methods employed toward this end provide the relevant definition of knowledge: empirically confirmed and logically consistent statements of regularities (which are, in effect, predictions). The institutional imperatives (mores) derive from the goal and the methods. The entire structure of technical and moral norms implements the final objective. The technical norms of empirical evidence, adequate, valid, and reliable, is a prerequisite for sustained true prediction; the technical norm of logical consistency, a prerequisite for systematic and valid prediction. The mores of science possess a methodological rationale but they are binding, not only because they are procedurally efficient, but

---

22. This was the title given to it when it appeared as a chapter in Merton 1968. Originally it was entitled "A Note on Science and Democracy." Both versions of the title point to the origin of the essay in the antifascist debates of the late 1930s about the role of science. The setting is explained by Robert Merton as follows:

> . . . A tower of ivory becomes untenable when its walls are under prolonged assault. After a long period of relative security, during which the pursuit and diffusion of knowledge had risen to a leading place if indeed not to the first rank in the scale of cultural values, scientists are compelled to vindicate the ways of science to man. Thus they have come full circle to the point of the reemergence of science in the modern world. Three centuries ago, when the institution of science could claim little independent warrant for social support, natural philosophers were likewise led to justify science as a means to the culturally validated ends of economic utility and the glorification of God. With the unending flow of achievement, however, the instrumental was transformed into the terminal, the means into the end. Thus fortified, the scientist came to regard himself as independent of society and to consider science as a self-validating enterprise which was in society but not of it. A frontal assault on the autonomy of science was required to convert this sanguine isolationism into realistic participation in the revolutionary conflict of cultures. The joining of the issue has led to a clarification of the ethos of modern science.

The reasons for public distrust of science have evidently changed since 1942. The ecological crisis, for instance, was still below the horizon, and the involvement of social science in the management of minds through scientifically designed advertisement and propaganda were not yet conspicuous (although it *had* begun as early as the 1920s, and Aldous Huxley had made it a prominent theme in his *Brave New World* from 1932). But the phenomenon of public distrust remains, for which reason clarification of the actual ethos of science is still important—not primarily for purposes of self-defense but rather as a basis for self-critical reflection.

> because they are believed right and good. They are moral as well as technical prescriptions.[23]

The final sentence has a clearly Kantian ring. "Moral prescriptions" are those which (are held to) have absolute character, while "technical prescriptions" are merely tools for instrumental rationality. But the actual understanding of the nature of norms comes close to the one proposed above. "Institutional imperatives," in fact, are understood as *norms which at least to a certain degree must be respected if the institution is going to fulfil its presumed role—in casu,* the production of "certified" (as opposed, for example, to *revealed*) knowledge.

These imperatives are not codified explicitly in any catechism for future scientists; they become visible, as Merton points out, in the "moral consensus of scientists as expressed in use and wont, in countless writings on the scientific spirit and in moral indignation directed toward contravention of the ethos."[24]

Merton himself, however, codified the system of institutional imperatives, finding four of them:

1. *Communism,* "in the non-technical and extended sense of common ownership of goods." Apart from eponymity expressing recognition ("Boyle's law," "Rorschach test"), nobody has property rights to scientific knowledge. Scientific results should be made public, firstly so that others may use them, secondly in order to be subjected to criticism (prerequisites for cumulativeness and certification, respectively). And further: "The communism of the scientific ethos is incompatible with the definition of technology as 'private property' in a capitalist economy. Current writings on the 'frustration of science' reflect this conflict. Patents proclaim exclusive rights for use and, often, nonuse. The suppression of invention denies the rationale of scientific production and diffusion."[25]

2. *Universalism,* which "finds immediate expression in the canon that truth-claims, whatever their source, are to be subjected to *preestablished impersonal criteria:*

---

23. Merton 1968/1942: 606f. Two terminological details should be taken note of: Merton speaks of "certified," not "certain" knowledge. And he speaks about empirical "confirmation," not "verification." Merton was never close to the logical empiricists, and what may remind vaguely of their idiom is nothing but a concise common sense description of the aims and methods of scientific work.

24. 1968/1942: 605f. As an aside on Merton's own "context of discovery," it can be told that Merton had intensive first-hand knowledge of these moral attitudes. He wrote his Ph.D. dissertation under the guidance of the historian of science George Sarton, who continually taught him about what and what not to do—as Merton told in a lecture at the George Sarton Centennial Conference, Ghent 1984.

25. Obviously, this norm has run into heavy weather since the Reagan-Thatcher era—patenting of results has become endemic in every discipline which offers the possibility, and financing authorities do what they can to accelerate the process.

consonant with observation and with previously confirmed knowledge. The acceptance or rejection of claims entering the lists of science is not to depend on the personal or social attributes of their protagonist: his race, nationality, religion, class and personal qualities are as such irrelevant." Universalism thus deals with knowledge, but no less with *persons*. The optimal progress of knowledge requires that nobody who is competent is excluded from the scientific institution.[26]

3. *Disinterestedness* is the norm which explains those features of scientific activity which traditional hagiography derives from the particular moral qualities of scientists (altruism, honesty, "objectivity"), or from their personal motives (curiosity, thirst for knowledge). Scientific disinterestedness requires that the scientist should not distort his science or his results in order to gain personal advantage or in the service of particular interests (in the terminology introduced on p. 331, disinterestedness thus imposes the elimination of particularism). Transgressions in this field are probably the ones which are most severely punished. As a rule, the scientist who has been caught in deliberate fraud can start looking around for a different career.[27] The same thing, of course, will happen to an accountant who has betrayed his employer. Other professions, in contrast, have quite different norms. The strictness with which the accountant is treated can be compared with the lenience with which his employer is handled when caught in insider trade at the stock market; the bad luck of the cheating scientist can be compared to the praise bestowed upon the fraudulently imaginative journalist.[28]

---

26. This was evidently, when written in 1942, a reference to the expulsion of Jews and Social Democrats from German universities. The effect had been described bitingly by the old David Hilbert (too much of a Nestor of mathematics to be maltreated by the Nazis at that date), as NS-Reichsminister Bernhard Rust asked him in 1934 whether it was really true that the Mathematical Institute in Göttingen (the most prestigious of the kind in the world until 1933, inspiration for the Princeton Institute of Advanced Studies) had suffered from the expulsion of Jews and their friends: "Jelitten? Dat has nicht jelitten, herr Minister. Dat jibt es doch janich mehr!" ("Suffered? That has not suffered, your excellence. After all, it does not exist any longer!" (quoted from Neuenschwander & Burmann 1987: 25).

27. In one Danish case from the 1950s, the Rector Magnificus of Copenhagen University resigned from his office, not because he had committed fraud himself, but because he had been unwilling to believe evidence that his son-in-law had done so. The son-in-law, when discovered, gave up his scientific career, changed his too characteristic name, and settled down as a practicing physician; his wife asked for divorce. A number of more recent cases from the U.S. are described in Broad and Wade 1982.

28. One example: Some years ago the Danish journalist Jan Stage was forced to admit in court that he had invented himself an interview with Bület Ecevit, endangering thus this Turkish social democratic politician, whom the military government had forbidden to make any public announcement. Shortly afterwards, Stage's employer, the newspaper *Politiken,* ran an advertisement campaign featuring precisely Stage—much better known and much better suited for advertisement purposes than his more honest colleagues, it seems.

Disinterestedness does not prohibit (and no norm can prevent) misunderstandings of experiments, blindness to adverse results, and overly dogmatic trust in established theories. What it proscribes are cases like the physics professor from Copenhagen going to Thule in January 1968 "in order to prove" that the crash of a B-52 carrying a number of H-bombs had produced no radioactive pollution. Jørgen Koch's slip (occurring during an interview in the Danish radio news) demonstrates that the norm is not universally respected. But the start one feels when hearing statements like this demonstrates the existence of the norm *as a norm*—and overt admissions of the kind can only undermine the public trust in the scientific institution. The rule thus *is* an institutional imperative: rampant disrespect endangers the institution.

4. *Organized scepticism* is the claim of the scientific institution that it should not be subjected to the interests of other institutions or submitted to their control, nor be bound by prevailing opinion and prejudice.[29] This norm is certainly useful for the cognitive functioning of science by offering moral support to scientists who risk conflict with those in power by staying loyal to what they (suppose to) know, and by censuring opportunism; but the attitude is one which will easily bring scientists into conflict with their surrounding society:[30]

There are good reasons that the scientist and the accountant are more strictly regimented than most other professions: in both cases, controls are almost automatically applied; and in both cases, the fraud undermines the very *raison-d'être* of the profession. The journalist, on his part, is rarely paid solely for telling the truth; entertainment value is quite as important for newspapers getting an appreciable part of their income from advertisement and the rest from customers paying rather for entertainment and relaxation and for the subjective impression of being informed than for information itself (in the case of papers which get support from political parties, employers' or trade unions, etc., other reasons ask at least for a specific perspective on truth).

Professional honesty thus depends on the situation and the rationale of the profession. So does the particular character of that honesty. The accountant has to be honest about money. The scientist who mixes up private and institutional money may be rebuked, fined, or perhaps dismissed—but may in even the worst of cases hope to get another job within the profession. Public trust in science, and the confidence with which others may use his scientific results, are not undermined.

29. It will be remembered from chapter 18 that this question was central to one of the important differences between Popper$_1$ and Popper$_2$/Lakatos: if followed, Popper$_1$'s methodological conventionalism would tend to make scientists bend to conventional wisdom; Lakatos's methodology of research programs, on the contrary, will protect research challenging accepted opinions as long as it remains fruitful.

30. Maintaining this ideal in spite of pressure is thus, in the isolated instance, contrary to act utilitarianism (with the "general happiness of the scientific institution" as the gauge). It would be much more remunerative for the institution to agree with government officials, newspaper magnates, and so on, in spite of what science itself appears to tell. But doing so may

> . . . Most institutions demand unqualified faith; but the institution of science makes scepticism a virtue. Every institution involves, in this sense, a sacred area that is resistant to profane examination in terms of scientific observation and logic. The institution of science itself involves emotional adherence to certain values. But whether it be the sacred sphere of political convictions or religious faith or economic rights, the scientific investigator does not conduct himself in the prescribed uncritical and ritualistic fashion. He does not preserve the cleavage between the sacred and the profane, between that which requires uncritical respect and that which can be objectively analyzed.

To be more precise: his professional ethos tells him that he does not need to preserve it. As we know, scientists are not only scientists but also members of society, and many of them split their allegiance between the norms of their profession and those of society (or their social group) at large. Obviously, the Atlantic allegiance of the physics professor mentioned above outweighed not only his allegiance to the norms of his profession but even his awareness that there might be a problem (and, as it turns out now that the U.S. archives have been opened, his scientific honesty).

Organized skepticism, it should be noted, has nothing to do with the customary concept of philosophical skepticism. It does not imply skepticism toward the possibility of obtaining reliable scientific knowledge; on the contrary, the latter kind of skepticism is often promoted by those who wish to domesticate the provocative self-assurance of science encroaching on the sacred domains of other institutions or to wash their hands when convicted of having acted in bad faith.[31] It is therefore totally mistaken to cite as evidence against the norm of organized skepticism five scientists out of seventeen who would not accept reports of flying-saucers "no matter who made the observations";[32] precisely these five, indeed, illustrate the norm, being so sure about the assertions of their science regarding the impossibility of interstellar travel and being so knowledgeable about the susceptibility of even fellow scientists to mass illusion that they felt entitled to contradict every report, be it published in

---

be prescribed by rule utilitarianism: a science which has bent too obviously to the desires or requests of authority tends to be decried at the next turn of the road (in Spring 1990, the very concept of 'social science' was officially abolished in the late German Democratic Republic, as a consequence of the too apologetic behavior of too many social scientists!).

Much hagiographic history of science probably serves the purpose of mediating between act and rule utilitarianism at precisely this point: if science can be shown to have been right in resisting heroically the now defamed authorities of former times, it might well be right in continuing to defy authority.

31. Thus the psychologist H. J. Eysenck (1979), when it turned out that the research on which he drew for proving that intelligence was determined by inheritance alone was one immense fraud, started a campaign to prove (Popperian methodology at hand) that astrology might be as good a science as any other.

32. Sklair 1973: 154, quoting a study from 1960.

the *New York Times* or made by the president of the National Association for the Advancement of Science. "Organized skepticism" does not contradict Kuhn's findings about "dogmatism"; the two are, in fact, sides of the same coin.

THEORETICAL VERSUS APPLIED SCIENCE

Merton's article became a Kuhnian paradigm for a generation of American sociology of science, though mostly in a sadly trivialized reading. As a reaction, a later generation has been eager to show that it is all wrong. Many of the objections are irrelevant in our present context, but one point of the discussion is not.

If scientific activity is regulated by the Mertonian norms, then scientists in general, or at least the majority, should subscribe to these norms when interviewed by sociologists—thus an alluring start of the argument. Most of those whom the sociologists regard as scientists, however, are active in industrial and other applied sciences; a survey which is meant to be representative will hence be dominated by applied scientists; some surveys have indeed looked solely at industrial scientists. The outcome of such studies has been that many industrial and similar scientists do not subscribe to the Mertonian norm system, in particular not to the communist imperative.

Though often represented as counter-evidence,[33] this is actually an essential underpinning of Merton's argument, which connects it to the above discussion. Industrial and other applied scientists do *not* work inside an institution whose primary "institutional goal . . . is the extension of certified knowledge." The aim of their work is the adaptation and application of new or (often) old knowledge (cf. chapter 21); they are paid for producing knowledge which can end up as privately owned technology.[34] That they do not follow all Mertonian norms (or do so only in a restrictive interpretation (cf. below) merely illustrates that these are *institutional* imperatives: they have little to do with the personal character and history of scientists, and they do not belong to the corpus of already existing scientific knowledge as an inseparable attribute; they crystallize within an institution, that is, a network of social interactions organized around a particular core value (as it was called above, see p. 339, n. 9): "the extension of certified knowledge" in a collective process.

It can still be objected that even many scientists working within the institutions of theoretical science (as well as whole institutions) do not obey the norms. The Ancient Testament, however, also abounds with stories about members of the

---

33. Thus still by Leslie Sklair (1973: 154).

34. These formulations fit only industrial scientists to the letter. Applied social scientists, for instance, rarely produce artefacts that may be bought and sold; their employer, however (be it the state or a private institution), will control their work and use its results in much the same way as an industrial corporation controls and uses the work of an industrial scientist.

Twelve Tribes killing each other, which does not invalidate the norm "Thou shalt not kill" (Judg. 19–21 tells a nauseating story of treason, group rape, and massacres culminating in the virtual extermination of one of the tribes at the hands of the remaining eleven). An analogue of the Popperian argument from p. 280 will be more justified in the present context than was the original version: the more often and the more strongly the norms are broken, the less efficient is the work of the institution, and the more likely is it to run into blind alleys. Keeping *all* knowledge as private property (that is, secret or encoded in a way which impedes others from building on it) would mean the end of science; prohibiting the teaching of "Jewish" physics in Germany (1933–1945) or (supposedly "bourgeois") genetics in the Soviet Union (1948–c. 1955) delayed research significantly in both countries.

FURTHER NORMS, CONTRADICTIONS, CONTRADICTORY INTERPRETATIONS

In another respect, the Mertonian norms seem to agree less well with the above analysis of the nature of norm systems: the scheme seems too simple, clear-cut, and free of inner contradictions. It looks more like a "code of ethics" for the profession laid down by a supervisory committee than a piece of real-life morality.[35]

That the normative regulation of scientific practice is indeed much more equivocal was already pointed out by Robert Merton in a humorous and less widely read paper on "the ambiguities of scientists." He lists nine pairs of mutually contradictory norms, beginning as follows (1963: 78f):

> — The scientist must be ready to make his new-found knowledge available to his peers as soon as possible, BUT he must avoid an undue tendency to rush into print. . . . .
>
> — The scientist should not allow himself to be victimized by intellectual fads, those modish ideas that rise for a while and are doomed to

---

35. The difference between such precisely stated "ethical rules" (for journalists, advertisement firms, etc., *not* to be confused with the general concept of 'ethics') and morality is that morality tells (more or less unambiguously) *how one should behave.* "Ethical rules" tend—in particular in cases where they do not derive from the immediate aim of the profession but go against it (mediating between this aim and general social norms, thus protecting the profession against suspicion of institutional immorality), as in the case of, for example, advertisement ethics—to state (like law) the limits of the permissible, ultimately thus telling *how far one may deviate from decent behavior without risk of condemnation or penalty*—or, if "positive" and noncommittal, they tend like Popper's methodology to advertise the honesty and altruism of the profession.

(The trivialized reading of) Merton's scheme has indeed inspired the ethical rules for scientific behaviour which authorities in the United States apply since long, and which are now spreading in Europe.

> disappear, BUT he must remain flexible, receptive to the promising new idea and avoid becoming ossified under the guise of responsibly maintaining intellectual traditions.
>
> — New scientific knowledge should be greatly esteemed, BUT the scientist should work without regard for the esteem of others.
>
> — The scientist must not advance claims to new knowledge until they are beyond reasonable dispute, BUT he should defend his new ideas and findings, no matter how great the opposition. . . . .
>
> — The scientist should make every effort to know the work of predecessors and contemporaries in the field, BUT too much reading and erudition will only stultify creative work. . . . .

"and so, on and on," as stated at the end of the list of "moral dilemmas." That a normative system may function even though ridden with such contradictions (and all practicing scientists know both the dilemmas and how to deal with them in single cases) hinges on the tacit knowledge of the participants in the social pattern that is regulated by the norms. Much of this tacit knowledge is part of the paradigm which governs work within a discipline; familiarity with this paradigm (and awareness of the number, the character, and the severity of the anomalies with which it is confronted) allows the worker to decide whether a new suggestion should, for instance, be dismissed as a mere "intellectual fad" or hailed as a "promising new idea."[36] The rest (inasmuch as two segments of tacit knowledge can be distinguished) structures the "merely social" interactions within the profession (for example, whom to honor, and how).

Even the four "institutional imperatives" are ambiguous and not interpreted in the same way by everybody. To some, for instance, the "communist" norm only means that final results should be made publicly known at some adequate moment without becoming thereby public property which can be used freely by everybody.[37] To others, the "communism of the scientific ethos" *is* indeed "incompatible with the definition of technology as 'private property' in a capitalist economy," as claimed by Merton.[38]

---

36. That not all workers make the same choice is one of the points in an activity governed by norms and not by rules—cf. Kuhn 1970: 186 as quoted on p. 334.

37. In recent years, this attitude has been demonstrated in glaring and dismal dimensions by the behavior of the molecular biologist Robert Gallo, purportedly co-discoverer of the HIV-virus. Within the humanities it is widespread in all fields where a monopoly on the right to publish findings can be upheld—e.g., in archaeology and Assyriology.

38. As the General Union of Danish Students fought its battle against contract research at universities in the early 1970s, it took care to appeal to both versions of the norm. Arguments that research paid by contracts could not be published freely were correlated with the "narrow" interpretation, and meant to find allies at universities; the slogan "Research for the people, not for profit" was an appeal to the broad version directed at the nonacademic public.

How the institutional imperatives are interpreted also changes—and must change—from one discipline to the other. To see why, we may look at the implications of "universalism" and "disinterestedness." Both are rejections (in somewhat different terms) of *particularism,* and thus *objectivity norms.* Yet how science achieves actual objectivity (and thus implements the two imperatives) depends on the problems and methods of the single discipline. In those branches of medical research which test individual cures, double-blind testing functions excellently; but to claim that this is the only way to guarantee scientific objectivity is evidently preposterous (already in medical disciplines like epidemiology or preventive medicine, but *a fortiori* in sciences like astronomy, sociology, and historical research). In sociology, where the value system of scientific workers may overlap or be in conflict with the value system that regulates the social unit under investigation, it may make sense to claim that sociological science should be "value-free"; as demonstrated by numerous hilariously absurd discussions taking place in the late 1960s, it gives little or no meaning to claim that science in general (and not only sociology and like human and social sciences) is or is not, should be or should not be value-free.

Even though discussions about the responsibility and political involvement of the sciences are often expressed in different terms nowadays, the theme can be pursued with profit. The idea that sociology should be value-free was formulated by Max Weber; already in *his* writings the use of the concept is ambiguous, and later interpretations are no less divergent. The sociologist Alvin Gouldner (1973: 11–13), in a lucid and stylistically sparkling essay on the meaning and function of "the myth of a value-free sociology," starts by presenting the opportunities offered by the concept:

> [The value-free doctrine] enhanced a freedom from moral compulsiveness; it permitted a partial escape from the parochial prescriptions of the sociologist's local or native culture. Above all, effective internalization of the value-free principle has always encouraged at least a temporary suspension of the moralizing reflexes built into the sociologist by his own society. From one perspective, this of course has its dangers—a disorienting normlessness and moral indifference. From another standpoint, however, the value-free principle might also have provided a *moral* as well as an intellectual *opportunity.* . . . .
>
> The value-free doctrine thus had a paradoxical potentiality; it might enable men to make better value judgements rather than none. . . . .
>
> The value-free doctrine could have meant an opportunity for a more authentic morality. It could and sometimes did aid men in transcending the morality of their "tribe," . . . , and to see themselves and others from the standpoint of a wider range of significant cultures.

The value-free doctrine could thus, we see, push sociologists from particularism toward universalism in their value judgements and enhance the organized skepticism of the discipline. But Gouldner goes on with harsher words:

> But the value-free doctrine also had other, less fortunate results as well.
>
> . . . many . . . used the value-free postulate as an excuse for pursuing their private impulses to the neglect of their public responsibilities. . . . Insofar as the value-free doctrine failed to realize its potentialities it did so because its deepest impulses were . . . dualistic. . . . .
>
> On the negative side, it may be noted that the value-free doctrine is useful both to those who want to escape *from* the world and those who want to escape *into* it. It is useful to those . . . who live off sociology rather than for it, and who think of sociology as a way of getting ahead in the world by providing them with neutral techniques that may be sold on the open market to any buyer. The belief that it is not the business of a sociologist to make value judgements is taken, by some, to mean that the market on which they can vend their skills is unlimited. From such a standpoint, there is no reason why one cannot sell his knowledge to spread a disease just as freely as he can fight it. . . . .
>
> In still other cases, the image of the value-free sociology is the armour of the alienated sociologist's self. . . . . Self-doubt finds its anodyne in the image of a value-free sociology because this transforms [the sociologist's] alienation into an intellectual principle. . . . .
>
> There is one way in which those who desert the world and those who sell out to it have something in common. Neither group can adopt an openly critical stance toward society. . . . .

When the value-free principle is understood in these ways—not as a statement that sociology need not repeat or apply prevalent value judgements but as a claim that the activity of the sociologist is itself above moral judgement—it has thus become a disguise for breaches of the norms of disinterestedness, organized scepticism, and communism (in the broad interpretation which emphasizes the general social responsibility of the sciences). It is only in this variant (when no longer a norm of objectivity but only of marketability) that the value-free postulate can be generalized to all sciences.

Other fields possess their own more or less idiosyncratic specifications of the "objectivity norms," which we may pass over without further discussion. The examples of the double-blind technique and the value-free doctrine should suffice to make a general point: that norms at the level of Merton's "institutional imperatives" function not only directly as regulators of scientific behaviour but also as "core values" around which more specific norms crystallize during the practice of the single scientific field. Further, the conflicting interpretations of the value-free principle illustrate that the norms regulating scientific work (occasionally down to the specific norms of the paradigm—cf. p. 286, n. 22) are susceptible to interaction with the norms of general social life.

CHAPTER 21

The Theory of Interdisciplinary and Applied Science

The discussion of chapters 18 and 19 presupposed (numerous examples at hand) that scientific work is made within separate disciplines; chapter 20 assumed that science may be either "theoretical" or "applied." Both suppositions refer to commonsense knowledge; none the less, both are contradicted by many traditional philosophies of science, not least empiricism.[1] At best, they regard the division of science into disciplines and the split between theoretical and applied knowledge as purely pragmatic divisions of labor with no epistemological or philosophical basis: "Knowledge is one. Its division into subjects is a concession to human weakness."[2]

The aim of the present chapter is to show that the divisions *are* epistemologically founded, and to explore the relations between disciplinary and interdisciplinary science and between these and the pair "basic" and "applied" science. After having looked in the preceding chapters at the nature of (scientific and general) knowledge and at that specific practice which aims at the production of scientific knowledge, we shall thereby be brought to consider the production of scientific knowledge as an aspect of general social practice.

KNOW-HOW AND KNOW-WHY

That scientific insight ("know-why") can be used in instrumentally rational practice as "know-how" is both a commonplace and part of our daily experience—as illustrated by an advertisement slogan painted on a van belonging to a pharmaceutical

1. Even Habermas, while setting aside the humanities and emancipatory social science, follows Peirce in his identification of natural science as a mere means to produce technology—cf. pp. 183 and 189.

2. H. J. Mackinder, quoted from Mackay 1977: 99.

firm which I noticed in the street some years ago: "Today's theory—tomorrow's therapy." The commonplace is as old as the scientific revolution; both Thomas More and Francis Bacon claimed that natural philosophy had the double aim of producing useful technology and honoring God by studying his creation.[3]

The commonplace was translated into an analytical tool for research statistics around 1960. At that moment, the Organization for Economic Development and Cooperation (OECD) had come to consider the development of adequate theory a crucial fundament for technological and social progress (and thus for the supremacy of the "West" over the Soviet Union[4]). As a consequence, the OECD embarked upon a massive patronage of "science policy" programs in the member countries, and in order to monitor the programs and the distribution of research funds, a handbook containing prescriptions for the production of research statistics (known as the *Frascati Manual*) was prepared. The manual contains the following definitions of "basic" and "applied" research (*Measurement of Scientific and Technical Activities*, 19f):

> — Basic research is experimental or theoretical work undertaken primarily to acquire new knowledge of the underlying foundations of phenomena and observable facts, without any particular application in view.

---

3. On the other hand, the idea that technical practice is or should be derived from scientific theory is not much older in Europe than the late Renaissance (in the Islamic world it can be traced back to the eighth or ninth century CE). Aristotle, in particular, tells in the *Metaphysics* that *first* the practical arts were invented in order to provide us with the necessities of life. Later, theoretical science developed, which was honored more highly because it had *no* technical purpose (cf. p. 26). Aristotle's slogan would thus rather have been "yesterday's therapy, today's theory"—or, even more radically, and denying that the two have any inner connection, "Yesterday therapy. Today's progress: theory."

We may find Aristotle's point of view outdated and perhaps be repelled by the philosopher's disdain for those who provided him with food and shelter. Yet what he describes was historically true in his times and corresponds exactly to an institutional and epistemological split between practitioner's knowledge and theoretical "know-why" which only broke down definitively in the nineteenth century. Here, the typical Early Modern (More to Bacon) legitimation of science as adoration of the creator *and* civically useful should not mislead us: it refers to the use by practitioners of the *elementary* items from the theoretician's tool-kit.

4. As we now know, this proved right, in the sense that the socialist countries were unable to transform their massive investments in research and scientific education into productive and institutional innovation. In the capitalist world, on the contrary, the outcome of the first oil crisis, in which the need for technological change had imposed itself, was accelerated development on all levels.

Technological development, it is true, does not by itself create or destroy social systems. But unequal economic development provided the basis for what came to happen on the political and institutional levels.

> Basic research analyses properties, structures and relationships with a view to formulating and testing hypotheses, theories or laws.
>
> — Applied research is also original investigation undertaken in order to acquire new knowledge. It is, however, directed primarily towards a specific practical aim or objective.
>
> Applied research is undertaken either to determine possible uses for the findings of basic research or to determine new methods or ways of achieving some specific and pre-determined objectives. . . .[5]

In agreement with the commonplace, the OECD thus sees the realms of "theory" and "therapy," that is, basic research and practical technology, as belonging together, linked through applied research (and, in fact, through a further bond constituted by "experimental development," which makes science-policy makers speak of R&D, "research and development")—but still as clearly distinct activities. Why this has to be so is a question to which we shall return below (p. 365 onwards).

THE ACQUISITION OF THEORETICAL KNOWLEDGE

From a commonsense point of view, OECD's strategy may seem awkward. If technological progress and know-how is the aim, why then finance activities aiming at the acquisition of "new knowledge of the underlying foundations of phenomena and observable facts, without any particular application in view" (the kind of knowledge of which I speak as "theoretical") and not merely research looking for *the relevant kinds* of knowledge? Or, if basic research *is* relevant, why is there any need for a particular stage of "applied science"?

The answer is connected to the dynamics of the Kuhnian cycle as described above but may also be approached from the common dynamics of student projects as made at Roskilde University.

Such a project normally takes as its starting point an interesting *phenomenon* (Greek for "that which is seen") providing a first formulation of the problem and a first idea about the relevant research strategy. However, work guided by this formulation will normally set the original phenomenon in a new light, *through concepts which only appear to the participants in the project due to the first phase of the systematic investigation.* Then they redefine the problem (that is at least what they *should* do) and continue to work according to the new definition of the problem and the strategy. This spiral process may continue; in the end, the treatment of the problem (and indeed the problem itself) may have very little direct connection with the phenomenon which originally inspired the work, and much more with the internal but

5. It should be clear already from this definition that "applied research" may be a much more creative process than the trivial "application of science" assumed by Popper and others (cf. p. 309, including n. 57).

hidden structures which were revealed by the systematic investigation (the "underlying foundations of phenomena and observable facts" of the *Frascati Manual*).

The Kuhnian cycle of successive paradigms repeats the same structure in larger scale. In the preparadigmatic phase, scholars start investigating phenomena which present themselves directly and which are either intriguing or in some sense important—maybe the curious fact that glass, when rubbed by a woollen cloth, attracts small pieces of paper, maybe the effect of taxation principles on the productive activity and the distribution of wealth and poverty within a country. The initial investigation results in the discovery of new connections between phenomena, giving rise to new questions. In the phase that follows, phenomena from seemingly quite different domains (lightning!) may enter the field, whereas others from the original area of investigation are discarded from view as, after all, irrelevant with regard to the underlying regularities which have come into focus (as when Mesmerism changed from a supposedly magnetic phenomenon into hypnosis); in this phase (and more radically in later repetitions of the cycle) questions will come to the fore which could not even have been imagined when the process started—as we have seen, the basic problems of *Das Kapital* could only be asked when Adam Smith had split society into classes according to kind of income (cf. p. 284). But even though the questions could not be imagined directly from commonsense observation of the phenomena which constituted the starting point, the answers remain relevant for explaining these phenomena—the Marxian analysis of the economic process tells better than both physiocrats and Adam Smith were able to do the dynamics of English economic life in the decades preceding the appearance of *The Wealth of Nations.*

What begins as an investigation of a particular range of phenomena belonging concretely together thus ends up as an investigation of the underlying regularities or laws for everything which can be explained (e.g.) from competition and monopolization of resources or from Maxwell's equations for the electromagnetic field; what has begun as scrutiny of an arbitrary section of the world (whether delimited spatially or by association with a particular practice) is stepwise transformed into a general examination of reality under the particular *perspective* of the techniques and concepts developed by the discipline. In this way we are led to much more fundamental insights not only into the phenomena which set us on the track but also into others—but only into particular *aspects* of these phenomena.[6]

This contradicts the general way of speaking according to which the different sciences deal with different *sections* of reality: biology deals with living beings (zoology with animals, botany with plants); physics with non-living matter; economics deals with the economy; sociology with *our* society and anthropology with the

6. That empiricist philosophy of science is blind to the epistemological foundation for the splitting of knowledge into disciplines is thus a specific instance of its general blindness to the role of *theory.*

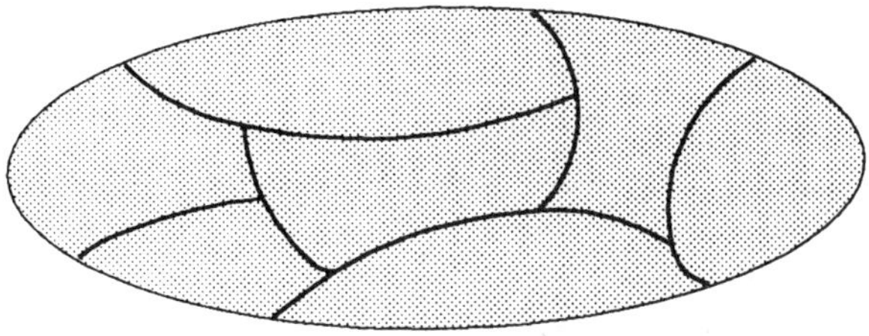

Separate sections of reality?

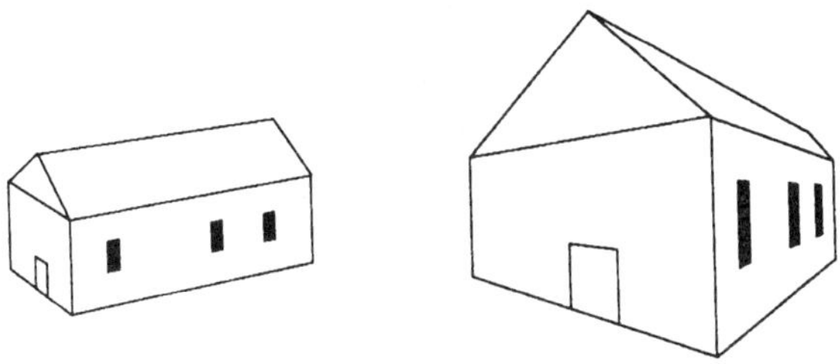

Or different perspectives on the same reality?

functioning of "primitive" societies;[7] electromagnetic theory of course deals with electricity and magnetism.

Yet electricity and magnetism are not specific sections of the physical world; they are all-pervasive. Similarly, no single human action is *solely* economic. When I buy a piece of cheese in the supermarket, my action can of course be analyzed as a step in the economic cycle. But how I move around when searching and paying may also be described by mechanical physics and biomechanics; that I feel a need to eat is a matter for the physiologist, and that I wish to eat cheese could give a cultural anthropologist the occasion for an explanation along quite different lines; my choice of a particular brand instead of another would certainly be of interest for the advertisement psychologist; and so on.

That we tend none the less to accept the common way of speaking illustrates to what extent we have come to see the world in which we live through the perspectives

7. The latter point, alas, is no parody in bad taste; it renders the objection actually raised by a fellow member of the profession when the anthropologist A. Shiloh wondered (1975; objection 1976: 554–555) that his discipline had done nothing to understand how the National Socialist system of extermination camps had been possible.

of the established sciences and to think in terms of their fundamental concepts. We have come to accept as self-evident, for instance, that society consists of an economic, a juridical, and a political sphere; only second thoughts allow us to rediscover that these spheres are wholly abstract.[8]

The simple Kuhnian cycle (as translated here into the metaphor of *perspectives in dynamic development*[9]) is not the whole truth about how disciplines develop. Above (p. 293), the "neo-classical" or "marginalist" revolution in academic economics was mentioned as a change of perspective that was inspired by political convenience rather than by the inner dynamics of classical political economy as based on the labor theory of value. Similar examples are numerous.[10] More cognitively productive, however, are two other processes: *branching* and *integration.*

Branching of disciplines can be explained at several levels. Since the later seventeenth century, the number of active scientists has doubled every fifteen years;[11] since it is impossible to follow in even modest detail the technical work of more than a few hundred colleagues, larger disciplines tend to split into specialties which loose contact (cf. p. 313 on the magnitude of the group sharing a paradigm in the narrow sense). But the process can also be explained at the epistemological level: every time new conceptualizations or new techniques emerge within a discipline (that is, at least when a new paradigm takes over), new open problems turn up, and specializations materialize when groups pursue different questions (and tend to loose contact

---

8. It may be objected that the three spheres are embodied in separate institutions: the stock exchange (etc.), the judicial system, and the parliament-and-party complex. But the transactions going on at the stock exchange are undertaken on legal conditions established by political authorities (and their very meaning indeed defined in commercial and corporation law); similarly in the case of other seemingly sphere-bound institutions. The spheres remain abstractions and analytical tools.

9. Where, if we continue the metaphor, each picture reveals where the camera should be moved next in order to take an even more detailed picture—a feature of the development of a discipline which was formulated on p. 331 as "progress toward greater objectivity of its subjectivity."

10. Without having researched the matter directly, it is for instance my definite feeling that the shift from Mertonian to anti-Mertonian sociology of science (see p. 352) is better classified as "academic patricide" than as a Kuhnian revolution. In disciplines without a firm inner structure (similar in many ways to Kuhn's preparadigmatic sciences), such patricides may be an easy way for the young wolves of a new generation (in the actual case *my* generation) to show their academic valor.

11. This rule of thumb was stated by Derek Price (1963). It is certainly only a rough approximation, and the process cannot go on in this pace for much longer (it seems indeed to have decelerated somewhat over the last twenty to thirty years), but it is not without value. Similar growth rates turn out to hold for the number of scientific publications appearing each year and for the population of scientific journals.

quickly if further conceptualizations and techniques developed around one question are not obviously relevant for other question types.[12]

Integration or convergence take place when the methods and concepts developed within one field turn out to be relevant for the understanding of another, and when two fields provide compatible explanations of a common object; they are furthered by the norm that scientific explanations should be consistent.[13] At times they lead to a genuine integration of separate disciplines into one, at times the process is better described as an assimilation (in a Piagetian sense) of a whole domain to the paradigm of another discipline (as when the neogrammarian sound shift laws were explained by structuralist linguistics—cf. p. 319); at times a new discipline emerges from the process alongside those which entered it; mostly (since the delimitation of disciplines is always blurred) the outcome is somewhere in between. In any case (as a colleague once summed up my view in a metaphor), the interplay between branching and integration assures that "the tree of science is no spruce but a banyan tree."

---

12. Explanations at still other levels can also be given, for example, through the specific recruitment structure of a scientific institution. One such explanation of the explosive development of new disciplines in nineteenth-century Germany was mentioned on p. 151, n. 5: Since there could be only one (ordinary) professor for each discipline at a university, the only way to argue for a new professorial position was the invention of a new subdiscipline.

The various explanations do not contradict each other. They relate to each other in much the same way as the pushing, the material and shape of the billiard ball, and the configuration of the table.

13. The ultimate consequence of this norm goes further than the integration of single disciplines. Since *reality* (or "physical reality" or "the reality of human life") is one, the descriptions provided by the different sciences ought to be compatible and in the final instance to fit together in a large coherent system of *unified science.* This was the dream of the Romanticists, and again of the logical empiricists (Kuhn's *Structure of Scientific Revolutions* was actually published in the "International Encyclopedia of Unified Science," a series directed by the major logical empiricists). The Neohumanist movement in early nineteenth-century Germany dreamt of unified *Geisteswissenschaft.* A rather recent formulation (concerning the social sciences alone) runs as follows:

> Economics and anthropology, and for that matter sociology and political science as well, are all—insofar as they are scientific—*ultimately the same science.* As economics broadens its horizons, it will increasingly seem to be invading realms of behavior that have in the past been reserved to other social disciplines. As other social sciences become more rigorous, they will increasingly grow to resemble economics (and, indeed, to *be* economics).
>
> (Hirschleifer 1977: 133)

The formulations are rather asymmetric, and social scientists outside economics will probably find them outright imperialist. But the principle remains the same.

THE "SCIENTIFIC-TECHNOLOGICAL REVOLUTION"

The spiral development of scientific disciplines leads, thus it was asserted above, "to much more fundamental insights not only into the phenomena which set us on the track but also into others—but only into particular *aspects* of these phenomena." But whoever wants to build a bridge or to improve the educational standard of the young generation is interested in the bridge or education *as functioning wholes,* not only in aspects. A bridge should be able to carry its load when built; but it should also be stable toward wind and perhaps (depending on its location) earthquakes, it should not corrode, and at best it should also correspond to certain aesthetic norms; an educational policy should involve *what* to teach to whom, but it does not function if teaching is badly made, if those who should be taught cannot afford participation, or if teachers cannot be recruited. Practice is concerned with many-sided *sections* of reality (by necessity and definition, if a "section of reality" is understood as what belongs together within a particular practice).

This difference between the orientation of "know-why" and "know-how" is a main reason that they were not only carried by separate professions but to a large extent by noncommunicating professions until the early nineteenth century: the higher levels of theory were carried by university scholars, members of scientific academies, and so on. Practitioners, on their part, were taught as apprentices by other practitioners. Most of their knowledge had originated in direct practice; what was borrowed from theoreticians consisted normally of specific techniques (for instance, the use of logarithms in navigation) and not of connected insights; typically these techniques would belong on the basic and not on the advanced levels of the sciences of the time.[14]

The first major change of this pattern was the appearance of the scientifically taught engineer in the early nineteenth century (cf. p. 149). Around the engineering schools, a particular "engineering science" developed (cf. Channell 1982), the aim of which was so to speak to translate and combine the knowledge of the "aspect-oriented" sciences into practically relevant information.

Efforts to integrate *theory developed with an eye on application* and *practice making use of actual research results* (and not just of the results that the engineer had been taught in his youth by a teacher who had learned about them during his early years in university) began in Germany around c. 1850, first in organic chemistry and soon in electrotechnology (Karl Wilhelm Siemens and the Siemens corporation) and metallurgy (Alfred Krupp and the Krupp Corporation). This step has been spoken of as the beginning of the "scientific-technological revolution." It was contemporary with parallel attempts to develop behavioral sciences for use in "social engineering,"

---

14. This is certainly only a rough approximation. It is, however, significantly less distorted than the reverse simplification: that the Technological Revolution taking place since the late Renaissance was derived from the Scientific Revolution.

the most important examples being probably Galton's eugenics and Lombroso's physiognomic criminology (cf. p. 161).

World War I, along with sonar, poison gas, and other new technologies developed by physical and chemical scientists on the basis of their theoretical insights and their best research techniques, gave rise (in alliance with contemporary industrial needs) to the development of "engineering psychology," that is, scientific investigation of how to design machinery in agreement with what psychology could know about the perception, discrimination, and reaction capabilities of the human operator (cf. Chapanis 1968).[15] Alfred Binet's and Cyril Burt's creation of the IQ-test technique (cf. p. 236) occurred in the same period and exemplifies the integration of psychological science with other divisions of general practice (*in casu* the educational system). It was followed in the interwar period (if we restrict the list to human, social, and organizational sciences) by the creation of industrial sociology, by mass media studies aiming at advertisement efficiency, and by welfare economics (John Maynard Keynes, Gunnar Myrdal); further on, during World War II, by operations research, enhanced propaganda studies, and by studies of the "cultural anthropology of the enemy" (undertaken in particular by Ruth Benedict; see Ember & Ember 1977: 42f). After 1950, the OEEC and its successor organization OECD promoted the generalization of science-based "policies" (the very use of the word in this sense is a postwar phenomenon): economic policy, educational policy, criminal policy, population policy, technology policy, science policy, and so on. The postwar era can hence be regarded as inaugurating the mature phase of the scientific-technological revolution.[16] This situation and its problems are what produced the *Frascati Manual;* "science policy" is indeed a metapolicy meant to produce the scientific knowledge needed in the other policies.

Some of the sciences created during the scientific-technological revolution started from a low level of theoretical knowledge and developed the necessary

---

15. The establishment of engineering psychology thus exemplifies that direct application of science in industry and warfare which provoked the Cambridge mathematician Godfrey Harold Hardy to formulate in 1915 that "a science is said to be useful if its development tends to accentuate the existing inequalities in the distribution of wealth, or more directly promotes the destruction of human life" (quoted in Hardy 1967: 120).

16. Evidently, speaking of this phase as "mature" does not imply that nothing new is going to happen. As a matter of fact, another phase shift is already taking place. The early integration of scientific knowledge into machinery and practical processes was, so to speak, put in once and for all: Burt constructed a number of tasks and put them together and then standardized the test; afterwards, the test was ready for use and would not be changed (only replaced by a new version when it proved inadequate). Increasingly since the 1960s, scientific knowledge is put into the machinery and the process itself so as to allow servo-control and other kinds of autoregulation (certainly with better success in car construction than in medical service and other social processes).

knowledge directly for the purpose of application. The whole field of mathematical statistics (which has had an splendid career since then) was founded by Francis Galton in this way as a tool for eugenics; Binet's IQ tests represent a similar instance (Pichot 1968). Both cases are characterized by the absence of developed theoretical disciplines which could have served.[17] The general experience has been, however, that the theoretical sciences were in possession of relevant knowledge of importance. If we refer to the questions formulated on p. 359 it is therefore clear why the science policy experts of the OECD would find it appropriate to invest in basic research.

It had also been common experience (since the development of "engineering science") that the knowledge possessed by the theoretical disciplines could not be used directly. If practice regards a section of reality, and theoretical disciplines only provide a particular perspective on this and other sections, no single theoretical discipline can do the job (whether we think of bridge building or the planning of an educational policy); as an engineers' saying goes, "the difference between theory and practice consists in condensed water."[18] Moreover, if communication between successive paradigms within the same field is only partial, the same holds by necessity in stronger form for the paradigms of different fields that must be combined in order to give a sufficiently complete understanding of the practical problem to be dealt with. The combination of several theoretical disciplines is therefore no easy process but one requiring active transformation of the conceptual frameworks involved, and active analysis of *how* the different frameworks relate to each other. This is the task of applied science. Often, applied science may have the further task to investigate questions left open by all theories; in the terms of the engineers' saying quoted above, when other disciplines have provided the knowledge needed to build a house which is stable and thermally and acoustically as well insulated as required, then the applied scientist has to find out how to modify their application in order to eliminate the unforeseen problem of condensed water.

A special task of applied science comes from its direct coupling to instrumentally rational action. Applied science should tell how to achieve certain effects by deliberate action (cf. the discussion on p. 325 of Humean causality as a formalization of

17. However, Piaget's epistemology, though chronologically secondary, shows the relation between the two types of knowledge, as does Chomsky's equally secondary transformational grammar in its relation to research in computer translation, cf. p. 196, n. 1.

18. A "real-life" example is the statement that anthropological investigation of societies that base their agriculture on artificial irrigation may well contribute to improvement of the social efficiency of irrigation systems—but only on the condition "that some anthropologists, some of the time, take their problems not from theories of social organization and social evolution but from the concerns of the bulk of mankind—problems of food production, productivity, income distribution, and employment—and work backwards from there" (R. Wade, commentary in Hunt & Hunt 1976: 405). Basic science can only function as applied science if it borrows the characteristics of applied science.

strategic action), and it should thus single out those factors which can be influenced by a human agent (the same as, or acting on behalf of, those who want the effect to be achieved). An applied educational science which concludes that pedagogical skill is a natural gift that cannot be taught is only of interest for educational authorities if it also tells how *they* may find the pedagogically gifted candidates; a science which asserts the skill to be teachable is only of interest if it also tells *how* to teach it.

An applied science, it should be clear, is no trivial collection of results from the theoretical sciences. It is no less an active process than theoretical science. It may also to some extent run through transformations similar to the Kuhnian cycle; but it cannot move from "section-" to "perspective-orientation" as long as it remains an applied science—its problems belong, so to speak, not to the science and its scientists but to those authorities, organizations, or corporations that want to apply their knowledge.[19]

INTERDISCIPLINARITY

Integration and convergence of theoretical disciplines has long been a familiar phenomenon, and might well have been spoken of as "interdisciplinarity" as long as "disciplines" in the modern sense have been known. As a matter of fact, however, nothing in that direction occurred—maybe because it was rarely clear whether genuine integration or cross-disciplinary inspiration was involved, maybe because the traditional normative ideal was "unified science" and not the unification of a few disciplines.

What goes nowadays under the name of *interdisciplinarity* evolved (without yet carrying the label) around the engineering and similar schools and the emergence of applied sciences. In a general sense, an "engineer" is a practitioner who has been taught, and makes use of, the results of actual science in his work (with some imprecision, an engineer is thus somebody who practices an applied science; actually, there is no one-to-one correspondence between separate engineering professions and existing applied sciences). The "engineers" of the nineteenth century were mostly engineers in the received, specific sense, and they were taught as engineers were still taught at technical high schools or technical universities a few decades ago (the changes since then are modest, but they are authentic); the curriculum consisted of a number of "basic disciplines"—mathematics, physics, and chemistry,

---

19. Cf. p. 352, n. 34 and the surrounding text. An informative discussion of the implications of this ownership is Schmid 1973, where it is argued that those who want to make "science for the people" should accept that "the people" as organized in unions or other organizations really formulate the problems (instead of devising their own "people" and determining what should be its problems), and the author's own objection (1981) that existing social science is not fit for that model—whether in a popular movement or in a firm or an organization, all the sociologist can honestly do is to offer participation as a critical intellectual.

more or less adapted to the particular needs of the profession[20]—and a number of "application disciplines."

During the twentieth century, many other professions have become "engineering-like" (e.g., nursing), and others have emerged as wholly new, still "engineering-like" professions (e.g., social work). Derek Price, alongside his doubling of the number of *scientists* every fifteen years since the later seventeenth century, suggests (1963: 6f) that the number of people who *apply scientific knowledge in their daily practice* ("engineers" in this generalized sense) has doubled every ten years since 1900. Until the 1960s, they were trained in much the same way as the classical engineers. Similarly, the new applied sciences that arose (e.g., communication studies, industrial sociology) followed the pattern of "engineering science," combining a fixed set of theoretical disciplines, adjusting and correcting their perspectives until condensed water and similar problems have been minimized. In neither context was any need felt to give the system a distinctive name. Instead we may speak of this particular kind of unacknowledged interdisciplinarity as the "engineering model," characterized by finite and well-defined multidimensionality.

The name *interdisciplinarity* only surfaced (soon to become high fashion) in the 1960s, as the finitist presupposition failed. It did so, on one hand, in the training of the new "engineering" professions (and, more cautiously, in the training of engineers stricto sensu). It was no longer possible to teach young people the disciplines and the particular knowledge from these disciplines they would need in the ensuing forty years of professional activity.

Two of the strategies which were invented to circumvent the problem are irrelevant to the epistemological issue: the production of human "dispensable items," trained for a very specific activity and dismissed when that activity was superseded (customary in the lower ranks of computer operation), and "lifelong education," primarily of members of the traditional professions by means of supplementary courses. A third strategy, championed by the policy makers of the OECD, was the creation of more flexible educational institutions encompassing both traditional university subjects and the training of members of the practical professions, based on "interdisciplinarity" and involving the students in "project work."[21] Such projects

20. Adapted at least through the selection of pertinent topics. When it came (and comes) to the adaptation of perspectives, the teachers of the basic disciplines often had (and have) difficulties. From my own experience at an engineering school, I remember two colleagues (*B,* a nuclear physicist, and *H,* trained as an engineer) producing a common course in electromagnetic theory for future building engineers. After a couple of years *H* observed that "*B* eliminates one Maxwell equation per year, but the outcome does not change."

21. A number of such teaching institutions connecting active basic research and the education of practitioners were indeed created in the late sixties and the early seventies. In Great Britain, in the wake of a reform plan "Education for the Future," a large number of teacher-training

should simulate or exemplify the confrontation with practical problems whose elucidation requires the integration of an array of disciplines that cannot be specified in advance and once for all.

But the finitist assumption also broke down in the applied sciences themselves, not only in teaching. Classical engineering science had been concerned with a particular and rather well-defined section of reality, and the formation of the early applied sciences was based on the assumption that practical reality *could* be cut up in pretty well-defined slices. A field like "educational studies," however, is not at all well-defined; as insights grow (and as unforeseen condensed water turns up), new approaches are included in the field, and old ones perhaps discarded as unsuccessful, at a pace which had not been known in the earlier phases of the scientific-technological revolution.

Worse perhaps than the indefinite number of disciplines that may be involved is their character. The classical engineering sciences drew on disciplines whose mutual relation was relatively clear: the Kuhnian incommensurability between the paradigms of (say) mechanical physics and metallurgical chemistry (both involved in our bridge) only becomes serious at their most advanced levels; on lower levels, it is normally not too difficult to establish who is responsible for condensed water. Even as complex a project as "Manhattan," the project that created the first atomic bombs and which involved in total some 250,000 collaborators, followed that model: mathematicians, physicists, and chemists made the research and the fundamental design, military people took care of secrecy, and large industrial corporations built and ran the factories.

The mature phase of the scientific-technological revolution, however, asked for scientific answers in realms where disciplines with much less well-defined perspectives were involved.[22] The many OECD-sponsored policies, *if* they were to build on

---

colleges were upgraded as "polytechnics" (now reclassified as "universities," but the substance remains the same). In West Germany, a number of such institutions were erected anew as *Gesamthochschulen,* while others of the same character were given the name of universities (e.g., Bremen and Bielefeld). Outside the OECD, but with similar aims, the GDR had its *Dritte Hochschulreform.* In Denmark, the new universities in Roskilde (1972) and Aalborg (1974) represented the new idea.

Not all institutions were equally organic in their interdisciplinarity. In many cases, the slogan covered realities not wholly different from those of the traditional engineering schools.

22. It is therefore somewhat paradoxical that participants in a research project on the *Living conditions in the industrial world* directed by Habermas formulated the thesis in 1973 that science was entering a "post-paradigmatic" phase of "finalization," where fundamental sciences could be oriented toward any practical problem where they were needed (Böhme et al 1973; cf. Schäfer (ed.) 1983 and Pfetsch 1979). The examples which were set forth in the argument were precisely basic sciences like classical mechanics whose perspective was well understood (because active research had stopped), and applied sciences like agricultural science which drew on such basic sciences.

scientific knowledge, would all involve economics, legal studies, organizational theory, and sociology, together with the sciences involved in their particular objective.[23]

"Global" problems of population growth, resource conservation, climate, and ecological balance also became urgent during the 1960s, and since neither traditional techniques nor common sense and *laissez-faire* had much to promise, scientific insight seemed to be necessary if anything was to be done about them. Even here, however, the mutual relations between the perspectives of the relevant disciplines were not clear; how much had to be presupposed, for example, about social, sexual and nutritional habits, about ploughing techniques in their interaction with the quality of the soil, and so on, if Ester Boserup's optimistic research results about the tolerability of population growth in Java were to be transferred to other contexts? In these applied social and human sciences[24] and in the scientific approach to the global problems, interdisciplinarity thus turned out to be a problem which had to have a name if it was to be discussed; very soon, the name of the problem got the status of a slogan which in itself was believed to procure the solution.

INTERDISCIPLINARITY IN BASIC RESEARCH

Very soon, the concept also spread to "non-applied science," and it may be here rather than in the genuine applied sciences that it really gained its spurs as a slogan. One current leading in this direction was that kind of "science for the people" which, in polemical formulation, tended to devise its own "people" and determine

---

23. The communication difficulties arising in these situations can be illustrated by a dialogue which took place in 1960 at one of the OECD-sponsored conferences which prepared the "new math" reform in mathematics education:

Hans Freudenthal (mathematician, but primarily a main authority on mathematics education): We could teach anything, drive the children in any direction. But there exist other things at school. We must see the whole together.

Jean Dieudonné (mathematician, and the leading figure in the formalist "Bourbaki" transformation of mid-twentieth-century mathematics): Non, nous parlons de mathématiques. Le reste je m'en fous. ("No, we are talking about mathematics. Fuck the rest!").

(Grubb & Kristensen, eds., 1960: 12f)

No wonder that the OECD was forced to discover the problem of interdisciplinarity.

24. Speaking of educational studies (etc.) as "applied social and human science" is in itself an illustration of the problem. Many of the policies erred because of a belief that everything could be planned from organizational theory, economics, and sociology alone. But any strategy to regulate our behavior (and the policies *are* such strategies) must take into account *both* our social existence and our existence as cultural beings, producing and reacting on meaningful communication in a historical context. On the level of applied science, the distinction between human and social science is absurd.

what should be its problems (cf. n. 19). In less polemical words: that widespread current in the radical scientific environment of the late sixties and the seventies that tried to develop knowledge of direct relevance for actual societal and political issues: peace research, women's studies, black studies, critical studies of education,[25] critical science policy studies, and others.

The *Frascati Manual* has a term for this kind of fundamental research (*Measurement of Scientific and Technical Activities,* 19f):

> ". . . in some instances basic research may be primarily oriented or directed toward some broad fields of general interest. Such research is sometimes called "oriented basic research."

Not all research belonging to this class is or was of course politically radical. Fields like "Soviet studies" and "China studies," whose aim it was to know *anything of relevance* for understanding the Soviet Union or China, were often sophisticated espionage in disguise,[26] and some of the scholars and institutions making peace and conflict research were more interested in how the United States might establish their own peace in Vietnam than in avoiding wars, or they were paid by institutions with such interests. Research laboratories financed by particular industries (Philips, Carlsberg, etc.) tend to ask for research connected to possible products ("research dealing with beer is in favor," as a colleague summed up what he had learned when examining the purportedly "basic" research of the Carlsberg Laboratory[27]).

There are strong institutional and financial reasons that China studies and beer studies retain their "orientation"; in this respect they show themselves to be more closely related to the applied sciences than to basic research. Fields defined by the personal engagement of the workers, on their part, have turned out to exhibit much of the dynamics of scientific disciplines in the preparadigmatic phase; if the vocabulary had been at hand, it would indeed have been possible for the economists of the eighteenth century (physiocrats as well as Smith) to speak about their science as "interdisciplinary studies of the problems of wealth," involving social statistics, political and social philosophy, and history. In some cases, for instance in women's studies, it has also been possible to observe something like a Kuhnian circle—not only in the sense that certain books (at one moment Germaine Greer's *Female*

---

25. Not only my knowledge of the debates surrounding the new math reform but also my familiarity with (and my particular interpretation of) Piagetian genetic epistemology goes back to engagement in critical studies of mathematics education.

26. The Danish Institute of Eastern European Studies, which was no cover for such intelligence work, had great difficulties in overcoming the suspicion of East European authorities (Andreas Jørgensen, personal communication).

27. "Det må gerne handle om øl" (Uffe Sæbye, personal communication). That was in 1977. Since then, the official aim of the laboratory has been redefined as applied research.

Eunuch) have acquired a paradigmatic status, only to be replaced by another exemplar after a while, but also through assimilation and accommodation. Thus, at a certain moment approaches which had been used in women's studies turned out to be relevant in various kinds of minority studies; thereby features of the female situation turned out to be specific instances of something more general, and many workers in women's studies began speaking about women as "a minority" regardless of actual numbers. In a commonsensical statistical interpretation, this is evidently absurd. Epistemologically, it is not: from that perspective it conveys the point that deeper work had shown the fundamental structure of the social minority situation not to depend on mere statistics. As it happened when Newton took the commonsense term *force* and gave it a specific interpretation (at odds with daily usage) and when Freud did the same to the "sexual drive," the concept of a 'minority' was reinterpreted so as to reveal deeper structures of (social) reality—structures that had only come to the fore through the dynamics of a theoretical discipline.

In the end, the interdisciplinary interests of the 1970s have resulted in that kind of processes which were spoken of above (p. 363) as "integration or convergence." Instead of being a universal solution, interdisciplinarity in the theoretical sciences (whether spoken of as such or not) has turned out to be one moment in the global dynamics of scientific knowledge, a complement of the Kuhnian cycle and no alternative—a mediation that takes care that the deeper knowledge produced in the development of theoretical science is never totally cut off from general practice, and that its production is always part of a contradictory process.[28] Dialectic, a fundamental feature of the individual acquisition of knowledge and in the development of a discipline, is also to be found at this level.

28. As formulated in a discussion about so-called theoretical archaeology, "growth in archaeological theory is usually obtained by optimally foraging the theoretical supermarkets of the human, biological, and physical sciences. It is worth emphasizing that there is nothing wrong with borrowing perspectives and approaches. However, there is everything wrong if such borrowing does not lead to the development of these theories, perspectives, or concepts—if only to adapt them to the analysis of *archaeological* phenomena. The fact remains that such development rarely takes place" (Murray 1995: 290).

CHAPTER 22

Art and Cognition

Further investigations of the sociology of the scientific-technological revolution might be fruitfully contrasted with the deliberations of chapter 20 concerning the role of norms in the regulation of the scientific process. It is clear, for instance, that "big science"—the activity of large, hierarchically organized research laboratories and organizations—leaves little space for individual, value-based decisions on the part of most participants. Decisions of importance are taken by the managers of the projects, and their primary loyalty is not necessarily directed toward the scientific value system. It is also clear that scientists who depend critically for the funding of their research (and ultimately for their living) on research councils or similar bodies may tend to let their research interests be determined not by the "prescriptions" of the paradigm as to what is important but from what they suppose the granting authority will appreciate. Quite often this authority can be safely supposed to favor some kind of "social" utility—relevance for export industry, not too critical understanding of social and cultural change. Scientists may therefore be driven toward presenting a more "finalized" picture of their scientific insights than warranted (cf. p. 369, n. 22), and work accordingly, that is, on levels where theoretical development has stopped.[1]

1. As *Danmarks Grundforskningsfond* (Danish Fund for Basic Research) distributed its first 800,000,000 DKr (roughly $120,000,000) to twenty-three research projects, critics pointed out that most projects were applied rather than basic; the chairman found no better defence than a claim that all "contained important elements of basic research," which seems to imply that all were at most *oriented basic research* (*Weekendavisen,* 21.5.1993 p. 5). The selection was also strongly biased toward "mainstream research." This might in itself seem a justifiable choice if we believe Kuhn's arguments in favor of normal science; but since oriented basic research tends either to be similar to preparadigmatic science (where "normal science" does not exist) or to be finalized, the cocktail is dubious.

Instead of pursuing such issues, however, we shall let these suggestions suffice and return to philosophical questions, examining what the epistemology developed so far has to tell us in relation to aesthetic theory; in other words, we shall address the relation between *art* and *cognition,* which is one of the central questions of aesthetics, though only one among several.

Raising this question may look like barking up the wrong tree, but the conclusion will lead us back to our main topic—it is indeed absurd (in Lenin's formulation) to deny the role of fantasy in even the most exact science.

KNOWING ABOUT ART

Let us imagine that we turn on the radio and hear the beginning of *Die schöne Magelone.* When encountering this or any other piece of art, we *know that it is there.* In the actual case, we perceive the sounds, we notice that they constitute music, we distinguish a piano and a voice; we may discover that the words are German, follow the words, perhaps we even recognize the work or at least the composer.

If instead the music is an ethnographic recording from Burundi in Central Africa, we may have greater difficulty in bringing the work from perception to classification. We may be unable to identify the instruments, and we may feel puzzled by its complex rhythm. But we certainly recognize that *it is there.*

In this sense, the problem of art and cognition is relatively trivial. We may also take Saussure's *Cours de linguistique générale* in our hands and notice that it is a book, that it is written in French, and so on. This is *not* what we mean by the cognitive dimension of a piece of scientific theory. What epistemology investigates is the relation between *the theory set forth in the book* and *the purported object of this theory* (language, in the actual case).

However, that a piece of art, if we are to understand it as such, has to "be there," is not quite as trivial as it would seem. This follows if we ponder the relation between two traditional conceptual pairs.

There is widespread agreement that the concepts of 'beauty' and 'art' are closely linked. Some define one from the other, others do not go so far but claim that "beauty" in some sense (at least through conspicuous, intentional and provocative absence, which we may characterize as a "negative aesthetic") distinguishes the work of art.[2] The explanation of 'beauty' with reference to (sensual and/or intellectual) 'pleasure' is also conventional. It is clear, however, that the pleasure of senses which have not been integrated in the "unified space" (gustation and olfaction, and the senses of pain, heat, and cold—cf. p. 342, n. 12) is never referred to as "beautiful" in what we feel to be the genuine sense. Sensations that cannot be apprehended as *sensation of something* cannot be "beautiful" however pleasant they are. But art, if

2. Useful general surveys are Dieckmann 1968 and Beardsley 1968.

associated with the category of beauty, must then be something which can be apprehended by the senses that give access to unified space.

However, since the beauty of a poem, whether apprehended through reading or through the ear, does not depend on its actual location, "being there" cannot itself be at stake. But then a necessary condition for something to be art must instead be that we grasp it by *that kind of intellect* which makes use of the senses of unified space, that is, that kind of intellect which sets itself apart from what it perceives, from the things that "are there" physically or conceptually.[3]

KNOWING IN ART

Though not fully trivial, this remains so far a modest conclusion, and we may still ask whether the work of art stands in a relation to something else which is similar to the way Saussure's book (theory) relates to language (reality). May we claim that a work of art encompasses or transmits knowledge about something?

At times the answer is an obvious *yes.* Many works have a clearly descriptive dimension (for which reason all aesthetic theory from Aristotle until the eighteenth century spoke of *mimesis* or "imitation" as an essential aspect of art). For instance Ivan Malinowski's poem "Kritik af kulden" ("Critique of frost", 1980: 5):[4]

Tidligt i marts vender vinteren tilbage	In early March winter returns
Og havens nybeskårne æbletræer	And the newly pruned apple trees
Svæver som pelsklædte spøgelser	Float like furred ghosts
I streng frost og fuldmåne. Herinde	In austere frost and full moon. In here
Blomstrer en gren.	A twig stands in bloom.

As it stands, this seems to be a naked description of a situation. But this situation is *of no interest in itself.* This *content* cannot be the reason that the poem is printed and sold (nor that Malinowski would spend his all too short life on writing poems).

In certain cases, analysis along these lines would even lead us to characterize the work as *a lie;* no wonder that the era which considered *mimesis* a central characteristic of art repeated time and again that Homer was the greatest of liars.[5] An illustration

---

3. It could be added that a work of art, *qua* work, that is, something which has been *produced* by somebody, must necessarily *be there.* But what has been produced and is indubitably somewhere may none the less be perceived without being grasped by the senses which locate in a *there*—as illustrated by the produce of the perfumer.

4. My translation, as usually (with apologies to Malinowski, and to all the poets whom I disfigure in the following).

5. Thus Aristotle, in *De poetica* 1460^{a}19f (trans. Bywater 1924): "Homer, more than any other has taught the rest of us the art of framing lies in the right way. I mean the use of paralogism."

That the mimesis of the poet does not have to depict an actual (or historical) state of affairs is what (etymologically) makes him *a poet,* somebody who *produces* something; we

of this point is provided by Cecil Bødker's poem "Generationer" ("Generations", 1964: 103)

Faderen stod på den bare jord og gjorde et hus med egne hænder.	The father stood on naked ground making a house with his own hands.
Sønnen steg op på faderens skuldre og satte nye etager på med andres hænder.	The son climbed his father's shoulders adding new stores with the hands of others.
Sønnesønnen lå på ryggen på tagterrassen og tog solbad.	The grandson lay on his back on the roof terrace sunbathing.

Evidently, the situation described has never existed. The poem, none the less, seems utterly meaningful—but this must be in an unfamiliar sense ("metaphorical," we would normally say, believing that a term solves the problem; it doesn't really before we have made clear(er) its relations to other terms and concepts).

At times, the content of the work is not a description but a message, a (moral or similar) *opinion* about the state of the world, directly expressed or implicit. Let us consider Otto Gelsted's "Døden" ("Death", 1957: 78):

Aldrig mere skal jeg se et træ —hvile i det svale bøgelæ.	Nevermore shall I see a tree —repose in the beeches' cool shelter.
Forårsskyer går i himlens blå, aldrig mere skal jeg se derpå.	Spring clouds drift in the blue of heaven, nevermore shall I watch them.
Timen kommer, som jeg ikke ved, hvor i glemsel alting synker ned.	The hour arrives that I do not know, where everything slides into oblivion.
Dækket til af jord som nu af nat ligger jeg af lys og liv forladt.	Covered by soil as now by night I lie, deserted by light and life.
Glemt er hver en drøm og hver en sang, intet er jeg som jeg var engang.	Forgotten every dream and every song, nothing I am which once I was.
Intet er den jord jeg dækkes af, ingen hvile er i ingen grav.	Nothing the soil that covers me, no rest is found in no grave.

---

might translate him literally into "a maker." As the term came to designate the maker of verse, it acquired connotations not too far from our idea of "creativity." However great the changes in the character, function, and understanding of art during the last 2500 years, it is hence also possible to point to continuities that make it meaningful to speak as generally about the artistic sphere as done in these pages.

Vilde, røde hjerte, alt du led,	Fierce red heart, all you suffered
intet er det i al evighed.	is nothing for ever and ever.

The poem may be read as a commentary to a famous Epicurean maxim (ed., French trans. Solovine 1965:139): "Death is nothing to us: since that which is dissolved is deprived of the ability to feel, and that which is deprived of the ability to feel is nothing for us." Since Gelsted was well versed in Greek literature and familiar with ancient philosophy, the poem probably *is* a commentary, and an objection. A work of art may thus also speak at the same level as a philosophical argument. This allows us to formulate with more precision the question which also follows from the observations that Malinowski's "facts" are uninteresting and Bødker's wrong: why is Gelsted's objection expressed in a poem, and not in another philosophical statement that "what I cannot accept about death is that my sufferings shall be forgotten"? Why will Malinowski tell in verse instead of prose that defeat is never complete and definitive? Why should Bødker express her critique of liberalist ideology as poetry? This question will be pivotal for the argument below.

At times the obvious answer to the question whether a work of art transmits knowledge about something is *no*. A generally accepted instance is Johann Sebastian Bach's *Musical Offering*, where neither description of things (not even in a metaphorical sense, whatever that is) nor moral or similar opinion is to be found.[6] Antonio Vivaldi's concert *The Spring* from *The four seasons*, however, is no less adequate in spite of its title and program. Only because we *know* that it should correspond to typical situations belonging to a particular season are we able to recognize them: they do not correspond to them in accordance with any code of more general validity—so far from it, indeed, that Vivaldi could use precisely the same music in his opera *Montezuma* for the scene where the victorious Spaniards celebrate their triumph over the Aztecs. Words, of course, are also different from what they describe (cf. the screwdriver and the screw of p. 270); but in this case correspondence follows from a more general code.

What makes Vivaldi differ from Bach is thus not presence versus absence of descriptive contents. It is rather that Vivaldi's style allows us to organize what we perceive with less mental effort, and that we are less disquieted by it. The *Musical Offering* forces us to concentrate. Anton Webern's reinstrumentation of its "Ricercare"

---

6. Except, of course, in the sense that a work of art which follows a certain stylistic canon "convincingly," can be taken as an indirect argument in favor of this canon, and hence also—inasmuch as the canon reflects norms and attitudes belonging in the domain of general culture—to express a view with general implications. In this sense, Bach's *Offering* to Friedrich II of Prussia can be seen as a justification of courtly formalism.

But, as when we spoke of the "metaphorical description": this statement opens up a problem and does not settle it. Namely, how does a stylistic canon reflect norms and attitudes? So far we may only conclude that the "obvious answer" is not necessarily the final word.

movement asks even more from us. Vivaldi's concert does not *force* us to concentrate but has sufficient content and density (again: whatever these metaphors may cover) to make attention and repetition rewarding.

FRESH EYES

The issue of concentration and attention brings us to one of the central twentieth-century views of art in relation to the problem of cognition: the role of art is to bring us beyond cognitive routine, to make the familiar unfamiliar in order to let us see it with fresh eyes. A radical yet quite plain example taken from the world of music is John Cage's *Credo in US,* which starts by flattering the listener's conventional musical understanding, quoting the beginning of Dvořák's Ninth Symphony *From the New World,* and then suddenly jumps to something which in this conventional understanding is totally cacophonous (and continues to jump back and forth between the two). But the point of view exists in many variants.

Very important among these is Brecht's theory of *Verfremdung* (to be translated "estrangement," "making unfamiliar," and *not* "alienation," which translates *Entfremdung*). When Brecht undertakes to shatter the illusions and the identification produced by naturalist stage play, his intention is that we shall not be allowed to indulge in trivially sentimental pity with the poor Mutter Courage as she loses all her children in the Thirty Years' War.[7] He wants us to recognize her role as co-responsible for what happens to herself as well as to her children—we should judge and hence learn and not only empathize. Similarly, when Peter Weiss lets the contradictory forces of the French Revolution be embodied by lunatics,[8] he prevents us from identifying with one of the parties and forces us to accept the dilemma—and having de Sade embody moral disengagement he prevents us from dodging the dilemma by taking *this* facile position.

Another exponent of the position that the role of art is to liquidate glib routine is *modernism.* Whereas Brecht and Weiss want to make us see *a state of affairs* of general importance with fresh eyes, it is asserted by much main-stream modernism (not by all! and the best modernism is rarely mainstream![9]), when it has to state a theoretical position, and in particular when it is explained by its academic advocates, that the aim of art is to force upon us the fresh eyesight abstractly, without

---

7. *Mutter Courage und ihre Kinder* (Gesammelte Werke, vol. 4).

8. Peter Weiss (1964), *Die Verfolgung und Ermordung Jean Paul Marats, dargestellt durch die Schauspielgruppe des Hospizes zu Charenton unter Anleitung des Herrn de Sade.*

9. One strong exception, borrowed from Paul Celan's *Schneepart,* which I shall abstain from translating:

> Ich höre, die Axt hat geblüht,
> ich höre, der Ort ist nicht nennbar,

any engagement in morally important real-life issues which might lure us into believing that this real life is what the work of art is "about." We might say that advocates of this view see Bach's *Musical Offering* as a model for art in general.

A theoretical formulation of the view was provided by the "Russian formalists" (c. 1915–1930), who influenced both Brecht and Sergey Eisenstein's montage technique. René Wellek (1968: 173b) sums up their views as follows:

> They had grown up in a revolutionary atmosphere which radically rejected the past, even in the arts. Their allies were the futurist poets. In contemporary Marxist art criticism art had lost all autonomy and was reduced to a passive reflection of social and economic change. The Formalists rejected this reduction of literature. But they could accept the Hegelian view of evolution: its basic principle of an immanent dialectical alteration of old into new and back again. They interpreted this for literature largely as a wearing-out or "automatization" of poetic conventions and then the "actualization" of such conventions by a new school using radically new and opposite procedures. Novelty became the only criterion of value. . . . .
>
> Jan Mukařovský (born in 1891), a follower of the Russian formalists in Czechoslovakia who developed their theories more coherently . . . formulated the theory very clearly: "A work of art will appear as positive value when it regroups the structure of the preceding period, it will appear as a negative value if it takes over the structure without changing it. " . . . In literary history there is only one criterion of interest: the degree of novelty.

Two things may be added. Firstly, that the view of the formalists emphasizes *the intellect* and the role of sober-minded reflection. Secondly, that it gives an explanation why artistic styles are worn out and have to be replaced when they themselves have become routine: if it is the style itself and not an irreconcilably contradictory tension between form and contents in the particular work (Bach!) that is supposed to crush standard expectations, a work cannot fulfil this role once its style has itself developed into a standard expectation.

Another example from the world of music may illustrate the point (and, at the same time, show its shortcomings). In the 1950s and early 1960s, Karlheinz Stockhausen and Pierre Boulez were recognized on a par as leading members of the Darmstadt School of serial music. As long as this extrapolation and automatization

---

ich höre, das Brot, das ihn ansieht,
heilt dem Erhängten,
das Brot, das ihm die Frau buk,

ich höre, sie nennen das Leben
die einzige Zuflucht.

(Celan 1975: II,342).

of Schönberg's dodecaphonic principles was unfamiliar, one (I!) would indeed find that the compositions of the two were equally forceful as sources for "fresh hearing." As time went on and the "conventions" became familiar, Boulez's music stopped triggering intense attention, and it came to be vaguely boring; Stockhausen conserved his interest, though now, one might say, as *music* and not as an enhancement of acoustical awareness.[10]

The formalist theory may also be applied to kinds of art (and quasi-art) which do not fulfil its requirements: trivial art, but perhaps in particular subgenres like pornography, horror and violence (in the following I shall speak in a generalized sense of these and their kin as "pornographic" genres). They draw on *what is already there* in the receiver's mind, preferably on strong dreams, drives, and prejudices, in ways which prevent rational deliberation and even the mere establishment of qualitatively new cognitive relations. The customary in customary amalgamation, in particularly that which is already so strongly felt that it is easily drawn upon and relatively resistent to reason and to integration in higher synthesis, is enhanced and stabilized.[11]

FORM VERSUS CONTENTS

In order to better understand the implications of the formalist view, we may return to chapter 16—more precisely to Piaget's concept of 'schemes' and to the dialectical relation between assimilation and accommodation.

The point of this dialectic was that actual knowledge "about" something (the *contents* of our cognition) is always organized in a specific *form,* namely the scheme and that cognitive structure in which the scheme participates. Formalism carries its

---

10. These observations and impressions are mine and of course are subjective; but they seem not to be quite private, since Boulez concentrated on a conductor's career while Stockhausen has remained a productive composer. The young Danish serial composer Thomas Koppel, to whom the serial technique was so much of an automatic and calculated technique that he made his compositions while listening to rock music, quit serial music altogether and became a rock musician.

11. Drawing preferentially on *what is already there* as strong dreams, drives and phobias, often to the exclusion of rational deliberation, also characterizes expressionist currents. But expressionist art (at its best) aims at creating new cognitive relations (another version of the "fresh eyes"), and therefore draws less on pre-established prejudice and on the customary and the customary blend than the pornographic genres.

But the difference may be subtle. The leap is short from certain kinds of expressionism to very efficient propaganda (another "pornographic" genre)—for instance from the use of the rats in Murnau's film *Nosferatu* from 1922 to certain scenes in the Nazi movie *The Eternal Jew* (already during World War I, Murnau had indeed participated in the production of propaganda films).

name because it sees the *form* of cognition as the aim of art,[12] not the creation of supplementary contents (who, in fact, is interested in knowing about the temperature in Malinowski's garden some forgotten day in March?).

Yet even if the Piagetian framework makes the central principle of formalism more clear, it also highlights its failings and problematic features.

A form, firstly, can no more be separated from content than content from form. You cannot, as Jean-Luc Godard parodically makes Ferdinand propose in the movie *Pierrot le Fou,* write a novel on "la vie simplement," "nothing but life." Novels telling about *life* must necessarily deal with "la vie des hommes," "the life of people," that is, it must necessarily be that novel which Ferdinand refuses to write. Piagetian schemes, in the same way, are only built up as generalizations of actions or interiorized actions, that is, as the form of something. If poetry is not allowed to deal with "morally important real-life issues," it will have to treat of *unimportant* issues. Or, as stated by Wassily Kandinsky, an innovator of artistic form if anybody ever was, in a strong attack on marketable *art for art's sake:* "Form without content is not a hand but an empty glove filled with air. An artist loves form passionately just as he loves his tools or the smell of turpentine, because they are powerful means in the service of content."[13]

Specifically, formalism possesses no instrument allowing us to understand the function of Gelsted's and Bødker's poems. These poems, indeed, do not force upon us a *new way* in which to see the use of words and phrases—their artistic form is quite conventional. Instead they convey a new understanding of *the world* (the implications of death, the reality behind the myth of "free initiative"). Formalism is unable to explain why these insights are better formulated as (traditionalist) poems than in thesis form.

More generally, formalism can be claimed to see innovation in *artistic* and in *cognitive* form as identical. Even though this may be warranted in many cases, it is no universal truth. In fact, becoming familiar with the rapidly changing artistic forms of the twentieth century has largely become an assimilative process: "Oh, music may also sound like this" / "Indeed, this is also an impressive painting." Like any assimilation, this involves accommodation, but mainly of our concepts of musical or visual beauty. Our interest in music, however, is not explainable through the observation that music changes our understanding of music.

---

12. Of course, one need not have read Piaget to use these terms, which do not even correspond to Piaget's own way to express himself.

One may also notice that the term *form* is used in aesthetics in various senses, not all of which are relevant for a discussion of "formalism." A useful discussion is Tatarkiewicz 1973, which however tends to distinguish mechanically (between "Form [concept] A," "Form B," . . . , "Form G") rather than pursuing the family likeness between the different meanings, and which also includes notions which *might* have been called "form" but which are actually labelled differently.

13. "L'arte d'oggi è più viva che mai," in Kandinsky 1976: II, 190; I borrow Tatarkiewicz's translation (1973: I, 221).

How are we then to understand the interest of art which does not live up to the formalist requirements? And how are we to understand the *actual* effect of trivial literature, pornography, and their like? We need an understanding of art which does not merely discard these genres but on the contrary understands the (empirically) different effects of "real art" and "trivial art."

GELSTED AND EPICUROS

The Epicurean maxim expresses a chain of purely analytical discourse. A familiar phenomenon is investigated stepwise: What does it mean to fear death? What does it mean to fear any*thing?* Since this peculiar "thing" is something which I cannot perceive, I cannot fear it.

It is easy to lose the thread of a complex analytical argument, and losing the thread deprives it of its power to convince emotionally. Epicuros's argument has therefore been summed up in an aphorism, "Why fear death? When I am here, death is absent. And when death is here, I am absent." This pulls together the main lines of the argument in a way which allows you to apprehend them in one glance. One reason that the aphorism can be grasped in this way is the doubly symmetric structure of the argument (presence/absence, I/ death), which, firstly, binds the two clauses constituting each of the last two periods together, and, secondly, joins these two periods; another reason that the aphorism is grasped and remembered is the artistic-humorous form. None the less, even the aphorism does not convince if one *is* scared of death.

Gelsted's poem is a protest against Epicuros's acceptance of death. Stated analytically, the argument might run something like this: what I cannot concede to death is that it reduces me to pure absence. I can only bear my existence if I am able to ascribe to it a meaning, if beauty is real, if suffering is real, etcetera. The fact that I shall die, that everything which in this moment is beautiful or bitter matters nothing under the perspective of Eternity, is unacceptable to me; what I fear is this fact of ultimate meaninglessness of *the present* and not an abstract future "thing" called death.

Yet instead of presenting this analytical string, which will easily be as existentially noncommittal as Epicuros's argument (although, as I discover, I actually put it into words and phrases which are more heavily loaded than a purely analytical argument should be), the poem *reminds* the readers of what *existence* is, and thus *suggests* how *absence* can be grasped. It draws upon the connotations of words and upon what readers can be supposed to know about the experience of spring under newly green beeches, upon their knowledge that "spring clouds drifting in the blue of heaven" imply sunshine, fragrance, and breeze but no storm; as a climax it draws upon the readers' own experience of suffering, and on the extra pain added to suffering if it is recognized to be meaningless.

All this is not put into any logical or analytical framework. The coherence of the poem is rhetorical and rhythmic, using the contrasts "nevermore/the beeches' cool shelter" and "spring clouds/nevermore" (a double contrast which keeps the two first stanzas together, enhancing their weight in the argument); then comes a middle part, strung together by "oblivion"/"forgotten" (*glemsel/glemt*) and leading to a final climax produced (at the rhetorical level) by a triple *nothing* (evidently, this structural analysis could be expanded). Just as the text draws upon the readers' total understanding of what it is *to live* in order to make it clear what it is *to die,* it is left to the readers to take bearing of the rhythmic and rhetorical structure in order to build up an ordered totality, an implicit synthetical argument. And it is, indeed, an ordered totality that is built up.

The words of the text do not serve in a sharply defined sense, as they would (ideally) do in a technical manual, where the role of the reader is reduced to understanding (or, perhaps, not understanding) the terms correctly. The words carry their whole, open-ended load of connotations, and should do so. By means of its rhythmic and rhetorical structure, however, the text puts these bundles of connotations into mutual relations which readers would not automatically produce on their own.[14] Assimilating these relations to their own understanding, readers will accomplish *an accommodation,* whereas the technical manual only gives information about familiar entities, that is, assimilation relatively free of accommodation. The manual, if well written, can be *used.* A work of art cannot: "using" it involves you in *co*-producing.

This is a general characteristic of the artistic product, and it explains why a work of art is not exhausted by a single reading (or whatever kind of reception is involved).[15] On one hand, assimilation normally presupposes repeated experience. On the other, one and the same work will be seen differently from different readers'

---

14. The minimal element of poetical language that achieves this effect is the *oxymoron,* consisting of two apparently contradictory terms whose juxtaposition forces the reader to reinterpret both. The standard example of dictionaries contains two of them in a single line, "faith unfaithful kept him falsely true" (Tennyson, "Lancelot and Elaine," verse 872)—even without knowledge of the actual poetical context enough to arouse sensitivity to the many dissimilar natures of faith and deceit.

15. Since even the forms of artistic expression carry a load of implied meanings and connotations derived from their use, it also explains why it is impossible to resurrect the styles and genres of former times. As expressed by Kandinsky in *Über das Geistige in der Kunst* (1976: II, 69):

> We cannot re-create the sensibility and the internal life of the ancient Greek; therefore, even if we try, for instance, to apply the Greek principles in sculpture, we shall create only shapes which are similar to the Greek ones, but the work will forever stay without spirit. Such imitation is similar to that of a monkey. Externally, the movements of the monkey are fully similar to those of a human.The monkey sits down, takes a book to its nose, turns over the leaves, seems to be fully absorbed; yet the internal meaning of these movements is completely absent.

perspectives and by the same reader in different moments. This is a simple consequence of the open-ended and nonoverlapping ranges of connotations (not only of words but also of rhythmic and other structures) produced in different readers and in different situations. Although one interpretation of a work can often be argued to be "better" than another, that is, to make better sense of more features of the work,[16] it is not possible to translate a work into a single definitive interpretation; even the artist will not be able to do so, since even the artist's range of connotations is open-ended.

On this account, the difference between *trivial art* and what we might call *complex art* (which is not the same as *complicated* art—complication may follow from confusion, but complexity not[17]) can be seen in a new perspective. Trivial art

---

16. This *may* mean that the better interpretation is also in better agreement with the intentions of the artist. But this need not be the case. Erich Maria Remarque intended to write a patriotic novel and believed to have done so. Only when the manuscript of *Im Westen nichts Neues,* after having been rejected by some ten patriotic publishing houses, was accepted by a left-wing publisher and became an immense success when read as a pacifist manifesto, did he discover that *his* patriotism was not that of the patriots (Nils Rickelt, personal communication).

17. The issue can be illustrated by Schönberg's comparison of Bach's style with that of his predecessors and contemporaries (1989: 89f):

> What is new in Bach's art can only be understood if, on one hand, one compares it with the style of the Netherlandian school and, on the other, with the art of Händel.
>
> The secrets of the Netherlanders, to which the uninitiated had absolutely no access, relied upon a complete knowledge of the possible contrapunctual relations between the seven tones of the diatonic scale. This allowed the initiates to produce combinations allowing many kinds of vertical and horizontal transposition and other similar transformations. But the remaining five tones were not included in these rules, and if they occurred at all then outside the contrapunctual combinations and as occasional variation.
>
> In contrast Bach, knowing more secrets than possessed by the Netherlanders, extended the rules in such measure that they included all twelve tones of the chromatic scale. . . . .
>
> If one has observed that the contrapunctual flexibility of Bach's themes in all probability depended on his instinctive thinking in multiple counterpoint giving room for additional voices, and if one compares his counterpoint with Händel's, then the counterpoint of the latter appears impoverished and simple, and his side voices are really mediocre.
>
> Even in other respects is Bach's art superior to Händel's. As a composer for the theatre, Händel was always able to start with a characteristic and often excellent theme. But then follow, apart from the repetition of the theme, a decline, . . . , empty and trivial, *étude*-like broken accords. In contrast, even Bach's connecting and subordinate passages are always full of character, invention, fantasy, and expression.

does not put things and concepts into unexpected relations and therefore does not produce accommodation. Yet what is utterly familiar for one person may be an unexpected and inciting discovery seen in another's perspective. Accommodation of the totally unfamiliar, furthermore, is not likely to occur (cf. the leaden ball of p. 237). Whether a particular work is "trivial art" or not is hence not merely to be determined from the work itself; it also depends on reception and on the capacities and preceding experience of the receiving mind.

On the same account, we may understand the frequent weakness of didactic art. "Leftist detective stories," "politically correct fairy tales," and "morally edifying versions" will all too often be reduced to one level of meaning. In order to make sure that the reader gets the "right" associations, the challenge of open-ended connotations is reduced to a minimum. Yet guiding the mind of the reader so that it performs no "wrong" movements, thus barring coproduction, prevents it from performing that autonomous activity which is a presupposition for accommodation. Preventing the occurrence of "wrong" new thoughts is tantamount to preventing the occurrence of new understanding.[18]

ART AS THOUGHT EXPERIMENTS

For the next step in our argument, we may return to the difference of opinion between Kuhn and Lakatos. For the latter, the hard core of a research program was precisely definable. For Kuhn, the paradigm was constituted in part by tacit knowledge, skills, and context-sensitive concepts. It is not possible to *define* exhaustively

---

18. Evidently, this is not the only reason that art that is morally or politically engaged is claimed to be banal or of bad quality. Often it simply means that the critic does not share and does not appreciate the values which are expressed.

A classical example of politically engaged art which is *not* banal art is provided by Eisenstein's *Potemkin.* There is no doubt about the political message. This, however, does not make the movie insipid. The political message, indeed, is not presented in homiletic one-dimensionality, but through highly complex use of the pictorial medium and the temporal organization. As also demonstrated by certain Baroque psalms, by the poetical introductory chapters to Augustine's *Confessions,* and even by certain homages from Racine's hand to Louis XIV (however hypocritical they are in content), the presence of a clear political or theological lesson does not *by necessity* entail banality.

As an (unusually transparent) example of a critic disguising political disagreement as art criticism one may cite a commentary to Shostakovitch's *Leningrad Symphony* written by Clive Bennett for a record edition (Decca D213D 2). If, as the composer's notes tell, the symphony is inspired by the war during which it was written, and if it is meant as a requiem for the victims of the Nazi atrocities, it is obviously nothing but vulgar propaganda, "a film score without a film." But "the symphony, if we believe [the] interpretation [set forth in a forged autobiography, *viz.* that the inspiration is rather Stalin's purges], becomes transformed from a propaganda work into a canvas of universal significance."

what, for example, a *text structure* is, nor to prescribe exactly a universal method for finding it. What you acquire when learning textual analysis is unsharply defined knowledge and skill in analyzing. Your experience within the field will be associatively connected, like Gelsted's "tree," "spring clouds" and "suffering."

As it was argued, this difference is what makes Kuhn's approach a more adequate description of the actual scientific process: scientific practice is to a considerable extent based upon intuitive knowledge, on knowledge organized in totality and in analogies.

This is the reason for the importance of the *thought experiment.* A prototype is the argument against the rotation of the Earth which was raised by the Paris philosopher Buridan in the fourteenth century:[19] as we know, the stars of heaven seem to circle around the Earth once every twenty-four hours. Wouldn't it be more economical if the Creator had made the Earth rotate and left the immense sphere of stars at rest? Perhaps, but hardly the case, Buridan explains. Reflect upon what happens if you shoot an arrow vertically upwards. It will fall down upon you own head. But if the Earth rotated, you would have moved a considerable distance (some 1500 feet per second) while the arrow was in the air, and it would fall to the ground far from where you are. This is evidently not the case.

According to later physics, Buridan's intuition is wrong. He is unaware of the law of inertia, according to which an arrow shot from a horizontally moving bow will receive a horizontal component of movement. This, however, is only important in so far as it shows that intuitive knowledge need not be correct. More central is the observation that scientific argumentation presupposes a certain measure of global knowledge (correct or incorrect) about the behavior of its object.

Works of art can also be regarded as thought experiments. In some cases this is obvious. If we consider a novel like George Orwell's *1984,* what makes the book influence us is the fact that it is psychologically plausible: the world which is depicted *could be* a world, with all the complexities of a real world, for all we know about social life and human beings. Evidently, "all we know" is a historical product (as was Buridan's knowledge); in the moment when our intuitive knowledge makes the world of the novel seem implausible, the novel itself will lose its force and actuality.[20]

---

19. In a discussion of a point in Aristotle's work *On the Heaven* (trans. Moody in Clagett 1959: 594–598).

20. When explaining in *De poetica* (1451^{a}36f: trans. Bywater 1924) the difference between the mimesis of the artist and that of the descriptive historian, Aristotle states that "the poet's function is to describe, not the thing that has happened, but a kind of thing that might happen, i.e. what is possible as being probable or necessary." And later (1454^{a}35ff), "whenever such-and-such a personage says or does such-and-such a thing, it shall be the necessary or probable outcome of his character; and whenever this incident follows on that, it shall be either the necessary or the probable outcome of it." Again (1460^{a}26), "a likely impossibility is always preferable to an unconvincing possibility"—*viz.* because a work of art does not function if its world appears to us as an implausible postulate.

Other works of art are only thought experiments in a transferred sense. A piece of pure music is neither plausible nor implausible with regard to our experienced daily world. Its "plausibility" depends on its inner coherence. But even a piece of pure music is a testing of possibilities and consequences within a space of plausible solutions (a form or style); the "transferred sense" thus *is* a sense.

The reason that Buridan needed his thought experiment (and that Niels Bohr and Einstein needed theirs!) is that analytical thought does not exhaust everything we know. Thought experiments allow us to gauge what we can formulate analytically against what we know tacitly (that is, to check those pieces of knowledge which we can isolate against the totality of the world as we have come to know it). Similarly, the genuine thought experiments of art allow us to gauge specifiable moral (and other) convictions, to see whether they will work acceptably in a specific situation created (fictionally) for that purpose. In this way, Shakespeare demonstrates in *Romeo and Juliet* that the morality of the family feud and honor are unacceptable. The thought experiment of the tragedy leads to better knowledge about how we should live, to superior practical knowledge. The fact that the questions explored by art (as questions about practical knowledge in general) are mostly less accessible to explicit analysis than those investigated by the sciences explains why thought experiments play a much more central role in art than in scientific discourse.[21]

"REALISM"

In mechanical interpretation, what was just said about *Romeo and Juliet* might be taken as an argument in favor of realism. Before we go on, this term has to be explained. *Realism* as I use it here is not identical with *naturalism*—it may be its opposite. The pair realism/naturalism is rather a transposition into aesthetics of the epistemological pair realism/nominalism. Nor is "realism" meant here as the antithesis of "embellishment" (as it sometimes is)—actually, the two concepts are close neighbors.[22]

---

21. As formulated provocatively by Richard Rorty (1982: xliii): "Physics is a way of trying to cope with various bits of the universe; ethics is trying to cope with other bits. Mathematics helps physics do its job; literature and the arts help ethics do its." Physics, to some extent, can replace thought experiments by mathematical argument; ethics cannot, however much utilitarianism argues as if it could.

22. The reason that these conceptual fences have to be erected is of course that the word is used in so many different senses. Taken as a synonym of naturalism, however, the term is superfluous; and taken as a token for the view that art should show the world as ugly or as cruel as it is instead of postulating a harmony which is not to be found in the real world, the term becomes epistemologically uninteresting (its relevance in political discussions of the responsibility of art and artists in a specific historical context notwithstanding).

Positively stated, realism (or *aesthetic realism* as I shall call it in the following, in order to avoid misunderstandings, and since we are dealing with art) is the view that a work of art should lead to understanding of the *essence* of things. It may do so by depicting phenomena naturalistically, but whether it does so or not, the important thing is that it should lead to insight in something more fundamental than these phenomena; in this sense, aesthetic "realism" is akin to philosophical realism—whether "strong," objectively idealist like Plato's variant, or "moderate" like Aristotle's.

An example of definitely non-naturalist realism is *Futurism.* Futurism did not use the realist label, but the label "Futurism" is itself a claim that art should tell the essence of the new world: speed, aggressiveness, and fight, breakdown of classical harmony. *Surrealism* (sometimes non-naturalist as with Max Ernst, sometimes deceitfully naturalist as with Salvador Dali), often inspired by psychoanalysis, aimed at showing the higher reality of the mind when uncensored by reason.

Within a philosophical context which itself tries to be different from classical (Platonic or Aristotelian) realism and its concepts of 'truth' and 'reality,'[23] a suggestion of aesthetic realism can still be found in Heidegger's "Ursprung des Kunstwerkes" (1977/1936). "Instating itself in work [of art], truth [which is 'the truth of being'] appears" (p. 69). Van Gogh's painting of a pair of peasant's shoes uncovers not only the shoes as things but the whole of the peasant's lived experience: the hard toil of monotonous ploughing, the hostile wind, the loneliness of the work in the fields, the anxiety for the daily bread, . . . (p. 19). After this innocuous prolegomenon, the essay involves itself in a cloud of home-brewed Greek etymologies and vague terms that may mean anything and nothing; but scattered in the mist a few explicit concepts turn up that anchor the text to a specified truth which has to reveal itself in the work if it is really a work of art: the very Platonic "truth of being" that is embodied in the true work of art coincides with the "simple and essential decisions in the destiny of a nation" (p. 35), and is concerned with victory and defeat, blessedness and curse, dominion and servility (*Sieg und Niederlage, Segen und Fluch, Herrschaft und Knechtschaft*—p. 50)—as it manifests itself (p. 49) in that deed of violence which establishes the state (*die Staatsgründende Tat*) and in the "essential sacrifice."

What I try to do here is to specify a sense which underlies some of the views on art which proclaim themselves "realist" (not least "Socialist Realism") and a number of others which use different banners.

23. Heidegger (1977/1936: 14f) indeed points out that the dichotomy "matter/form" is derived from tools created with a certain purpose; and concludes that seeing everything *that is* as the result of an imposition of *form* on *matter* presupposes that it is the outcome of an act of creation similar to the one by which men create their tools—and thus in the final instance, and however much theologians and philosophers try to deny the parallel—faith.

Later on in the same work (p. 51), however, the concept of a 'shape' (*Gestalt*) is used as a perfect stand-in for the concept which has just been rejected.

Everything in *true art* suggests that new beginning for the German nation which is deemed to be imminent (p. 66; the year is 1936).

What was said above regarding *Romeo and Juliet* also points to a "realist" interpretation, though much more open and with much less prophetic pretensions. The tragedy does not tell merely about a particular sequence of events arousing pity and fear (in agreement with Aristotle's view of the nature of the tragedy, cf. p. 42); it also tells what are the real consequences of the prevailing code of honor and the practice of family feuds—no less than van Gogh's shoes as *lived experience.* Aristotle moves on a comparable level when he asserts that "poetry is something more philosophic and of graver import than history, since its statements are of the nature rather of universals, whereas those of history are singulars" (*De poetica* 1451^{b}5–7, trans. Bywater 1924). In spite of what one might believe to be implied by the term *mimesis,* Aristotle's ideal for art is thus no naturalistic imitation of phenomena (cf. also the quotations in n. 20).

Aristotle's aesthetic realism is cautious and unpretentious in the sense that it is not coupled to a statement of what the underlying truth revealed or displayed by the work of art should be. Heidegger's "extreme aesthetic realism" does not share this restraint—nor does the "Socialist Realism" that was proclaimed as a program at the Congress of Soviet Writers in 1934,[24] and which may be the most outstanding example of a declared aesthetic realism from this century (Heidegger's program is cloaked by uncommitted metaphors and private grammar to such a degree that few readers discover that specific declarations are made).[25] It was coupled to a version of Marxism which already knew not only the past and the present essential conditions of class struggle but also its inescapable outcome;[26] art was therefore to reflect the movement of (this) history past and future and to lay bare how particular situations were explainable in terms of the general laws of history.

The way this program was coupled for a while to political power falls as much outside the scope of a discussion of the relation between art and cognition as does Heidegger's Party membership. What falls inside is the observation that the program,

24. "Socialist Realism" is thus wholly distinct from "social realism," a trend in American and Western European art going back to the same years, and which in the present terminology is rather a *naturalist* social criticism.

25. The program of Socialist Realism is certainly not the only example of political control of art from the twentieth century. Other instances of this phenomenon, however, have not been supported by a similarly elaborate aesthetic philosophy.

26. This (Stalinist) version of Marxism was thus in itself (like all the brands of Marxism which subscribe to strict economic determinism or to a closed Hegelian dialectic where "history has been but is no more") a brand of extreme philosophical realism or objective idealism—as caustically pointed out in Sartre's *Questions de méthode* (cf. p. 217).

if followed to the letter, makes the expression in artistic form superfluous.[27] If the essence of things is already known so precisely that it can be translated into prescriptions, then it can no less easily be explained as theory, and art which follows the prescriptions *becomes* theory (and ultimately ineffective as art, cf. the above observations on didactic art); what can be expressed, and what needs to be expressed as art is open-ended knowledge.

Aesthetic realism (not least this realism) is often, and justly, seen as antithetical to formalism. The reason is not that realist art does not care about form. Quite the contrary: that Madame Bovary kills herself by swallowing arsenic does not in itself tell anything about the obtuseness of provincial bourgeois society; if Flaubert's novel manages to relate this "essential truth" about bourgeois life, it is through the way it selects, orders, and conveys its material—that is, by means of *the form* of the novel. Aesthetic realisms (and not just this realism) are passionately absorbed in putting art into *the correct* form, the form which corresponds to their assumed underlying order of reality. What distinguishes realism from formalism is not the degree of absorption in the question of form; it is that formalism rejects the idea that any particular form should be correct and hence definitive, arguing instead that *every* artistic form is wrong when it has become customary or trivialized.

SYNTHETICAL UNDERSTANDING AND PRACTICAL KNOWLEDGE

The experience of artistic thought experiments and the analysis of Gelsted's poem suggested that the formalist understanding was insufficient, however much its "realist" counterpart exaggerates its own point. It remains, however, that a large class of artistic works refer to nothing outside themselves and the stylistic canon to which they relate, neither as thought experiments creating a world in agreement with our tacit knowledge and scrutinizing it, nor with connotations derived from our "lived experience" of spring clouds and suffering. If we are not satisfied by the

27. This "letter," it should be said to do justice to the better theoreticians of the movement (not least Georg Lukács), was not theirs explicitly—they had too much pragmatic sense and artistic feeling to reject everything which was valuable art, and could avoid doing so by moving imperceptibly between "realism" as here understood and "naturalism." What I do in these pages is to draw some consequences which the fathers of Socialist Realism had too much insight to draw, but which are none the less inherent in their ideas, as demonstrated in not a few of the works that came out of the program. Lukács himself (1969/1936: 78) notices their "monotony . . . . One has barely begun the reading of most of these novels before he knows everything that is going to happen: in a factory, vermin is at work; everything is chaotic, but finally the Party cell or the GPU discovers the nest of wrongdoers, and production flourishes . . . ." As far as predictability is concerned, no worse than much trivial art—but certainly not what a philosopher of art whose ideals were Tolstoy and Balzac would like to sponsor.

formalist explanation that they allow us to see with fresh eyes (and as we have seen, there are reasons not to stop at that point), what are we then to do about works that (like Bach's *Musical Offering*) are only thought experiments in a transferred sense?

Clearly they cannot gauge the validity of moral convictions (etc.), since such convictions are neither present nor represented in the work. What they do to us when we put their "plausibility" on trial—in terms of inner coherence and of the tension between form and material, stylistic canon and actual use—is rather to sharpen our ability to comprehend totalities, to perceive intuitively.[28] For instance: if we have become able to grasp (consciously or, just as well, subconsciously) the structure of Beethoven's piano sonatas, we shall have no difficulty in following the implicit prescriptions given by the rhythmic and rhetorical structure of Gelsted's poem; and we shall also have enhanced our chances to grasp a structure in complex real situations. Certainly not *the* structure: real-life situations are, no less than works of art with their indefinite range of connotations and resonances, infinitely complex, and we shall find no bottom if we dive into them; realist aesthetic theory fails on both accounts. But precisely therefore, training in grasping as much and as essentially as possible and concluding from that is crucial. In Brecht's words:[29]

> Was hilft zweifeln können dem
> der sich nicht entschließen kann!
> Falsch mag handeln
> wer sich mit zu wenigen Gründen begnügt
> aber untätig bleibt in der Gefahr
> wer zu viele braucht

Because no description of reality (no scientific description, nor certainly any other) can be transcendentally and exhaustively true (that is, identical with what it describes), translating one's apprehension of a situation into a formal system from which a conclusion can be drawn always involves a moment of *synthetical judgement* which integrates our analytical and tacit knowledge but cannot itself be argued exhaustively in analytical terms. Under which moral rule a certain act is to be counted, for instance, is itself not to be derived from rules alone (cf. above, p. 338). Alf Ross, the Danish legal philosopher, gave the strong formulation to this observation that the leap from

28. The following line of reasoning was originally inspired by Feinberg 1977. As it unfolded, it developed a certain (unforeseen) affinity with Kant's position in *Critik der Urtheilskraft*—cf. below.

29. From "Lob des Zweifels" (*Gesammelte Werke*, vol. 9, 626–628). Once more, copyright reasons restrict me to paraphrasing: Doubting is of little help unless you are able to conclude; being satisfied with too few reasons may lead you into error—but asking for too many will make you unfit in danger.

legal premises to action is *irrational.*[30] Knowing how to make this leap (and all the analogous leaps) is *practical knowledge about how to know.*

We may conclude that *if skill in logical inference and analytical thought is of any use in human life, then only if coupled to practical knowledge about how to use this skill, to a corresponding skill in synthetical or intuitive judgement.* We may then continue an argument begun on p. 255. There, the fundamental cognitive categories were suggested—like the conserved object—to allow more adequate action, and to have developed biologically for that reason. This, we now see, they could do only if they developed *together* with a faculty for synthetical or intuitive judgement.

Art, we have also seen, trains that faculty and allows its development into a genuine skill. Thereby, what looked initially as a side track returns us to our main theme. As it was argued, scientific practice is largely governed by paradigmatic knowledge, as social life is regulated by norm systems. Both presuppose for their application the wielding of intuitive and integrative knowledge. Art, by training the faculty for integrative thought, is thereby also a training of the very basis of scientific practice and social life. By being less bound than real-life moral reflection and scientific work to a specific content, it may provide a more thorough and comprehensive training than the two activities provide occasion for themselves.[31]

Paraphrasing what Gouldner says about the value-free doctrine (cf. p. 355), the free activity of art may enable us "to make *better* value judgements rather than none"—"better" in the sense of less parochial, better since based on a broader understanding of our total situation and the implications of our actions. This may be a first step in a justification (and, at the same time, a critique) of the idea that understanding

---

30. Cf. Jarvad 1993: 94. Evidently, the idea of rationality underlying the argument is what was spoken of above (p. 231) as the "absolutist ideal of the strict proof," not the open rationality of dialogue.

Elsewhere (p. 43), Jarvad sums up Ross's view in the words that "theories and analyses do not lead to decisions, do not designate certain decisions as correct and others as wrong, at most they pinpoint errors in the basis for the decision. Science does not lead to decision and action."

31. This is the point where the argument comes close to Kant's *Critik der Urtheilskraft.* That faculty of judgement, indeed, which we need in order to decide whether a particular act falls under a specific rule and, in general, to perceive *the particular* as a case of *the general* (cf. p. 338, n. 8)—a faculty which is a cornerstone of theoretical as well as moral reflection: that faculty is exerted in its purest form as judgement of art, and it is in the domain of aesthetic judgement that its particular character is exposed most clearly. This character Kant describes (A VII, B VII) technically as "want of a principle," where theoretical thought has the *concept of nature* (involving "the nature of things," regularity, causality, predictability, etc.) as its principle and moral thought the *concept of freedom,* that is, in terms that correspond to his conception of theoretical and practical reason. As argued forcefully by Garroni (1992), however, what is involved is the role of the faculty of judgement as a general guide to the intellect in its wielding of conceptual thought.

of art leads to moral improvement and that aesthetic education is the best moral education.

From a slightly different point of view, the discussion is connected to the role of metaphors in scientific thinking (of which a particular instance was discussed on p. 270, n. 19); understanding of such metaphors—which serve the *construction* of general concepts from particulars—relies on the same faculty as the understanding of the metaphors of art: no reasons of principle would prevent Cecil Bødker's sunbathing grandson from serving as the starting point for a theory about Torstein Veblen's "leisure class." Yet once again, the understanding of art may provide more abundant opportunity than theory to exert—whence train—that faculty.

Formulated in less aphoristic terms: Apart from its function in concrete intuitive reasoning and as a source for better practical knowledge, as discussed *inter alia* in connection with Gelsted's poem and the role of art as thought experiment, art (and here, abstract music no less than figurative painting or texts that tell a story) may also have a fundamental cognitive function as *training of that integrative or synthesizing competence without which the best analytical abilities are empty*—as training of practical cognition, the *faculty* for practical knowledge.[32]

This sounds like a conclusion to the question concerning art and cognition, and in a way it is. Yet it is important to remember that these functions of art as a way to practical knowledge and as training of practical cognition *per se* do not answer (or at least do not exhaust) the question why we engage *as individuals* in art, either as producers or as co-producers. The phenomenology of *beauty*—as told by two illustrious witnesses—may serve once more.

In his *Confessions* (X.xxx–xxxv, ed. trans. Trabucco 1960: 123–147), St. Augustine aligns indulgence in the pleasures of the senses with erotic concupiscence (and with interest in natural philosophy characterized as "vain curiosity"). Pleasant smells are no problem for him (they produce no beauty, we might say), but all the more is music. Even though he has improved, he still cannot ignore the musical beauty of a psalm melody sung by a beautiful voice or avoid rejoicing in it—and no better is *seeing* the sweetness of the world. Aristotle, when locating the essential beauty of the tragedy, points to its arousal of "pity and fear." Less immediately than the "pornographic" genres but no less truly, *all* impressions of beauty—and all engagement in artistic production or co-production—appear to be rooted in our affections. *How* it relates to them is a major question if we wish to understand what art is, and one might postulate that the experience of "beauty" results from some kind of unity of affection and integrative insight; but the problem does not belong within the present line of argumentation, and we shall leave the argument here.

---

32. This fits an observation made on p. 375, *viz.* that art must be grasped "by *that kind of intellect* which makes use of the senses of unified space." This is the kind of intellect which needs to be trained if analysis and synthesis shall work together.

With respect to its root in affections, of course, engagement in art only differs by *degree* and *degree of immediacy* and perhaps in the *kind of affections involved* from other kinds of human conscious activity. No such activity is undertaken without a motive (if only the pleasure of the activity itself), and motives are by definition rooted in the affective. The sphere of art may therefore be less absolutely separable from other spheres of life than presupposed for convenience in the preceding pages. Still, if large enough, differences in degree remain decisive, and art, if no absolutely separable sphere, remains a sphere of its own no less than science and morality.

ABBREVIATIONS AND BIBLIOGRAPHY

Achebe, Chinua, 1986. *Things Fall Apart.* London: Heinemann. [1]1958.

Aldridge, A. Owen, 1968. "Ancients and Moderns in the Eighteenth Century," *in* DHI I, 76–87.

Alexander, Peter (ed.), 1951. William Shakespeare, *The Complete Works.* London & Glasgow: Collins.

Alquié, Ferdinand (ed.), 1963. Descartes, *Œuvres philosophiques.* 3 tomes. Paris: Garnier, 1963–1973.

Anderson, John M., 1971. *The Grammar of Case. Towards a Localistic Theory.* Cambridge: Cambridge University Press.

Anderson, Perry, 1974. *Passages from Antiquity to Feudalism.* London: New Left Books.

Apel, Karl-Otto, et al, 1971. *Theorie-Diskussion. Hermeneutik und Ideologiekritik.* Frankfurt a.M.: Suhrkamp, 1971.

Aristotle, *Works.* Translated into English under the Editorship of W. D. Ross. 12 vols. Oxford: The Clarendon Press, 1908–1952.

Armstrong, A. H. (ed.), 1970. *The Cambridge History of Later Greek and Early Medieval Philosophy.* Cambridge: Cambridge University Press.

Austin, M. M., & P. Vidal-Naquet, 1977. *Economic and Social History of Ancient Greece: An Introduction.* London: Batsford. [1]1972 (in French).

Bacon, Francis, 1937. *Essays.* London: Oxford University Press.

Baczko, Bronislaw, 1978. *Lumières de l'utopie.* (Critique de la politique). Paris: Payot.

Baczko, Bronislaw (ed.), 1982. *Une Éducation pour la démocratie. Textes et projets de l'époque révolutionnaire.* (Les Classiques de la politique). Paris: Garnier.

Baldwin, John W., 1971. *The Scholastic Culture of the Middle Ages, 1000–1300.* (Civilization and Society. Studies in Social, Economic, and Cultural History). Lexington, Mass.: D. C. Heath.

Baldwin, Marshall W. (ed.), 1971. *Christianity through the Thirteenth Century.* (The Documentary History of Western Civilization). London: Macmillan.

Barber, William J., 1967. *A History of Economic Thought.* Harmondsworth, Middlesex: Penguin.

Barnes, Barry, 1979. "Thomas Kuhn," *in* Q. Skinner (ed.) 1985: 83–100.

Batkin, Leonid M., 1979. *Die historische Gesamtheit der italienischen Renaissance. Versuch einer Characterisierung eines Kulturtyps.* (Fundus-Bücher 66/67/68). Dresden: Verlag der Kunst. [1]1978 (in Russian).

Bautier, Robert-Henri, 1971. *The Economic Development of Medieval Europe.* London: Thames and Hudson.

Beardsley, Monroe C., 1968. "Theories of Beauty since the Mid-Nineteenth Century," *in* DHI I, 207–214.

Bec, Christian, 1967. *Les Marchands écrivains. Affaires et humanisme à Florence 1375–1434.* Thèse pour le Doctorat ès Lettres. Paris & Le Hague: Mouton.

Benac, Henri (ed.), 1962. Diderot, *Œuvres romanesques.* Paris: Garnier.

Berger, Peter L., 1973. *The Social Reality of Religion.* Harmondsworth, Middlesex. [1]1967 (U.S., as *The Sacred Canopy*).

Berlin, Isaiah (ed.), 1970. *The Age of Enlightenment. The Eighteenth Century Philosophers.* Selected, with Introduction and Interpretative Commentary. (The Great Ages of Western Philosophy, IV). New York: Books for Libraries. [1]1956.

Berlin, Isaiah, 1973. "The Counter-Enlightenment," *in* DHI II, 100–112.

Bickerton, Derek, 1983. "Creole Languages." *Scientific American* 249:1, 108–115 (European edition).

Bloch, Marc, 1965. *Feudal Society.* I. *The Growth of Ties of Dependence.* II. *Social Classes and Political Organization.* London: Routledge & Kegan Paul, 1965. [1]1940 (in French).

Bloom, Alfred H., 1979. "The Impact of Chinese Linguistic Structure on Cognitive Style." *Current Anthropology* 20, 585–586.

Bødker, Cecil, 1964. *Samlede digte.* København: Hasselbalch.

Böhme, Gernot, Wolfgang van den Daele & Wolfgang Krohn, 1973. "Die Finalisierung der Wissenschaft." *Zeitschrift für Soziologie* 2, 128–144.

Boncompagni, Baldassare, 1851. "Della vita e delle opere di Gherardo cremonese, traduttore des secolo duodecimo, e di Gherardo da Sabbionetta astronomo del secolo decimoterzo." *Atti dell'Accademia pontificia de' Nuovi Lincei* 4 (1850–51), 387–493.

Bonfantini, Mario (ed.), 1991. Niccolò Machiavelli, *La Mandragora, Belfagor, Lettere.* Nota critica di Gabriella Mezzanotte. Milano: Mondadori. [1]1954.

Bonner, S. F., 1977. *Education in Ancient Rome. From the Elder Cato to the Younger Pliny.* London: Methuen.

Borzi, Italo (intr), Giovanni Fallani, Nicola Maggi & Silvio Zennaro (commenti), 1993. Dante, *Tutte le opere.* Roma: Newton Compton.

Bos, Henk J. M., 1993. "On the Interpretation of Exactness," pp. 23–44 *in* J. Czermak (ed.), *Philosophie der Mathematik. Akten des 15. Internationalen Wittgenstein-Symposiums, Kirchberg am Wechsel, 1992,* 1. (Schriftenreihe der Wittgenstein-Gesellschaft, 20/I). Wien: Hölder-Pichler-Tempsky.

Bottomore, T. B., & Maximilien Rubel (eds), 1963. Karl Marx, *Selected Writings in Sociology and Social Philosophy.* Harmondsworth, Middlesex: Penguin. [1]1956.

Bowen, James, 1972. *A History of Western Education.* I. *The Ancient World.* II. *Civilization of Europe, Sixth to Sixteenth Century.* III. *The Modern West. Europe and the New World.* London: Methuen, 1972, 1975, 1981.

Brecht, Bertolt, 1967. *Gesammelte Werke.* 20 vols. Frankfurt a.M.: Suhrkamp.

Brentjes, Burchard, 1966. "Einige Quellen zur Geschichte der Klassenkämpfe im Alten Orient." *Klio* 46, 27–44.

Brinkmann, Richard (ed.), 1969. *Begriffsbestimmung des literarischen Realismus.* (Wege der Forschung, CCXII). Darmstadt: Wissenschaftliche Buchgesellschaft.

Broad, William, & Nicholas Wade, 1982. *Betrayers of the Truth.* London: Century Publishing.

Brooke, Christopher, 1973. *The Twelfth Century Renaissance.* (History of European Civilization Library). New York: Harcourt, Brace & World. [1]1969.

Brown, Peter, 1971. *The World of Late Antiquity.* London: Thames and Hudson.

Brown, Peter, 1992. *Power and Persuasion in Late Antiquity. Towards a Christian Empire.* Madison, Wisconsin: University of Wisconsin Press.

Buhr, Manfred (ed.), 1989. Johann Gottlieb Fichte, *Schriften zur Französischen Revolution.* Köln: Pahl-Rugenstein.

Burchell, Graham, Colin Gordon & Peter Miller (eds), 1991. *The Foucault Effect: Studies in Governmentality.* With Two Lectures by and an Interview with Michel Foucault. London: Harvester Wheatsheaf / Chicago: University of Chicago Press, 1991.

Burke, Peter, 1972. *Culture and Society in Renaissance Italy, 1420–1540.* London: B.T. Batsford.

Burkert, Walter, 1962. *Weisheit und Wissenschaft. Studien zu Pythagoras, Philolaos und Platon.* (Erlanger Beiträge zur Sprach- und Kunstwissenschaft, 10). Nürnberg: Hans Carl.

Burkert, Walter, 1972. *Lore and Science in Ancient Pythagoreanism.* Cambridge, Mass.: Harvard University Press.

Burling, Robbins, 1993. "Primate Calls, Human Language, and Nonverbal Communication." *Current Anthropology* 34, 25–37, discussion 37–53, 165–170.

Butler, H. E. (ed., trans.), 1920. Quintilian, *Institutio Oratoria.* (Loeb Classsical Library 124–127). Cambridge, Mass.: Harvard University Press / London: Heinemann, 1920–22.

Bywater, Ingram (ed., trans.), 1924. Aristotle, *De poetica,* in Aristotle, *Works,* vol. XI.

Camesasca, Ettore (ed.), 1985. Benvenuto Cellini, *Vita.* Milano: Rizzoli, 1985.

Capshew, James H., 1993. "Engineering Behavior: Project Pigeon, World War II, and the Conditioning of B. F. Skinner." *Technology and Culture* 34, 835–857.

Casini, P., 1988. "Newton's *Principia* and the Philosophers of the Enlightenment." *Notes and Records of the Royal Society of London* 42, 35–52.

Cassirer, Ernst, Paul Oskar Kristeller & John Herman Randall, Jr. (eds), 1956. *The Renaissance Philosophy of Man.* Selections in Translation. Chicago & London: University of Chicago Press.

Celan, Paul, 1975. *Gedichte.* 2 vols. Frankfurt a.M.: Suhrkamp.

Chamisso, Adalbert von. *Werke.* 4 vols. Leipzig: Max Hesse, n.d.

Channell, David F., 1982. "The Harmony of Theory and Practice: The Engineering Science of W. J. M. Rankine." *Technology and Culture* 23, 39–52.

Chapanis, Alphonse, 1968. "Engineering Psychology," pp. 81–87 *in* IESS 5. New York: Macmillan and The Free Press.

Chenu, Marie-Dominique, O. P., 1966. *La théologie au douzième siècle.* 2. édition. Paris: Vrin. [1]1957.

Chenu, Marie-Dominique, 1977. "Praxis historique et relation Église-société." *La Pensée* 192 (Avril 1977), 5–11.

Chevalier, Jacques (ed.), 1954. Pascal, *Oeuvres complètes.* (Bibliothèque de la Pléiade, 34). Paris: Gallimard.

Chomarat, Jacques, 1988. "Le 'De reciprocatione sui et suus' de Lorenzo Valla," pp. 283–292 *in* I. Rosier (ed.), *L'Héritage des grammariens latins de l'antiquité aux lumières.* Paris: Société pour l'Information grammaticale.

Chomsky, Noam, 1979. *Language and Responsibility.* Hassocks, Sussex: Harvester. [1]1977.

Cifoletti, Giovanna, 1996. "The Creation of the History of Algebra in the Sixteenth Century", pp. 121–142 *in* Catherine Goldstein, Jeremy Gray & Jim Ritter (eds), *L'Europe mathématique. Mathematical Europe.* Paris: Éditions de la Maison des sciences de l'homme, 1996.

Claessen, Henri J. M., & Peter Skalník (eds), 1978. *The Early State.* (New Babylon. Studies in the Social Sciences, 32). Hague: Mouton.

Clagett, Marshall, 1959. *The Science of Mechanics in the Middle Ages.* Madison, Wisconsin: University of Wisconsin Press.

Clagett, Marshall, 1970. "Archimedes," *in* DSB I, 213–231. New York: Scribner.

Clarke, M. L., 1971. *Higher Education in the Ancient World.* London: Routledge & Kegan Paul.

Cobban, A. B., 1975. *The Medieval Universities, their Development and Organization.* London: Methuen.

Cole, Michael, & Sylvia Scribner, 1974. *Culture and Thought. A Psychological Introduction.* New York: Wiley.

Cooter, Roger, 1984. *The Cultural Meaning of Popular Science: Phrenology and the Organization of Consent in Nineteenth-Century Britain.* (Cambridge History of Medicine). Cambridge: Cambridge University Press.

Craik, T. W., & R. J. Craik (eds), 1986. John Donne, *Selected Poetry and Prose.* London & New York: Methuen.

Crombie, Alistair C. (ed.), 1963. *Scientific Change.* London: Heinemann.

Daiches, David, Peter Jones & Jean Jones (eds), 1986. *A Hotbed of Genius: The Scottish Enlightenment, 1730–1790.* Edinburgh: Edinburgh University Press.

Darnton, Robert, 1968. *Mesmerism and the End of the Enlightenment in France.* Cambridge, Mass.: Harvard University Press.

Daston, Lorraine, 1992. "Objectivity and the Escape from Perspective." *Social Studies of Science* 22, 597–618.

de Waal, Frans, 1982. *Chimpanzee Politics. Power and Sex among Apes.* London: Jonathan Cape, 1982.

Debus, Allen G., 1978. *Man and Nature in the Renaissance.* (Cambridge History of Science Series). Cambridge: Cambridge University Press.

Denifle, Heinrich, 1885. "Die Constitutionen des Prediger-Ordens vom Jahre 1228." *Archiv für Literatur- und Kirchengeschichte des Mittelalters* 1, 165–227.

Derathé, Robert (ed.), 1973. Montesquieu, *De l'Esprit des lois.* 2 vols. Paris: Garnier.

DHI = *Dictionary of the History of Ideas.*

Diakonoff, I. M. (ed.), 1969. *Ancient Mesopotamia, Socio-Economic History.* A Collection of Studies by Soviet Scholars. Moscow: 'Nauka' Publishing House, Central Department of Oriental Literature.

Dictionary of Scientific Biography. 16 vols. New York: Scribner, 1970–1980.

Dictionary of the History of Ideas. Studies in Selected Pivotal Ideas. 5 vols. New York: Scribner, 1968–74.

Diderichsen, Paul, 1971. *Elementær dansk grammatik.* 3. udg., 5. oplag. København: Gyldendal.

Diderichsen, Paul, 1974. "The Foundation of Comparative Linguistics," *in* D. Hymes (ed.) 1974: 277–306.

Dieckmann, Herbert, 1968. "Theories of Beauty to the Mid-Nineteenth Century," *in* DHI I, 195–206.

Dilthey, Wilhelm, 1924. *Die Geistige Welt. Einleitung in die Philosophie des Lebens.* Erste Hälfte. *Abhandlungen zur Grundlegung der Geisteswissenschaften.* (Gesammelte Schriften, V. Band). Leipzig: Teubner.

Dilthey, Wilhelm, 1988. *Introduction to the Human Sciences: An Attempt to Lay a Foundation for the Study of Society and History.* Translated with an Introductory Essay by Ramon J. Betanzos. London: Harvester-Wheatsheaf.

Drake, Stillman (ed., trans.), 1957. *Discoveries and Opinions of Galileo.* New York: Doubleday.

Drake, Stillman, & C. D. O'Malley (eds., trans.), 1960. *The Controversy of the Comets of 1618. Galileo Galilei, Horatio Grassi, Mario Guiducci, Johann Kepler.* Philadelphia: University of Pensylvania Press.

Drake, Stillman, 1970. "Early Science and the Printed Book: The Spread of Science Beyond the Universities." *Renaissance and Reformation* 6, 43–52.

DSB = *Dictionary of Scientific Biography.*

Duby, Georges, 1973. *Guerriers et paysans, VII–XIIe siècle. Premier essor de l'économie européenne.* (Bibliothèque des histoires). Paris: Gallimard. [1]1969 (in English).

Dutton, Paul Edward, 1983. "*Illustre civitatis et populi exemplum:* Plato's *Timaeus* and the Transmission from Calcidius to the end of the Twelfth Century of a Tripartite Scheme of Society." *Mediaeval Studies* 45, 79–119.

Eamon, William, 1994. *Science and the Secrets of Nature. Books of Secrets in Medieval and Early Modern Culture.* Princeton: Princeton University Press.

Eddy, William Alfred (ed.), 1932. Swift, *Satires and Personal Writings.* London: Oxford University Press, 1973.

Einstein, Albert, 1921. "Geometrie und Erfahrung." Offprint from *Sitzungsberichte der Preussischen Akademie der Wissenschaften* 5.

Eisenstein, Elizabeth L., 1979. *The Printing Press as an Agent of Change. Communications and Cultural Transformations in Early-Modern Europe.* 2 vols. Cambridge: Cambridge University Press.

Ellegård, Alvar, 1973. "The Study of Language," *in* DHI II, 659–673.

Ember, Carol R., & Melvin Ember, 1977. *Cultural Anthropology.* 2nd edition. Englewood Cliffs, New Jersey: Prentice-Hall.

EncBrit: *Encyclopaedia Britannica.* (1)+X+19 vols. 15th Edition. Chicago: Encyclopaedia Britannica, 1979.

Eysenck, H. J., 1979. "Astrology—Science or Superstition." *Encounter* 53:6 (December 1979), 85–90.

Fanfani, Amintore, 1951. "La préparation intellectuelle à l'activité économique, en Italie, du XIVe au XVIe siècle." *Le Moyen Age* 57, 327–346.

Farrington, Benjamin, 1965. *Science and Politics in the Ancient World.* 2nd ed. London: Allen & Unwin. [1]1939.

Farrington, Benjamin, 1969. *Greek Science.* Harmondsworth, Middlesex: Penguin. [1]1944–49.

Febvre, Lucien, 1948. "Sorcellerie, sottise ou révolution mentale?" *Annales. Économies—Sociétés—Civilisations* 3, 9–15.

Feinberg, Evgeny L., 1977. "Art and Cognition." *Soviet Studies in Philosophy* 15:4, 62–91. Russian original: *Voprosy filosofij* 1976 no. 7.

Feyerabend, Paul, 1975. *Against Method.* London: New Left Books.

Field, J. V., 1984. "A Lutheran Astrologer: Johannes Kepler." [Includes a translation of Kepler's *De fundamentis atrologiae certioribus*]. *Archive for History of Exact Sciences* 31 (1984), 189–272.

Finley, Moses I. (ed.), 1968. *Slavery in Classical Antiquity. Views and Controversies.* Cambridge: Heffer / New York: Barnes & Noble. [1]1960.

Finley, Moses I., 1973. *Democracy Ancient and Modern.* London: Chatto & Windus.

Foucault, Michel, 1966. *Les Mots et les choses. Une archéologie des sciences humaines.* Paris: Gallimard.

Fowler, Harold North (ed., trans.), 1977. Plato, *Euthyphro, Apology, Crito, Phaedo, Phaedrus.* Cambridge, Mass.: Harvard University Press / London: Heinemann, 1977. [1]1914.

Franci, Raffaella, 1988. "L'insegnamento della matematica in Italia nel Tre-Quattrocento." *Archimede* 4, 182–194.

Fraser, Alexander Campbell (ed.), 1959. John Locke, *An Essay Concerning Human Understanding.* 2 vols. New York: Dover. [1]Oxford: Oxford University Press, 1894.

Fried, Morton, 1967. *The Evolution of Political Society. An Essay in Political Anthropology.* New York: Random House, 1967.

Garin, Eugenio, 1965. *Italian Humanism. Philosophy and Civic Life in the Renaissance.* Oxford: Blackwell / New York: Harper & Row. Reprint Westport, Connecticutt: Greenwood Press, 1975.

Garroni, Emilio, 1992. *Estetica. Uno sguardo-attraverso.* (Strumenti di studio). Milano: Garzanti.

Gay, P., 1967. *The Enlightenment: An Interpretation.* I. *The Rise of Modern Paganism.* II. *The Science of Freedom.* New York: Alfred A. Knopf, 1967–69.

Gelsted, Otto, 1957. *Udvalgte digte.* København: Gyldendal.

Gertz, M. Cl. (ed., trans.), 1967. *Sven Aggesøns historiske skrifter.* Oversatte efter den på grundlag af Codex Arnamagnæanus 33,4[to] restituerede nye text. København: Rosenkilde og Bagger. [1]1916–17.

Gewirtz, Alan, 1941. "Experience and the Non-Mathematical in the Cartesian Method." *Journal of the History of Ideas* 2, 183–210.

Gibson, Quentin, 1980. "Social Technology and Praxis." *Philosophy of the Social Sciences* 10, 25–37.

Giddens, Anthony, 1976. *New Rules of Sociological Method.* London: Hutchinson.

Giddens, Anthony, 1985. "Jürgen Habermas," *in* Q. Skinner 1985: 121–139.

Gillispie, Charles Coulston, 1980. *Science and Polity in France at the End of the Old Regime.* Princeton, N.J.: Princeton University Press.

Gilson, Étienne, 1955. *History of Christian Philosophy in the Middle Ages.* London: Sheed and Ward.

Gledhill, John, Barbara Bender & Mogens Trolle Larsen (eds), 1988. *State and Society. The Emergence and Development of Social Hierarchy and Political Centralization.* (One World Archaeology, 4). London: Unwin Hyman.

Goethe, Johann Wolfgang von, n.d.. *Werke.* 15 vols. Leipzig & Wien: Bibliographisches Institut.

Gombrich, E. H., 1971. *Norm and Form. Studies in the Art of the Renaissance.* 2nd ed. London: Phaidon. [1]1966.

Goody, Jack (ed.), 1968. *Literacy in traditional Societies.* Cambridge: Cambridge University Press.

Goody, Jack, 1977. *The Domestication of the Savage Mind.* Cambridge: Cambridge University Press.

Goody, Jack, 1986. *The Logic of Writing and the Organization of Society.* Cambridge: Cambridge University Press.

Goody, Jack, 1987. *The Interface between the Oral and the Written.* Cambridge: Cambridge University Press.

Gould, James L., & Peter Marler, 1987. "Learning by Instinct." *Scientific American* 256:1 (January 1987), 62–73.

Gouldner, Alvin W., 1973. *For Sociology. Renewal and Critique in Sociology Today.* London: Allen Lane.

Grabmann, Martin, 1941. *I divieti ecclesiastici di Aristotele sotto Innocenzo III e Gregorio IX.* (Miscellanea Historiae Pontificiae. Vol. V:i). Roma: Saler.

Grafton, Anthony, & Lisa Jardine, 1986. *From Humanism to the Humanities. Education and Liberal Arts in Fifteenth- and Sixteenth-Century Europe.* London: Duckworth.

Grafton, Anthony, 1992. *New Worlds, Ancient Texts: The Power of Tradition and the Shock of Discovery.* Cambridge, Mass.: Harvard University Press.

Gravelle, Sarah Stever, 1988. "The Latin-Vernacular Question and the Humanist Theory of Language and Culture." *Journal of the History of Ideas* 49, 367–386.

Gregory, Tullio, 1975. "La nouvelle idée de nature et de savoir scientifique au XIIe siècle," *in* Murdoch & Sylla (eds) 1975: 193–218.

Grmek, M. D., 1970. "Bernard, Claude," *in* DSB II, 24–34.

Grubb, Gerd, & Erik Kristensen, 1960. "First Draft of a Summary of the Discussions Following the Lectures." ICMI-Seminar, University of Aarhus, May 30 to June 2, 1960 *Mimeo,* Mathematical Institute, University of Aarhus.

Guevara, Ernesto Che, 1967. *Obra revolucionaria.* Prólogo y selección de Roberto Fernández Retamar. Altos, Mexico: Ediciones ERA, 1967.

[Gurevič =] Gurevitj, Aron J., 1979. *Feodalismens uppkomst i Västeuropa.* Stockholm: Tidens Förlag. [1]1970 (in Russian).

[Gurevič =] Gurjewitsch, Aaron J., 1986. *Mittelalterliche Volkskultur. Probleme zur Forschung.* Dresden: Verlag der Kunst. [1]1981 (in Russian).

Gustafsson, Bo, 1968. "Klassicism, marxism och marginalism." *Häften för Kritiska Studier* 1:1–2, 3–16.

Guthrie, W. K. C., 1962. *A History of Greek Philosophy.* I. *The Earlier Presocratics and the Pythagoreans.* II. *The Presocratic Tradition from Parmenides to Democritus.* III. *The Fifth-Century Enlightenment.* IV. *Plato. The Man and his Dialogues: Earlier Period.* V. *The Later Plato and the Academy.* VI. *Aristotle. An Encounter.* Cambridge: Cambridge University Press, 1962–1981.

Guthrie, W. K. C., 1967. *The Greek Philosophers from Thales to Aristotle.* London: Methuen. [1]1950.

Habermas, Jürgen, 1968. *Strukturwandel der Öffentlichkeit. Untersuchungen zu einer Kategorie der bürgerlichen Gesellschaft.* 3. Auflage. Neuwied & Berlin: Luchterhand. [1]1962.

Habermas, Jürgen, 1973. *Erkenntnis und Interesse.* 2. Auflage. Frankfurt a.M.: Suhrkamp. [1]1968.

Hahn, Roger, 1971. *The Anatomy of a Scientific Institution: The Paris Academy of Sciences, 1666–1803.* Berkeley: University of California Press.

Hankins, Thomas L., 1985. *Science and the Enlightenment.* Cambridge: Cambridge University Press.

Hardie, R. P., & R. K. Gaye (eds., trans.), 1930. Aristotle, *Physica, in* Aristotle, *Works,* vol. II.

Hardy, Godfrey Harold, 1967. *A Mathematician's Apology.* With a Foreword by C. P. Snow. Cambridge: Cambridge University Press.

Harris, William V., 1989. *Ancient Literacy.* Cambridge, Mass.: Princeton University Press.

Haskins, Charles Homer, 1924. *Studies in the History of Mediaeval Science.* Cambridge, Mass: Harvard University Press.

Haskins, Charles Homer, 1976. *The Renaissance of the Twelfth Century.* Cambridge, Mass., & London: Harvard University Press. [1]1927.

Hauser, Arnold, 1962. *The Social History of Art.* 4 vols. London: Routledge & Kegan Paul. [1]1951.

Havelock, Eric A., 1976. *Origins of Western Literacy.* (Monograph Series/14). Toronto: The Ontario Institute for Studies in Education.

Haydn, Hiram, 1950. *The Counter-Renaissance.* New York: Scribner.

Heidegger, Martin, 1977/1936. "Ursprung des Kunstwerkes," pp. 1–74 *in* Heidegger, *Holzwege* (Gesamtausgabe, 5). Frankfurt a.M.: Klostermann.

Heidelberger, Michael, & Sigrun Thiessen, 1981. *Natur und Erfahrung: Von der mittelalterlichen zur neuzeitlichen naturwissenschaft.* Hamburg: Rowohlt Taschenbuch Verlag.

Henningsen, Gustav, 1980. "The Greatest Witch-Trial of All: Navarre, 1609–1614." *History Today* 30:11, 36–39.

Herlihy, David (ed.), 1968. *Medieval Culture and Society.* (The Documentary History of Western Civilization). New York: Harper & Row.

Herlihy, David (ed.), 1971. *The History of Feudalism.* (The Documentary History of Western Civilization). London: Macmillan.

Herrmann, Joachim, & Jens Köhn (eds.), 1988. *Familie, Staat und Gesellschaftsformation. Grundprobleme vorkapitalistischer Epochen einhundert Jahre nach Friedrich Engels' Werk "Der Ursprung der Familie, des Privateigentums und des Staats."* Berlin: Akademie-Verlag.

Hicks, R. D. (ed., trans.), 1925. Diogenes Laertius, *Lives of Eminent Philosophers.* 2 vols. Cambridge, Mass.: Harvard University Press / London: Heinemann.

Hirschleifer, J., 1977. "Further Comments on Economics and Anthropology." *Current Anthropology* 18, 133–134.

Holtz, Louis, 1981. *Donat et la tradition de l'enseignement grammatical.* Paris: Centre National de Recherche Scientifique.

Holy Bible, Authorized King James Version. London & New York: Collins, n.d. (c. 1946).

Hooykaas, R., 1958. *Humanisme, science et réforme: Pierre de la Ramée (1515–1572).* Leiden: Brill.

Horkheimer, Max, & Theodor W. Adorno, 1972. *Dialectic of Enlightenment.* New York: Herder & Herder, 1972. [1]1944.

Høyrup, Jens, 1988. "Jordanus de Nemore, 13th Century Mathematical Innovator: an Essay on Intellectual Context, Achievement, and Failure." *Archive for History of Exact Sciences* 38, 307–363.

Høyrup, Jens, 1992. "Archimedism, not Platonism: on a Malleable Ideology of Renaissance Mathematicians (1400 to 1600), and on its Role in the Formation of Seventeenth-Century Philosophies of Science." Pp. 81–110 *in* C. Dollo (ed.), *Archimede. Mito Tradizione Scienza.* (Biblioteca di Nuncius. Studi e testi IV). Firenze: Leo S. Olschki.

Høyrup, Jens, 1994. *In Measure, Number, and Weight. Studies in Mathematics and Culture.* New York: State University of New York Press.

Høyrup, Jens, 1997. "Reflections on the Baroque in the History of Science". *Physis* 34 (1997), 675–694.

Huizinga, Johan, 1955. *The Waning of the Middle Ages. A Study of the Forms of Life, Thought, and Art in France and the Netherlands in the Fourteenth and Fifteenth Centuries.* Harmondsworth, Middlesex: Penguin. [1]1924.

Hunt, R. C., & E. Hunt, 1976. "Canal Irrigation and Local Organization" *Current Anthropology* 17, 389–398, discussion 398–411, 18 (1977), 116.

Huppert, George, 1970. *The Idea of Perfect History: Historical Erudition and Historical Philosophy in Renaissance Florence.* Urbana: University of Illinois Press.

Hymes, Dell (ed.), 1974. *Studies in the History of Linguistics.* Bloomington & London: Indiana University Press.

IESS = *International Encyclopedia of the Social Sciences.*

Iggers, Georg G., 1973. "Historicism," *in* DHI II, 456–464.

International Encyclopedia of the Social Sciences. 17 vols. New York: Macmillan & The Free Press, 1968.

Jacobsen, Thorkild, 1988. "Sumerian Grammar Today." *Journal of the American Oriental Society* 108, 123–133.

Jaeger, Werner, 1973. *Paideia. Die Formung des griechischen Menschen.* Ungekürzter photomechanischer Nachdruck in einem Band. Berlin & New York: Walter de Gruyter. [1]1933, 1944, 1947.

Jahnke, Hans Niels, & Michael Otte (eds.), 1981. *Epistemological and Social Problems of the Sciences of the Early Nineteenth Century.* Dordrecht: Reidel.

Jarvad, Ib Martin, 1993. *Ret og Stat. Tværvidenskab.* København: Christian Ejlers.

Jerison, Harry J., 1973. *Evolution of the Brain and Intelligence.* New York & London: Academic Press.

Jerison, Harry J., 1976. "Paleoneurology and the Evolution of the Mind." *Scientific American* 234:1, 90–101 (European pagination).

Jones, W. H. S. (ed., trans.), 1972. Hippocrates. 4 vols. (Loeb Classical Library). London: Heinemann / Cambridge, Mass.: Harvard University Press, 1972–1979. [1]1923–1931.

Jowett, Benjamin (ed., trans.), 1921. Aristotle, *Politica, in* Aristotle, *Works* (ed. W. D. Ross), vol. X. Oxford: The Clarendon Press.

Jürß, Fritz, 1977. *Von Thales zu Demokrit. Frühe griechische Denker.* Leipzig: Urania Verlag.

Jürß, Fritz, & Dietrich Ehlers, 1982. *Aristoteles.* (Biographien hervorragender Naturwissenschaftler, Techniker und Mediziner, 60). Leipzig: Teubner.

Kalikow, Theodora J., 1980. "Die ethologische Theorie von Konrad Lorenz: Erklärung und Ideologie, 1938 bis 1943." Pp. 189–214 *in* H. Mehrtens & S. Richter (eds.), *Naturwissenschaft, Technik und NS-Ideologie. Beiträge zur Wissenschaftsgeschichte des Dritten Reiches.* Frankfurt a.M.: Suhrkamp.

Kandel, Eric R., & Robert D. Hawkins, 1992. "The Biological Basis of Learning and Individuality." *Scientific American* 267:3, 53–60 (European pagination).

Kandinsky, Wassily, 1976. *Tutti gli scritti.* 2 vols. Milano: Feltrinelli.

Kant, Immanuel. *Werke.* 6 vols. Wiesbaden: Insel Verlag, 1956–1964.

Kelley, Donald R., 1970. *Foundations of Modern Historical Scholarship: Language, Law, and History in the French Renaissance.* New York & London: Columbia University Press.

Kelley, Donald R., 1988. "The Theory of History," *in* C. B. Schmitt et al (eds) 1988: 746–761.

Kemp, Martin, 1977. "From 'Mimesis' to 'Fantasia': The Quattrocento Vocabulary of Creation, Inspiration and Genius in the Visual Arts." *Viator* 8, 347–397, + Plates 1–7.

Kiernan, C., 1973. *The Enlightenment and Science in Eighteenth Century France.* (Studies on Voltaire and the Eighteenth Century, LIX A). 2nd ed. Banbury, Oxfordshire: The Voltaire Foundation.

Kinser, S., 1981. "*Annaliste* Paradigm? The Geohistorical Structuralism of Fernand Braudel." *American Historical Review* 86, 63–105.

Kircher, Athanasius, 1650. *Musurgia universalis.* Roma.

Kirk, G. S., & J. E. Raven, 1957. *The Pre-Socratic Philosophers.* Cambridge: Cambridge University Press.

Kirk, G. S., J. E. Raven & M. Schofield, 1983. *The Presocratic Philosophers. A Critical History with a Selection of Texts.* 2nd Edition. Cambridge: Cambridge University Press.

Klengel, Horst, 1980. *Hammurapi von Babylon und seine Zeit.* 4., überarbeitete Auflage. Berlin: Deutscher Verlag der Wissenschaften.

Koch, Ludovica, 1994. "Écho, sa raison et ses rimes," pp. 89–96 in Inge Degn, Jens Høyrup & Jan Scheel (eds), *Michelanea. Humanisme, litteratur og kommunikation.* Festskrift til Michel Olsen. Aalborg: Center for Sprog og Interkulturelle Studier, Aalborg Universitetsforlag.

Koenigsberger, H. G., 1979. "Republics and Courts in Italian and European Culture in the Sixteenth and Seventeenth Centuries." *Past & Present* 83, 32–56.

König, René, 1968. "Comte, Auguste," *in* IESS III, 201–206.

Kretzmann, Norman, Anthony Kenny & Jan Pinborg (eds), 1982. *The Cambridge History of Later Medieval Philosophy. From the Rediscovery of Aristotle to the Disintegration of Scholasticism, 1100–1600.* Cambridge etc.: Cambridge University Press.

Kristeller, Paul Oskar, 1956. *Studies in Renaissance Thought and Letters.* Roma: Edizioni di Storia e Letteratura.

Kristeller, Paul Oskar, 1961. *Renaissance Thought. The Classic, Scholastic, and Humanistic Strains.* New York: Harper & Brothers.

Kristeller, Paul Oskar, 1965. *Renaissance Thought.* II. *Papers on Humanism and the Arts.* New York: Harper & Brothers.

Kuhn, Thomas S., 1963. "The Function of Dogma in Scientific Research," *in* A. C. Crombie (ed.) 1963: 347–369.

Kuhn, Thomas S., 1970. *The Structure of Scientific Revolutions.* (International Encyclopedia of Unified Science, Volume 2, Number 2). 2nd ed. Chicago: University of Chicago Press. [1]1962.

Kuhn, Thomas S., 1974. "Reflections on my Critics," *in* Lakatos & Musgrave 1974: 231–278.

Kuhn, Thomas S., 1978. *Black-body Theory and the Quantum Discontinuity, 1894–1912.* New York: Oxford University Press.

Kuhn, Thomas S., 1992. *The Trouble with the Historical Philosophy of Science: Robert and Maurine Rotschild Distinguished Lecture, 19 November 1991.* Cambridge, Mass.: History of Science Department, Harvard University.

Kurdziałek, Marian, 1976. "David von Dinant als Auslager der aristotelischen Naturphilosophie," pp. 181–192 *in* Albert Zimmermann (ed.), *Die Auseinandersetzungen an der Pariser Universität im XIII. Jahrhundert.* (Miscellanea mediaevalia, 10). Berlin & New York: Walter de Gruyter.

Laistner, M. L. W., 1957. *Thought and Letters in Western Europe, A.D. 500–900.* Revised edition. London: Methuen / Ithaca, N.Y.: Cornell University Press. [1]1931.

Lakatos, Imre, & Alan Musgrave (eds.), 1974. *Criticism and the Growth of Knowledge.* Corrected reprint. Cambridge: Cambridge University Press. [1]1970.

Lakatos, Imre, 1974a. "Falsification and the Methodology of Scientific Research Programmes," *in* Lakatos & Musgrave (eds.) 1974: 91–196.

Lattin, Harriet Pratt (ed., trans.), 1961. *The Letters of Gerbert with his Papal Privileges as Sylvester II.* (Records of Civilization, 60). New York: Columbia University Press.

Lea, Kathleen Marguerite, 1974. "Bacon, Francis." *EncBrit,* vol. 2, 561a–567a.

Leavitt, Judith Walzer, 1992. "'Typhoid Mary' Strikes back. Bacteriological Theory and Practice in Early Twentieth-Century Public Health." *Isis* 83, 608–629.

Leff, Gordon, 1976. *The Dissolution of the Medieval Outlook. An Essay on the Intellectual and Spiritual Change in the Fourteenth Century.* New York: Harper & Row.

[Lenin =] Lénine, V., 1973. *Cahiers philosophiques.* Paris: Éditions Sociales / Moscou: Éditions du Progrès.

Lévy, Jacques R., 1973. "Le Verrier, Urbain Jean Joseph," in DSB VIII, 276–279.

Ljungdal, Arnold, 1967. *Marxismens världsbild.* Stockholm: Pan/Nordstedts. [1]1947.

Lloyd, G. E. R., 1979. *Magic, Reason and Experience. Studies in the Origin and Development of Greek Science.* Cambridge: Cambridge University Press.

Longrigg, James, 1989. "Presocratic Philosophy and Hippocratic medicine." *History of Science* 27, 1–39.

Longrigg, James, 1993. *Greek Rational Medicine: Philosophy and Medicine from Alcmaeon to the Alexandrians.* London: Routledge.

Lorenz, Konrad, 1943. "Die angeborenen Formen möglicher Erfahrung." *Zeitschrift für Tierpsychologie* 5, 235–409.

Losee, John, 1972. *A Historical Introduction to the Philosophy of Science.* London: Oxford University Press.

Lucas, Christopher J., 1979. "The Scribal Tablet-House in Ancient Mesopotamia." *History of Education Quarterly* 19, 305–332.

Lukács, Georg, 1969/1936. "Erzählen oder Beschreiben? Zur Diskussion über Naturalismus und Formalismus," *in* R. Brinkmann (ed.) 1969: 33–85.

Luria, Aleksandr R., 1976. *Cognitive Development. Its Cultural and Social Foundations.* Edited by Michael Cole. Cambridge, Mass., & London: Harvard University Press. [1]1974 (in Russian).

Mackay, A. L., 1977. *The Harvest of a Quiet Eye. A Selection of Scientific Quotations.* Bristol & London: The Institute of Physics.

MacKenna, Stephen (ed., trans.), 1969. Plotinus, *The Enneads.* Fourth edition revised by B. S. Page. London: Faber and Faber.

MacPherson, Crawford Brough (ed.), 1968. Thomas Hobbes, *Leviathan.* Harmondsworth, Middlesex: Penguin, 1968.

Mahoney, Michael S., 1975. "Ramus," *in* DSB XI, 286–290.

Malinowski, Bronislaw, 1955. *Magic, Science and Religion, and Other Essays.* Garden City, New York: Doubleday. [1]1948.

Malinowski, Bronislaw, 1955/1925. "Magic, Science and Religion," *in* B. Malinowski 1955: 9–148.

Malinowski, Ivan, 1980. *Vinterens hjerte.* København: Borgen.

Mandonnet, Pierre, O.P., 1914. "La crise scolaire au début du XIIIe siècle et la fondation de l'ordre des Frères-Prêcheurs." *Revue d'histoire ecclésiastique* 15, 35–49.

Manuel, Frank E., 1965. *The Prophets of Paris. Turgot, Condorcet, Saint-Simon, Fourier, Comte.* New York: Harper & Row. [1]1962.

Marrou, Henri-Irénée, 1956. *A History of Education in Antiquity.* London: Sheed and Ward.

Martines, Lauro, 1983. *Power and Imagination. City-States in Renaissance Italy.* Harmondsworth, Middlesex: Penguin. [1]1979.

Martzloff, Jean-Claude, 1988. *Historie des mathématiques chinoises.* Paris: Masson.

McCluskey, Stephen, 1993. "Astronomies in the Latin West from the Fifth to the Ninth Centuries." Pp. 139–160 *in* P. L. Butzer & D. Lohrman (eds.), *Science in Western and Eastern Civilization in Carolingian Times.* Basel: Birkhäuser.

McGarry, Daniel D. (ed., trans.), 1971. The *Metalogicon* of John of Salisbury. *A Twelfth-Century Defence of the Verbal and Logical Arts of the Trivium.* Gloucester, Mass.: Peter Smith.

McKitterich, Rosamond (ed.), 1990. *The Uses of Literacy in Early Medieval Europe.* Cambridge: Cambridge University Press.

McKitterick, Rosamond (ed.), 1994. *Carolingian Culture: Emulation and Innovation.* Cambridge: Cambridge University Press, 1994.

McCracken, D. S. et al. (eds., trans.), 1957. Saint Augustine, *The City of God against the Pagans.* 7 vols. (Loeb Classical Library, 411–417) London: Heinemann/ Cambridge, Mass: Harvard University Press, 1957–1972.

Measurement of Scientific and Technical Activities. Proposed Standard Practice for Surveys of Research and Experimental Development. *"Frascati Manual."* Third edition. Paris: OECD, 1976.

Merton, Robert K., 1963. "The Ambivalence of Scientists." *Bulletin of the Johns Hopkins Hospital* 112, 77–97.

Merton, Robert K., 1968. *Social Theory and Social Structure.* Enlarged Edition. New York & London: The Free Press.

Merton, Robert K., 1968/1942. "Science and Democratic Social Structure," *in* Merton 1968: 604–615. First published in *Journal of Legal and Political Sociology* 1 (1942).

Metcalf, George J., 1974. "The Indo-European Hypothesis in the 16th and 17th Centuries," *in* D. Hymes (ed.) 1974: 233–257.

MEW: Karl Marx, Friedrich Engels. *Werke.* 39 vols. Berlin: Dietz Verlag, 1965–1972.

Middleton, W. E. Knowles, 1971. *The Experimenters: A Study of the Accademia del cimento.* Baltimore: John Hopkins Press.

Mollat, Michel, & Philippe Wolff, 1970. *Ongles bleus, Jacques et Ciompis. Les révolutions populaires en Europe aux XIVe et XVe siècles.* Paris: Calmann-Lévy,1970.

Montagu, Ashley (ed.), 1980. *Sociobiology examined.* London: Oxford University Press.

Monter, E. William, 1980. "French and Italian Witchcraft." *History Today* 30:11, 31–35.

Mossner, Ernest C. (ed.), 1969. David Hume, *A Treatise of Human Nature.* Harmondsworth, Middlesex: Penguin.

Murdoch, John Emery, & Edith Dudley Sylla (eds.), 1975. *The Cultural Context of Medieval Learning.* (Boston Studies in the Philosophy of Science, 26). Dordrecht & Boston: Reidel.

Murray, Tim, 1995. "On Klejn's Agenda for Theoretical Archaeology." *Current Anthropology* 36 (1995), 290–292.

Naves, Raymond, & Julien Benda (eds.), 1967. Voltaire, *Dictionnaire philosophique.* Paris: Garnier.

Neuenschwander, Erwin, & Hans-Wilhelm Burmann, 1987. "Die Entwicklung der Mathematik an der Universität Göttingen." *Georgia Augusta. Nachrichten der Universität Göttingen,* November 1987, 17–28.

Nietzsche, Friedrich, 1982. *Die fröhliche Wissenschaft.* Frankfurt a.M.: Insel Verlag.

Nissen, Hans J., Peter Damerow & Robert Englund, 1994. *Archaic Bookkeeping: Writing and Techniques of Economic Administration in the Ancient Near East.* Chicago: Chicago University Press.

O'Brien, Conor Cruise (ed.), 1969. Edmund Burke, *Reflections on the Revolution in France and on the Proceedings in Certain Societies in London Relative to that Event.* Harmondsworth, Middlesex: Penguin.

Ong, Walter J., 1958. *Ramus, Method, and the Decay of Dialogue. From the Art of Discourse to the Art of Reason.* Cambridge, Mass., & London: Harvard University Press.

Ong, Walter J., 1967. *The Presence of the Word. Some Prolegomena for Cultural and Religious History.* New Haven & London: Yale University Press.

Ong, Walter J., 1982. *Orality and Literacy. The Technologizing of the World.* London & New York: Methuen.

Ovitt, George, Jr., 1986. *Perfection: Labor and Technology in Medieval Culture.* New Brunswick: Rutgers University Press.

Panofsky, Erwin, 1960. *Renaissance and Renascenses in Western Art.* I. *Text.* II. *Plates.* (The Gottesman Lectures, Uppsala University, VII). Stockholm: Almquist & Wiksell.

Pappe, Helmut O., 1973. "Enlightenment," *in* DHI II, 89–100.

Parker, Geoffrey, 1980. "The European Witchcraze Revisited. Introduction." *History Today* 30:11, 23–24.

Partington, J.R., 1961. *A History of Chemistry.* 4 vols. London: Macmillan, 1970, 1961, 1962, 1964.

Pasquini, Emilio (ed.), 1975. Francesco Guicciardini, *Ricordi.* Milano: Garzanti.

Paton, W. R. (ed., trans), 1922. Polybius, *The Histories.* 6 vols. Cambridge, Mass.: Harvard University Press / London: Heinemann, 1921–1927.

Pedersen, Olaf, 1979. *Studium generale. De europæiske universiteters tilblivelse.* Copenhagen: Gyldendal.

Perrault, Charles, 1688. *Paralelle des Anciens et des Modernes.* I. *En ce qui regarde les Arts et les Sciences.* II. *En ce qui regarde l'Eloquence.* II. *En ce qui regarde la poesie.* IV. *Ou il est traitté de l'Astronomie, de la Geographie, de la Navigation, de la Guerrre, de la Philosophie, de la Musique, de la Medecine, etc.* Paris: Jean Baptiste Coignard, 1688–1697. Facsimile Munich: Eidos Verlag, 1964.

Perrin, Bernadotte (ed., trans.), 1967. Plutarch's *Lives.* III. *Pericles and Fabius Maximus. Nicias and Crassus.* London: Heinemann / Cambridge, Mass.: Harvard University Press. [1]1916.

Pevsner, Nikolaus, 1940. *Academies of Art, Past and Present.* Cambridge: Cambridge University Press.

Pfetsch, Frank R., 1979. "The 'Finalization' Debate in Germany: Some Comments and Explanations." *Social Studies of Science* 9, 115–124.

Piaget, Jean, 1950. *The Psychology of Intelligence.* London: Routledge & Kegan Paul.

Piaget, Jean, 1967. *Six Psychological Studies.* New York: Random House.

Piaget, Jean, 1972. *Psychology and Epistemology.* London: Allen Lane.

Piaget, Jean, 1973. *The Child and Reality. Problems of Genetic Psychology.* New York: Grossman.

Piaget, Jean, 1976. *The Grasp of Consciousness: Action and Concept in the Young Child.* Cambridge, Mass.: Harvard University Press.

Pichot, Pierre, 1968. "Binet, Alfred", pp. 74–78 *in International Encyclopedia of the Social Sciences,* vol. II. New York: Macmillan & The Free Press.

Piltz, Anders, 1981. *The World of Medieval Learning.* Oxford: Blackwell. [1]1978 (in Swedish).

Popkin, Jeremy D., 1991. "Periodical Communication and the Nature of Knowledge in Eighteenth-Century Europe", pp. 203–213 *in* Donald R. Kelley & Richard H. Popkin (eds.), *The Shapes of Knowledge from the Renaissance to the Enlightenment.* Dordrecht: Kluwer.

Popper, Karl R., 1972. *Conjectures and Refutations. The Growth of Scientific Knowledge.* 4th, Revised Edition. London: Routledge & Kegan Paul.

Popper, Karl R., 1972a. *The Logic of Scientific Discovery.* 6th Revised Impression. London: Hutchinson. [1]Wien, 1935.

Popper, Karl R., 1973. *Objective Knowledge. An Evolutionary Approach.* 2nd Revised Printing. London: Oxford University Press.

Popper, Karl R., 1974. "Normal Science and its Dangers," *in* Lakatos & Musgrave 1974: 51–58.

Price, Derek J. de Solla, 1963. *Little Science, Big Science.* New York & London: Columbia University Press.

Propp, Vladimir, 1968. *Morphology of the Folktale.* 2nd ed. Austin & London: University of Texas Press.

Propp, Vladimir, 1984. *Theory and History of Folklore.* Edited, with and Introduction and Notes, by Anatoly Liberman. (Theory and History of Literature, 5). Manchester: Manchester University Press.

Quine, Willard Van Orman, 1963. *From a Logical Point of View. Logico-Philosophical Essays.* Second Edition, Revised. New York: Harper & Row.

Quine, Willard Van Orman, 1963/1951. "Two Dogmas of Empiricism," *in* Quine 1963: 20–46. First published in *Philosophical Review,* January 1951.

Ramus, Petrus, 1569. *Scholarum mathematicarum libri unus et triginta.* Basel: Eusebius Episcopius.

Rand, Edward Kennard, 1957. *Founders of the Middle Ages.* New York: Dover. [1]1928.

Randall, John Herman, 1960. *Aristotle.* New York & London: Columbia University Press.

Randall, John Herman, 1962. *The Career of Philosophy from the Middle Ages to the Enlightenment.* New York & London: Columbia University Press.

Rappaport, Roy A., 1986. "Desecrating the Holy Woman: Derek Freeman's Attack on Margaret Mead." *The American Scholar* 55, 313–347.

Rashdall, Hastings, 1936. *The Universities of Europe in the Middle Ages.* A New Edition in Three Volumes Edited by F. M. Powicke and A. B. Emden. Oxford: The Clarendon Press. [1]1895.

Riché, Pierre, 1962. *Éducation et culture dans l'Occident barbare, vi[e]–viiie siècles.* Thèse pour le Doctorat ès Lettres. Paris: Seuil.

Riché, Pierre, 1976. *Education and Culture in the Barbarian West, Sixth through Eighth Centuries.* University of South Carolina Press. [1]1962 (in French).

Riché, Pierre, 1979. *Les Écoles de l'Occident chrétien de la fin du Ve siècle au milieu du XIe siècle.* Paris: Aubier Montaigne.

Richter, Michael, 1994. *The Formation of the Medieval West. Studies in the Oral Culture of the Barbarians.* Dublin: Four Courts.

Rieux, Jacques, et al. (eds., trans.), 1975. Antoine Arnauld & Claude Lancelot, *General and Rational Grammar. The Port-Royal Grammar.* (Janua linguarum. Series minor, 208). De Haag & Paris: Mouton.

Robins, R. H., 1971. *General Linguistics: An Introductory Survey.* 2nd edition. London: Longman.

Robinson, I. S., 1978. *Authority and Resistance in the Investiture Contest: The Polemical Literature of the Late Eleventh Century.* Manchester: Manchester University Press / New York: Holmes and Meier.

Robinson, Joan, & John Eatwell, 1973. *An Introduction to Modern Economics.* London etc.: McGraw-Hill.

Rorty, Richard, 1982. *Consequences of Pragmatism.* (Essays: 1972–1980). Brighton: Harvester, 1982.

Ross, James Bruce, & Mary Martin McLauchlin (eds.), 1977. *The Portable Renaissance Reader.* Harmondsworth, Middlesex: Penguin. [1]New York: Viking Press, 1953.

Ross, James Bruce, & Mary Martin McLauchlin (eds.), 1977a. *The Portable Medieval Reader.* Harmondsworth, Middlesex: Penguin. [1]New York: Viking Press, 1949.

Ross, W. D. (ed., trans.), 1928. Aristotle, *Metaphysica, in* Aristotle, *Works,* vol. VIII (second edition). Oxford: Clarendon Press.

Rotman, Brian, 1977. *Jean Piaget: Psychologist of the Real.* Hassocks, Sussex: Harvester.

Roubinet, P. (ed.), 1964. *Logique de Port-Royal.* Facsimile reprint of [A. Arnauld & P. Nicole,] *La Logique ou L'Art de penser.* (Cinquième édition, 1683). (Publications de la Faculté des Lettres et Sciences Humaines de l'Université de Lille, 12). Lille: René Giard.

Sainte-Beuve, Charles Augustin (ed.), 1868. Boileau, *Oeuvres.* Paris: Furne, Jouvet et Cie.

Sajo, Géza (ed.), 1964. Boetii de Dacia *Tractatus de eternitate mundi.* (Quellen und Studien zur Geschichte der Philosophie, IV). Berlin: De Gruyter.

Sartre, Jean-Paul, 1946. *L'Existentialisme est un humanisme.* Paris.

Sartre, Jean-Paul, 1960. *Critique de la raison dialectique* précédé de *Questions de méthode.* Tôme I, *Théories des ensembles pratiques.* Paris: Gallimard.

Sartre, Jean-Paul, 1964. *Les Mots.* Paris: Gallimard.

Sartre, Jean-Paul, 1976. *L'Etre et le néant. Essai d'ontologie phénoménologique.* Paris: Gallimard. [1]1943.

Saussure, Ferdinand de, 1972. *Cours de linguistique générale.* Publié par Charles Bally et Albert Sechehaye. Paris: Payot. [1]1922.

Schäfer, Wolf (ed.), 1983. *Finalization in Science: The Social Orientation of Scientific Progress.* (Boston Studies in the Philosophy of Science, 77). Dordrecht etc: D. Reidel.

Schmeidler, Felix (ed.), 1972. Joannis Regiomontani *Opera collectanea.* Faksimiledrucke von neun Schriften Regiomontans und einer von ihm gedruckten Schrift seines Lehrers Purbach. (Milliaria X,2). Osnabrück: Otto Zeller.

Schmid, Herman, 1973. "Om forskning för folket." *Nordisk Forum* 8, 291–305.

Schmid, Herman, 1981. "Tillämpad forskning som praktik." *Institut for Samfundsøkonomi og Planlægning, Roskilde Universitetscenter. Arbejdspapir* nr. 16/1981.

Schmitt, Charles B., Quentin Skinner & Eckhard Keßler (eds.) 1988. *Cambridge History of Renaissance Philosophy.* Cambridge: Cambridge University Press.

Schmitz, Rudolf, & Fritz Krafft (eds.), 1980. *Humanismus und Naturwissenschaften.* (Beiträge zur Humanismusforschung, Band VI). Boppard: Harald Boldt.

Schönberg, Arnold, 1989. *Stil und Gedanke.* Leipzig: Reclam. [1]1950.

Seidel, Helmut, 1988. *Aristoteles und der Ausgang der antiken Philosophie. Vorlesungen zur Geschichte der Philosophie.* 2., ergänzte Auflage. Berlin: Dietz Verlag.

Seidel, Helmut, 1989. *Von Thales bis Platon. Vorlesungen zur Geschichte der Philosophie.* 5. Auflage. Berlin: Dietz Verlag.

Selby-Bigge, L. A. (ed.), 1975. David Hume, *Enquiries Concerning Human Understanding and Concerning the Principles of Morals.* Reprinted from the Posthumous Edition of 1777 and Edited with Introduction [etc.]. Third Edition with text Revised and Notes by P. H. Nidditch. Oxford: Oxford University Press.

Serres, Michel, François Dagognet & Allal Sinaceur (eds.), 1975. Auguste Comte, *Philosophie première. Cours de philosophie positive, leçons 1 à 45.* Paris: Hermann.

Service, Elman R., 1975. *Origins of the State and Civilization. The Process of Cultural Evolution.* New York: W. W. Norton.

Shiloh, Ailon, 1975. "Psychological Anthropology: A Case Study in Culture Blindness." *Current Anthropology* 16, 618–620; discussion 17 (1976), 326f, 349f, 554f.

Sjöberg, Åke W., 1976. "The Old Babylonian Eduba," pp. 159–179 *in Sumerological Studies in Honor of Thorkild Jacobsen.* (Assyriological Studies, 20). Chicago & London: University of Chicago Press.

Skinner, B. F., 1973. *Beyond Freedom and Dignity.* Harmondsworth, Middlesex: Penguin. [1]1971.

Skinner, Quentin, 1985. *The Return of Grand Theory in the Human Sciences.* Cambridge: Cambridge University Press, 1985.

Sklair, Leslie, 1973. *Organized Knowledge.* St Albans Herts: Paladin.

Smith, Adam, 1974. *The Wealth of Nations.* Harmondsworth, Middlesex: Penguin.

Snow, C. P., 1964. *The Two Cultures: And a Second Look. An Expanded Version of the Two Cultures and the Scientific Revolution.* London: Cambridge University Press.

Solovine, Maurice (ed., trad.), 1965. Epicure, *Doctrines et maximes.* Paris: Hermann.

Southern, R. W., 1953. *The Making of the Middle Ages.* London: Hutchinson.

Southern, R. W., 1970. *Western Society and the Church in the Middle Ages.* Harmondsworth, Middlesex: Penguin.

Southern, R. W., 1970a. *Medieval Humanism and Other Studies.* Oxford: Blackwell.

Stone, Lawrence (ed.), 1974. *The University in Society.* I. *Oxford and Cambridge from the 14th to the Early 19th Century.* II. *Europe, Scotland, and the United States from the 16th to the 20th Century.* Princeton, N.J.: Princeton University Press.

Störig, Hans Joachim, 1957. *Kleine Weltgeschichte der Wissenschaft.* Salzburg: Europäischer Buchklub. [1]1954.

Strohmaier, Gotthard, 1991. "Zur Erfindung der Vokalbuchstaben durch die Griechen." *Philologus* 135, 38–44.

Sulloway, Frank J., 1992. *Freud, Biologist of the Mind: Beyond the Psychoanalytic Legend.* [2]Cambridge, Mass., & London: Harvard University Press.

Summers, Montague (ed., trans.), 1971. Heinrich Kramer & Jacob Sprenger, *Malleus maleficarum.* New York: Dover. [1]London: John Rodker, 1928.

Surtz, Edward, S. J., & J. H. Hexter (eds., trans.), 1965. Thomas More, *Utopia.* (The Yale Edition of the Complete Works of St. Thomas More, 4). New Haven & London: Yale University Press.

Taine, Hippolyte, 1877. *Histoire de la littérature anglaise.* Quatrième édition. 5 vols. Paris: Hachette, 1877– 78.

Tainter, J. A., 1988. *The Collapse of Complex Societies.* (New Studies in Archaeology). Cambridge: Cambridge University Press.

Tatarkiewicz, W., 1973. "Form in the History of Aesthetics," *in* DHI II, 216–225.

Taylor, Alan B. H., 1994. "An Episode with May-Dew." *History of Science* 32, 163–184.

Taylor, Jerome (ed., trans.), 1961. The *Didascalicon* of Hugh of St. Victor. *A Medieval Guide to the Arts.* (Records of Civilization. Sources and Studies, 64). New York & London: Columbia University Press.

Thibaudet, Albert, & Maurice Rat (eds), 1962. Montaigne, *Oeuvres complètes.* Paris: Gallimard.

Thomas, Keith, 1971. *Religion and the Decline of Magic in Sixteenth and Seventeenth Century England.* London: Weidenfeld & Nicholson.

Thompson, C. R., J. K. Sowards & A. H. T. Levi (ed., trans.) 1978. Erasmus, *Literary and Educational Writings.* 6 vols. (Collected Works of Erasmus, 23–28). Toronto: Toronto University Press, 1978–86.

Thomson, George, 1961. *Studies in Ancient Greek Society. The First Philosophers.* Second edition. London: Lawrence & Wishart. ^{1}1955.

Thomson, Robert, 1968. *The Pelican History of Psychology.* Harmondsworth, Middlesex: Penguin, 1968.

Thorndike, Lynn, 1923. *A History of Magic and Experimental Science.* 8 vols. New York & London: Columbia University Press, 1923–58.

Thorndike, Lynn, 1944. *University Records and Life in the Middle Ages.* (Records of Civilization). New York: Columbia University Press.

Thorpe, Lewis (ed., trans.), 1969. Einhard and Notker the Stammerer, *Two Lives of Charlemagne.* Harmondsworth, Middlesex: Penguin.

Thuillier, Pierre, 1985. "La sociobiologie au service du féminisme?" *La Recherche* 16, 226–230.

Trabucco, Joseph (ed., trad.), 1960. Saint Augustin, *Les Confessions.* Paris: Garnier.

Turner, R. Steven, 1971. "The Growth of Professorial Research in Prussia 1818–48—Causes and Context." *Historical Studies in the Physical Sciences* 3, 137–182.

Turner, R. Steven, 1974. "University Reformers and Professorial Scholarship in Germany 1760–1806," *in* L. Stone 1974: II, 495–531.

Turner, R. Steven, 1981. "The Prussian Professorate and the Research Imperative, 1790–1840," *in* Jahnke & Otte 1981: 109–121.

Twain, Mark, 1986. *The Adventures of Tom Sawyer—The Adventures of Huckleberry Finn—Mark Twain's Sketches—[etc.]* New York: Chatham River Press.

Ullmann, Walter, 1970. *A History of Political Thought: The Middle Ages.* Harmondsworth, Middlesex: Penguin. ^{1}1965.

van Steenberghen, Fernand, 1955. *The Philosophical Movement in the Thirteenth Century.* Lectures Given under the Auspices of the Department of Scholastic Philosophy, The Queen's University, Belfast. Edinburgh: Nelson.

van Steenberghen, Fernand, 1966. *La Philosophie au XIIIe siècle.* (Philosophes Médiévaux, tôme IX). Louvain: Publications Universitaires / Paris: Béatrice-Nauwelaerts.

Vartanian, Aram, 1973. "Man-Machine from the Greeks to the Computer," *in* DHI III, 131–146.

Venturi, F., 1971. *Utopia and Reform in the Enlightenment.* Cambridge: Cambridge University Press.

Vernant, Jean-Pierre, 1982. *The Origins of Greek Thought.* Ithaca, New York: Cornell University Press. ^{1}1962 (in French).

Vernières, Paul (ed.), 1964. Diderot, *Oeuvres philosophiques.* Paris: Garnier Frères.

Vloebergh, Assomption, 1984. "Comment les rats évitent les aliments empoisonnés." *La Recherche* 15, 1280–1281.

von Clausewitz, Carl, 1973. *Vom Kriege.* Bonn: Dümmler.

Wade, Ira O., 1977. *The Structure and Form of the French Enlightenment.* I. *Esprit Philosophique.* II. *Esprit Révolutionnaire.* Princeton, N. J.: Princeton University Press.

Watkins, J. W. N., 1974. "Against 'Normal Science'," *in* Lakatos & Musgrave 1974: 25–37.

Weber, Max, 1947. *The Theory of Social and Economic Organization.* Edited with an Introduction by Talcott Parsons. New York: Oxford University Press.

Weber, Max, 1984. *Die protestantische Ethik.* I. *Eine Aufsatzsammlung.* II. *Kritiken und Antikritiken.* Herausgegeben von Johannes Winckelmann. Gütersloh: Gerd Mohn: 1984, 1987.

Weiss, Peter, 1964. *Die Verfolgung und Ermordung Jean Paul Marats dargestellt durch die Schauspielgruppe des Hospizes zu Charenton unter Anleitung des Herrn de Sade.* Frankfurt a.M.: Suhrkamp.

Wellek, René, 1968. "Evolution of Literature," *in* DHI II, 169–174.

Wellek, René, 1973. "Baroque in Literature," *in* DHI I, 189–195.

Werner, Ernst, 1976. "Stadtluft macht Frei: Frühscholastik und bürgerliche Emancipation in der ersten Hälfte des 12. Jahrhunderts." *Sitzungsberichte der Sächsischen Akademie der Wissenschaften zu Leipzig, Philologisch-historische Klasse* 118, Heft 5.

Westman, Robert S., 1980. "The Astronomer's Role in the Sixteenth Century: A Preliminary Study." *History of Science* 18, 105–147.

White, Lynn, 1962. *Medieval Technology and Social Change.* Oxford: Clarendon Press.

Wilson, Edward O., 1975. *Sociobiology.* Cambridge: Belknap Press.

Winkle, Stefan, 1983. *Johann Friedrich Struensee: Arzt—Aufklärer—Staatsmann. Beitrag zur Kultur-, Medizin- und Seuchengeschichte der Aufklärungszeit.* Stuttgart: Gustav Fischer.

Winter, Alison, 1994. "Mesmerism and Popular Culture in Early Victorian England." *History of Science* 32, 317–343.

Wittfogel, Karl, 1957. *Oriental Despotism: A Comparative Study of Total Power.* New Haven: Yale University Press.

Wittgenstein, Ludwig, 1968. *Philosophical Investigations.* Translated by G. E. M. Anscombe. Oxford: Basil Blackwell.

Wolfgang, Marvin E., 1968. "Lombroso, Cesare," *in* IESS IX, 471b–473a.

Wrenn, C. L., 1967. *A Study of Old English Literature.* London: Harrap.

Yates, Frances, 1964. *Giordano Bruno and the Hermetic Tradition.* London: Routledge & Kegan Paul.

Zilsel, Edgar, 1976. *Die sozialen Ursprünge der neuzeitlichen Wissenschaft.* Herausgegeben von Wolfgang Krohn. Frankfurt a.M.: Suhrkamp.

Zimmermann, Albert (ed.), 1976. *Die Auseinandersetzungen an der Pariser Universität im XIII. Jahrhundert.* (Miscellanea mediaevalia, 10). Berlin & New York: Walter de Gruyter.

Name Index

Subject Index